QUALITATIVE
RESEARCH
METHODS
—— IN ——
HUMAN
GEOGRAPHY

QUALITATIVE RESEARCH METHODS

— IN —

HUMAN GEOGRAPHY

THIRD EDITION

IAIN HAY

OXFORD

UNIVERSITY PRESS

OXFORD
UNIVERSITY PRESS

8 Sampson Mews, Suite 204, Don Mills, Ontario M3C 0H5
www.oupcanada.com

Oxford University Press is a department of the University of Oxford.
It furthers the University's objective of excellence in research, scholarship,
and education by publishing worldwide in

Oxford   New York

Auckland   Cape Town   Dar es Salaam   Hong Kong   Karachi
Kuala Lumpur   Madrid   Melbourne   Mexico City   Nairobi
New Delhi   Shanghai   Taipei   Toronto

With offices in
Argentina   Austria   Brazil   Chile   Czech Republic   France   Greece
Guatemala   Hungary   Italy   Japan   Poland   Portugal   Singapore
South Korea   Switzerland   Thailand   Turkey   Ukraine   Vietnam

Oxford is a trade mark of Oxford University Press
in the UK and in certain other countries

Published in Canada
by Oxford University Press

Library and Archives Canada Cataloguing in Publication

Qualitative research methods in human geography / [edited by] Iain Hay. — 3rd ed.

Includes bibliographical references and index.
ISBN 978-0-19-543015-8

1. Human geography—Methodology—Textbooks. 2. Qualitative research—Textbooks. I. Hay, Iain

GF21.Q83 2010   304.2072   C2009-906164-3

Cover image: OJO Images Photography/Veer

Printed on permanent acid-free paper ∞.

Printed and bound in Canada.

2 3 4 — 14 13 12

Contents

3 Cross-Cultural Research: Ethics, Methods, and Relationships   40

Richie Howitt and Stan Stevens

4 Qualitative Research Design and Rigour   69

Matt Bradshaw and Elaine Stratford

7   Oral History and Human Geography   139

Karen George and Elaine Stratford

8   Focusing on the Focus Group   152

Jenny Cameron

9   Historical Research and Archival Sources   173

Michael Roche

10   Using Questionnaires in Qualitative Human Geography   191

Pauline M. McGuirk and Phillip O'Neill

11 Doing Foucauldian Discourse Analysis— Revealing Social Identities 217

Gordon Waitt

12 Seeing with Clarity: Undertaking Observational Research 241

Robin A. Kearns

16 Writing a Compelling Research Proposal
Janice Monk and Richard Bedford

17 Writing Qualitative Geographies, Constructing Geographical Knowledges
Juliana Mansvelt and Lawrence D. Berg

18 From Personal to Public: Communicating Qualitative Research for Public Consumption 356

Dydia DeLyser and Eric Pawson

List of Boxes and Figures

Boxes

Figures

Notes on Contributors

Jamie Baxter, BA (Hons) (Queen's) 1989, MA (McMaster) 1992, PhD (McMaster) 1997, is an Associate Professor of Geography at the University of Calgary and Graduate Chair at the University of Western Ontario. He has also worked as an associate professor of geography at the University of Calgary. His research interests involve environmental risks from hazards, community responses to technological hazards, geography of health, noxious facility siting, and methodology. His recent projects have involved case studies of risk perception in communities living with toxic waste and urban pesticides. He is currently principal investigator for a study entitled Environmental Inequity in Canada. Jamie has published in a wide array of journals, including *Transactions of the Institute of British Geographers, Journal of Risk Research, Risk Analysis, Environmental Planning and Management, Social Science and Medicine,* and *The Canadian Geographer.*

Richard Bedford, BA (Auckland) 1965, MA (Auckland) 1967, PhD (Australian National University) 1972, is Professor of Population Geography and Director of the Population Studies Centre at the University of Waikato, Hamilton, New Zealand. Prior to taking up the chair of geography at the University of Waikato in 1989, he was on the staff of the University of Canterbury. His research interests are mainly in the field of population dynamics, especially migration, in the Asia-Pacific region. He is a Fellow of the Royal Society of New Zealand and in 2008 was made a Companion of the Queen's Service Order in recognition of his services to geography. He has published widely on population movement and development in the Pacific Islands, including a book co-authored with Harold Brookfield, Tim Bayliss-Smith, and Marc Latham in the Cambridge Human Geography series on the colonial and post-colonial experience of eastern Fiji (1989). His articles have appeared in journals such as *Asia-Pacific Viewpoint, Asian and Pacific Migration Journal, Contemporary Pacific, Espaces Populations Sociétés, GeoJournal, Journal de la Société des Oceanistes, Journal of International Migration and Integration, International Migration, New Zealand Geographer, New Zealand Population Review,* and *Oceania.*

Lawrence D. Berg, BA (Dist.) (Victoria) 1988, MA (Victoria) 1991, DPhil (Waikato) 1996, began his career as a lecturer in the School of Global Studies at Massey

University, Aotearoa/New Zealand. Lawrence now works in Canada at the University of British Columbia where he recently completed his term as a Canada research chair and is now Co-director of the Centre for Social, Spatial and Economic Justice. He has a diverse range of academic interests in radical geography, focusing especially on issues relating to analyses of identity politics and place and the emplaced cultural politics of knowledge production. Lawrence has been editor of both *SITES* and *Canadian Geographer,* and he is one of the founding editors of *ACME: An International E-Journal for Critical Geographies.* He is active in the open-access movement. His current research involves a wide range of participatory action research projects aimed at contesting and disrupting various taken-forgranted forms of everyday social relations that contribute to oppression in a range of spaces. Lawrence has published extensively in journals such as *Antipode, Geoforum, Gender Place and Culture, Progress in Human Geography, Social and Cultural Geography,* and *Society and Space.*

Matt Bradshaw, BA (Hons) (Tasmania) 1989, MEnvStudies (Tasmania) 1992, PhD (Tasmania) 2000, is an Honorary Associate in the Department of Geography and Environmental Studies, University of Tasmania. He is also a social research consultant to the Irish Sea Fisheries Board, the Hobart City Council, and the Tasmanian Aquaculture and Fisheries Institute, among others. His current academic interests include economic geography, the social impact assessment of fisheries, and community involvement in local government planning. He has published in *Antipode, Applied Geography, Area, Australian Geographical Studies, Environment and Planning A, Fisheries Research,* and *New Zealand Geographer,* among others.

Jenny Cameron, DipTeach (Brisbane College of Advanced Education) 1983, BAppSc (Queensland University of Technology) 1990, MA (Sydney) 1992, PhD (Monash) 1998, is Associate Professor in the Discipline of Geography and Environmental Studies at the University of Newcastle. Jenny has conducted focus groups and facilitated workshops as part of her participatory action research working with economically marginalized groups to develop community economic projects. She has also collaborated with the Queensland Department of Local Government and Planning to evaluate, through action research, public participation in regional planning. Her most recent work has been with grassroots community enterprises, particularly documenting the issues faced by these enterprises. She has published articles in journals that include *Australian Geographer, Geoforum, Gender, Place and Culture, Rethinking Marxism, Socialist Review,* and *Urban Policy and Research.* She is also committed to communicating research outcomes to a general audience and has produced a documentary on asset-based community economic development and several resource kits.

Meghan Cope, AB (Vassar) 1989, MA (Colorado) 1992, PhD (Colorado) 1995, is Associate Professor and Chair of the Department of Geography at the University of Vermont. She recently completed a four-year project funded by the National Science Foundation (US) on 'children's urban geographies' in which she worked directly with low-income children with diverse ethnic and racial backgrounds in Buffalo, New York. Meghan is now focusing her research on young people's mobility and access to public space in Vermont towns in the context of a larger interest in young people, place, and identity across Europe and North America. Her work has always blended qualitative research with a sustained interest in 'the everyday'. From this work and her love of walking and cycling has grown a new activist interest in fostering 'youth-friendly cities' through design, outreach, and education. Meghan's experiences have also led her to develop work on blending qualitative research with geographic information systems (GIS), and she recently co-edited a book with Sarah Elwood on this topic (*Qualitative GIS*, Sage, 2009). She has published more than 20 articles and chapters on methods, children's geographies, urban post-industrial decline, feminist geography, and participatory research and pedagogy.

Dydia DeLyser, BA (California, Los Angeles) 1992, MA (Syracuse) 1996, PhD (Syracuse) 1998, is Associate Professor of Geography at Louisiana State University. Her qualitative research—largely on tourism, landscape, and social memory in the American West—has been both ethnographic and archival. She teaches cultural geography, qualitative methods, and writing at the graduate and undergraduate levels. Dydia has published more than a dozen articles and chapters on these topics in such journals as the *Annals of the Association of American Geographers, Journal of Geography in Higher Education, Journal of Historical Geography, Cultural Geographies*, and *Social and Cultural Geography* as well as a book, *Ramona Memories: Tourism and the Shaping of Southern California* (University of Minnesota Press, 2005). She serves as North American editor of *Cultural Geographies*.

Robyn Dowling, BEc (Hons) (Sydney) 1988, MA (British Columbia) 1991, PhD (British Columbia) 1995, is Associate Professor in the Department of Environment and Geography at Macquarie University, NSW, Australia. Her broad research interests lie at the intersection of cultural and urban geography. Past research explored the material and cultural geographies of home in Australian suburbia and gendered aspects of transport and mobility, while her current research focuses on the changing governance and social patterns of suburban Australia. Robyn's publications include papers in *Antipode, Environment and Planning D, Australian Geographical Studies, Housing Studies, Social and Cultural Geography*, and *Urban Geography* as well as a book with Alison Blunt entitled *Home* (Routledge, 2006).

Kevin Dunn, BA (Hons) (Wollongong) 1990, PhD (Newcastle) 1999, is Professor in Human Geography and Urban Studies at the University of Western Sydney. Like other contributors to this volume, Kevin has a broad range of academic interests, including geographies of racism, transnationalism and migrant settlement, identity and place, media representations of place and people, and the politics of heritage and memorial landscapes. His PhD focused on opposition to mosques in Sydney. He co-authored *Introducing Human Geography* (Longman-Pearson Education Australia, 2000), which won the Australian Award for Excellence in Publishing, and *Landscapes* (Pearson, 2003) and has published more than 50 chapters and articles in various books and in journals including *Australian Geographer*, *Australian Geographical Studies*, *Environment and Planning A* and *D*, *Social and Cultural Geography*, and *Urban Studies*.

Karen George, BA (Hons) (Adelaide) 1984, MA (Australian National University) 1988, PhD (Adelaide) 1994, is a consultant historian and writer. She is sole proprietor of Historically Speaking, a business she has operated since 1993. From 1993 to 2001, she was oral historian for the Corporation of the City of Adelaide. She is a past-president of the South Australian branch of the Oral History Association of Australia and currently teaches oral history workshops for the branch. She is the author of *A Place of Their Own: The Men and Women of War Service Land Settlement at Loxton after World War II* (Wakefield Press, Adelaide, 1999) and *City Memory: A Guide and Index to the City of Adelaide Oral History Collection 1978–1998* (Corporation of the City of Adelaide, 1999) and was a contributor to Doreen Mellor and Ann Haebich, eds, *Many Voices: Reflections on Experiences of Indigenous Child Separation* (National Library of Australia, 2002). In 2005, she completed a major research project for SA Link-Up, identifying and describing surviving records of homes and institutions into which Aboriginal children were placed. This work led her into the role of research historian for the Children in State Care Commission of Inquiry in South Australia (2005–8). As a consultant historian, she specializes in oral history and has worked on numerous projects ranging over many subject areas. She is currently involved in exhibition work with community museums and a major oral history project on the apple and pear industry in Australia for the National Library of Australia.

Iain Hay, BSc (Hons) (Canterbury) 1982, MA (Massey) 1985, PhD (Washington) 1989, GradCertTertEd (Flinders) 1995, MEdMgmt (Flinders) 2004, LittD (Canterbury) 2009, is Professor of Geography in the School of Geography, Population and Environmental Management at Flinders University, South Australia. His principal research interests revolve around geographies of domination and oppression. He is the author or editor of eight books, several of which have gone to second and third editions, and author of more than 75 scholarly chapters and papers in journals such as *Geografiska Annaler B*, *Political Geography*, *Professional Geographer*, and *Progress in Human Geography*. In 2006, Iain was named Australian University Teacher of the Year and in 2008 was admitted as a Senior Fellow of the Higher Education Academy. He is currently Vice-President (President-Elect) of the Institute of Australian Geographers.

Richie Howitt, BA (Hons) Dip Ed (Newcastle) 1978, PhD (New South Wales) 1986, is Professor and Discipline Leader of Human Geography in the Department of Environment and Geography, Macquarie University, NSW, Australia. He is author of *Rethinking Resource Management* (Routledge, 2001) and co-editor of *Resources, Nations and Indigenous Peoples* (Oxford University Press, 1996, with J. Connell and P. Hirsch). Richie teaches and researches in areas of resource management, social impact assessment, social justice, and geographical education. He received the Australian Award for University Teaching (Social Science) in 1999 and is a Distinguished Fellow of the Institute of Australian Geographers.

Robin Kearns, BA (Auckland) 1981, MA (Hons) (Auckland) 1983, PhD (McMaster) 1988, is Professor in the School of Environment at the University of Auckland. He has published *Putting Health into Place: Landscape, Identity and Well-being* (Syracuse University Press, 1998) and *Culture/Place/Health* (Routledge, 2002) with Wilbert Gesler, as well as many refereed articles and book chapters. His current research interests include neighbourhood design and physical activity, activism in the voluntary sector, geographies of mental health, aging, and place, and the social dynamics of coastal settlements. Robin is an editor of *Health and Place* (Elsevier) and *Health and Social Care in the Community* (Blackwell). He maintains a range of involvements within the wider research community, including ministerial appointments to the National Health Committee and the Public Health Advisory Committee.

Sara Kindon, BA (Hons) (Durham) 1989, MA (Waterloo) 1993, DPhil (Waikato) in progress, is Senior Lecturer in Human Geography and Development Studies at Victoria University of Wellington, New Zealand. Sara is interested in feminist praxis, participatory research, and community development as applied to understanding place–identity relationships, particularly within indigenous and refugee-background communities. Her thesis work thinks through aspects of complicity, power, and desire in a participatory video research project with a North Island *iwi*. She also co-ordinates participatory research with refugee-background youth in Wellington. In 2007, she published *Participatory Action Research Approaches and Methods: Connecting People, Participation and Place* (Routledge, edited with Rachel Pain and Mike Kesby). She has also published in a number of journals, including *Transactions of the Institute of British Geographers* (as mrs kinpaisby, with Rachel Pain and Mike Kesby), and contributed chapters to books, including *The Handbook of Social Geography* (Sage, 2009), *The International Encyclopaedia of Human Geography* (Elsevier, 2009), *Development Fieldwork: A Practical Guide* (Sage, 2003), and *The Myth of Community: Gender Issues in Participatory Development* (Intermediate Technology Publications, 1998).

Pauline M. McGuirk, BA (Hons) (Dublin) 1986, Dip Ed (Dublin) 1987, PhD (Dublin) 1992, is Professor of Human Geography at the University of Newcastle, NSW,

Australia, where she is Director of the university's Centre for Urban and Regional Studies. Pauline's research focuses on the interaction of structure, politics, and practice in reconfiguring urban governance, urban policy, and urban space. She has published widely in leading international journals, with more than 50 refereed publications. Pauline is on the editorial boards of *Irish Geography, Geographical Research*, and *Geographical Compass*. She was awarded Fellowship of the Geographical Society of New South Wales in 2008.

Juliana Mansvelt, BA (Hons) (Massey) 1989, PhD (Sheffield) 1994, is currently employed as a Senior Lecturer in Geography in the School of People, Environment and Planning, Massey University, New Zealand. Her teaching and research interests include geographies of leisure, aging, and consumption, and she has recently conducted research for the New Zealand Ministry of Education on professional development and e-learning. Her current research involves interviews and participant observation to examine the socialities, subjectivities, and spatialities of elders' experiences of shopping for, purchasing, and using commodities. She is author of *Geographies of Consumption* (Sage, 2005) and has contributed to a number of books and journals on topics related to learning and teaching, qualitative research, aging and consumption. In 2006, Juliana was awarded a New Zealand National Tertiary Teaching Award for sustained excellence.

Janice Monk, BA (Hons) (Sydney) 1958, MA (Illinois at Urbana-Champaign) 1963, PhD (Illinois at Urbana-Champaign) 1972, is Professor in the Scool of Geography and Development, at the University of Arionza. For more than two decades, she served as executive director of the Southwest Institute for Research on Women at the University of Arizona where she was responsible for funded research and educational projects supported by an array of governmental agencies and private foundations. She was president of the Association of American Geographers (2001–2) and has been honoured by the Association of American Geographers, the Institute of Australian Geographers, the National Council for Geographical Education, and the Royal Geographical Society (with the Institute of British Geographers). Janice's interests include gender studies, the history of women in American geography, and geography in higher education. Among her numerous publications are *The Desert Is No Lady* (Yale University Press, 1987, 1989; University of Arizona Press, 1997, with V. Norwood), *Full Circles* (Routledge, 1993, with C. Katz), *Women of the European Union* (Routledge, 1996, with M.D. Garcia-Ramon), *Encompassing Gender* (Feminist Press, 2002, with M.M. Lay and D.S. Rosenfelt), and *Aspiring Academics* (Pearson Education, 2009, with Michael Solem and Kenneth Foote). Her articles appear in such journals as *Annals of the Association of American Geographers, Journal of Geography in Higher Education, Journal of Geography*, and *Professional Geographer*.

Phillip O'Neill, BA (Hons) Dip Ed (Macquarie) 1975, MA (Hons) (Macquarie) 1986, PhD (Macquarie) 1995, is Professor and Director at the Urban Research Centre,

University of Western Sydney. Phillip is editor-in-chief of *Geographical Review* and a member of the international editorial boards of the *Journal of Economic Geography*, *Progress in Human Geography*, *Human Geography*, and *Geography Compass*. He is a regular newspaper columnist and a frequent national media commentator and public speaker on economic change and regional development issues. Phillip is widely published in the fields of industrial, urban, and regional economic change and has a keen interest in the geographical behaviours of large corporations.

Eric Pawson, MA, DPhil (Oxford) 1975, is Professor of Geography at the University of Canterbury, New Zealand. He teaches at all levels in the curriculum, from large first-year classes to PhD students, while his research interests include environmental history and landscape and memory. He chaired the Advisory Board of the *New Zealand Historical Atlas* (Department of Internal Affairs, 1997, with David Bateman) and co-edited *Environmental Histories of New Zealand* (Oxford University Press, 2002). He is managing editor of the *New Zealand Geographer*, a holder of the Distinguished New Zealand Geographer Medal, and received a National Tertiary Teaching Excellence Award in 2009.

Robin Peace, DipTchg (Christchurch College of Education) 1978, BA (Canterbury) 1993, BSocSci (Hons) (Waikato) 1993, DPhil (Waikato) 1999, is an Associate Professor in the School of People, Environment and Planning at Massey University (based at the Wellington campus). After completing a PhD in human geography at the University of Waikato, she spent five years at the Centre for Social Research and Evaluation in New Zealand's Ministry of Social Development in Wellington before returning to an academic life centred on migration research, methodological interests, and post-graduate teaching focused on the development of theory and practice in the field of evaluation.

Michael Roche, MA, PhD (Canterbury) 1983, is Professor of Geography in the School of People, Environment and Planning at Massey University at Palmerston North in New Zealand. He has written on historical and contemporary aspects of forestry and agriculture in New Zealand. His publications based on archival research have appeared in *Historical Geography*, *Journal of Historical Geography*, *New Zealand Forestry*, *New Zealand Geographer*, *South Australian Geographical Journal*, the *Historical Atlas of New Zealand*, and 'Te Ara the online encyclopedia of New Zealand'. He is a life member of the New Zealand Geographical Society, a member of the New Zealand Geographic Board Ngā Pou Taunaha o Aotearoa, and one of three academic editors of the *New Zealand Geographer*.

Matthew W. Rofe, BEd (Avondale College) 1993, BA (Hons) (Newcastle) 1996, PhD (Newcastle) 2002, is Lecturer and Program Director of Postgraduate Studies at the School of Natural and Built Environments (Urban and Regional Planning) at the University of South Australia. Matthew has worked at a number of

Australian universities, including the University of Newcastle, Sydney and Adelaide. The central theme of his research agenda involves unravelling the complexity of human landscapes and the often conflicting meanings that are attached to place. Currently, Matthew is researching globalization and the emergence of international educating cities, urban revitalization projects and entrepreneurial governance, and media constructions of landscape of serial killings.

Stan Stevens, BA (California, Berkeley) 1983, MA (California, Berkeley) 1986, PhD (California, Berkeley) 1989, is Senior Lecturer in Geography in the Department of Geosciences at the University of Massachusetts, Amherst. His research in political ecology, cultural ecology, environmental history, protected areas, and tourism has been based on post-colonially informed ethnographic, collaborative fieldwork with more than 30 Sharwa (Sherpa) co-researchers. He is particularly concerned with issues of indigenous peoples' land use, community-based resource management, conservation values and practices, and struggles for sovereignty and self-determination. These projects have involved fieldwork in the Chomolungma (Mount Everest) region during 20 of the past 28 years and in more than 60 villages. He is also active in promoting indigenous rights recognition through the International Union for the Conservation of Nature and Natural Resources (IUCN). Stan is author of *Claiming the High Ground: Sherpas, Subsistence, and Environmental Change in the Highest Himalaya* (University of California Press, 1993) and editor and contributor to *Conservation through Cultural Survival: Indigenous Peoples and Protected Areas* (Island Press, 1997). He has also contributed chapters to edited books and published articles in journals such as *Journal of Cultural Geography, The Geographical Journal*, and *Geographical Review*.

Elaine Stratford, BA (Flinders) 1984, BA (Hons) (Flinders) 1986, PhD (Adelaide) 1996, is Associate Professor and Head, School of Geography and Environmental Studies, University of Tasmania. Her research centres on various theoretical and empirical place-based, political and ecological puzzles related to environmental planning, community, cultural expression, and governance in island places. Elaine teaches undergraduate human geography and graduate level environmental planning. She has published widely in leading journals such as *Transactions of the Institute of British Geogra Environment and Planning A* and *B, Health and Place*, and *Political Geography* as well as in numerous edited collections.

Bettina van Hoven, BA (Plymouth, UK) 1996, PhD (Plymouth, UK) 1999, is Lecturer and Researcher at the Department of Cultural Geography at the University of Groningen in the Netherlands. Her interest has long been focused on eastern Europe, gender issues, and identity. Recently, a four-year research grant shifted her main interest to the geographies of prison. She has published in Area and Antipode and edited the books Europe: Lives in Transition (Pearson, 2003) and Spaces of Masculinities (Routledge, 2004).

Gordon Waitt, MA (Hons Soc Sci) (Edinburgh) 1985, PhD (Edinburgh) 1988, is Associate Professor in the School of Earth and Environmental Sciences at the University of Wollongong, NSW, Australia. He has published widely on the geographies of tourism, sexuality, gender, sports, and festivals. He co-authored *Introducing Human Geography* (Longman-Pearson Education Australia, 2000) and *Gay Tourism: Culture and Context* (Howarth Publications, 2006). He has written numerous articles in various journals, including the *Annals of the Association of American Geographers*, *Annals of Tourism Research*, *Australian Geographer*, *Environment and Planning D*, *Gender, Place and Culture*, *Social and Cultural Geography*, *Transactions of the Institute of British Geographers*, and *Urban Studies*. Gordon is actively involved in the professional activities of the Geographical Society of New South Wales.

Hilary P.M. Winchester, BA (Hons) (Oxford) 1974, MA (Oxford) 1977, DPhil (Oxford) 1980, is Pro Vice Chancellor and Vice President: Organisational Strategy and Planning at the University of South Australia. She joined UniSA in 2003 from Flinders University, where she was pro vice chancellor (academic) for two years, after a decade at the University of Newcastle, NSW, during which time she held various positions including head of the Department of Geography and Environmental Science and president of the Academic Senate. Hilary has a wide range of interests in population, social, and cultural geography, specifically including marginal social groups (especially one-parent families), the social construction of place, the geography of gender, and urban social and landscape planning. She is co-author of *Landscapes: Ways of Imagining the World* (Pearson, 2003) and author of numerous articles and book chapters.

Preface

Qualitative Research Methods in Human Geography provides a succinctly written, comprehensive, and accessible guide to thinking about and practising qualitative research methods. This third edition has been revised and expanded to include one new chapter on case studies and to reflect some of the influence of new technologies on qualitative research while at the same time maintaining all of the previous editions' strengths. The book is aimed primarily at an audience of upper-level undergraduate students, but feedback on ways in which the previous editions have been used indicates that the volume is of considerable interest to people commencing honours theses and post-graduate study, and indeed, some professional geographers with many decades of research experience have let me know—quietly in some instances—that they have learned a great deal from the previous editions. Chapters have been written with the dual intent of providing novice researchers with clear ideas on how they might go about conducting their own qualitative research thoroughly and successfully and offering university academics a teaching-and-learning framework around which additional materials and exercises on research methods can be developed. The text maintains its dedication to the provision of practical guidance on methods of qualitative research in geography.

Without realizing it at the time, I started work on this book in 1992 when I was asked to teach Research Methods in Geography at Flinders University in South Australia. I developed lectures and extensive sets of notes for my classes. I also referred students to helpful texts of the time, such as Bernard (1988), Kellehear (1993), Patton (1990), Sarantakos (1993), and Sayer (1992). All of these books came from disciplines other than geography or referred to the social sciences in general. It disturbed me that despite the renewed emphasis on qualitative research and the teaching of qualitative methods in the discipline, no geographer had produced an accessible text on the day-to-day practice of qualitative research.

In the meantime, during their regular visits to my office, publishers' representatives asked what sorts of books might be useful in my teaching. I repeatedly mentioned the need for a good text that dealt with qualitative research methods in geography. That message came back to haunt me when I was asked whether I would like to write a book on qualitative methods for Oxford University Press. I declined but offered instead to try to draw together the expertise and energies of a group of active and exciting geographers from Australia and Aotearoa/New Zealand to produce an edited collection.

Of course, during the period when I planned the first edition of this book and fulfilled the terms of the publishing contract, a new group of research methods texts emerged! This group included Flowerdew and Martin (1997), Kitchin and Tate (2000), Lindsay (1997), and Robinson (1998). Despite their various merits, none of these books deals exclusively, or as comprehensively, with qualitative methods as the first edition of this volume. While other excellent volumes on research methods in geography have now emerged (e.g., Clifford and Valentine 2009; DeLyser et al. 2009; Moss 2002), *Qualitative Research Methods in Human Geography* continues to be received very well, not only in Australia and Aotearoa/ New Zealand but also much farther afield. Indeed, the book's growing popularity in North America has led to its shift from Oxford University Press, Australia, to Oxford University Press, Canada.

This third edition is revised and expanded to respond to constructive comments made about its predecessors. For instance, across the book the well-received emphasis on reflexivity and ethical practice has been maintained, and several chapters have been modified substantially in recognition of the ways in which the Internet and other electronic technologies are affecting the conduct of qualitative research.

In overview, *Qualitative Research Methods in Human Geography* is subdivided loosely into three main sections: 'introducing' qualitative research in human geography, 'doing' qualitative research, and 'interpreting and communicating' qualitative research. Issues considered within these broad sections are also raised, where appropriate, within chapters: matters of ethical practice, data reduction, and communicating research results are among the more prominent. The third edition includes contributions to Chapter 1 by South Australian geographer Matthew Rofe and one new chapter by Canadian scholar Jamie Baxter on the place of case studies in qualitative research. The addition of Jamie to the team of contributors continues a shift made in the second edition to include more authors from beyond Australia and Aotearoa/New Zealand. Despite this shift, the book maintains some emphasis on examples drawn from the Asia–Pacific region. Notwithstanding that regional focus, the examples should still inform the learning, teaching, and research work of 'northern' readers. Indeed, it seems from the success of the first and second editions that just as teachers in Australia, Aotearoa/New Zealand, Southeast Asia, and the Pacific elaborate on British and North American texts by referring to examples from their local region, colleagues in 'our' antipodes have been willing to supplement this volume's examples with local illustrations of their own.

As I note above, I have subdivided the book into three sections: 'introducing', 'doing', and 'interpreting and communicating'. To some extent, it may be imprudent to impose such separations. I acknowledge, for instance, the ways in which writing—which might be seen superficially as a part of communication—is embedded in the process of 'doing' qualitative research in human geography. Similarly, 'doing' qualitative research typically implies ongoing interpretation and reinterpretation. The ordering of the book's contents is the product of the medium in which the contents are communicated (a book of linear structure). It is a device to help make more quickly comprehensible a book of 18 chapters. And it follows the organizing sequence used in many research methods classes—such as the one that gave rise to this book—for which this might serve as a textbook.

The first section of the book deals with 'introducing' qualitative research methods in human geography. In Chapter 1, Hilary Winchester and Matthew Rofe situate qualitative research methods within the context of geographic inquiry. They outline the range of qualitative techniques commonly used in human geography and explore the relationship between those methods and the recent history of geographic thought. On this foundation, Robyn Dowling in Chapter 2 builds a review of some critical issues associated with qualitative research. These issues include power relationships between researchers and their co-researchers, questions of subjectivity and intersubjectivity, and points of ethical regulation and concern. Similar issues are taken up in further detail by Richie Howitt and Stan Stevens who—drawing from their considerable experience working with indigenous communities in Australia and in Nepal respectively—present in Chapter 3 an illuminating and engaging conversation on ethics, methods, and relationships in cross-cultural qualitative research. Chapter 4 by Matt Bradshaw and Elaine Stratford examines the difficult but vitally important matters of design and rigour in qualitative research. There follows a strong addition to this volume, a new chapter that examines in more detail some of the methodological issues raised in Chapter 4 and explores research design issues associated with case studies in detail. In this new chapter, Jamie Baxter sets out the value of case study methodology for understanding specific situations as well as for developing theory.

The second section of the book focuses on the 'conduct' of qualitative research methods in human geography. Each of the eight chapters offers concise yet comprehensive, similarly structured outlines of good practice in some of the main forms of qualitative research practice employed in geography. I have grouped together three chapters on research that focus on spoken testimony, three that emphasize 'texts', or inscription, and two that emphasize worlds of observation and participation.

First, Kevin Dunn draws on his extensive experience of interviewing in cross-cultural settings to provide a complete and valuable outline of interviewing practice. Acknowledging some of the effects of the Internet on qualitative research practice, this long but rich chapter includes substantial new material on computer-mediated communications (CMC) interviewing. Karen George and Elaine Stratford then set out the basic scope of oral history, discussing the ways in which its practice differs from interviewing and describing how oral history can enhance our understandings of space, place, region, landscape, and environment. In Chapter 8, Jenny Cameron considers the research potential of focus groups in geography, outlines key issues to take into account when planning and conducting successful focus groups, and provides an overview of strategies for analyzing and presenting the results. New to this edition is material on on-line focus groups.

The book then turns to three chapters on 'texts'. They deal, in turn, with archival research, the use of questionnaires in qualitative human geography, and discourse analysis. Historical geographer Mike Roche provides a helpful chapter on the use of archival resources, a sometimes neglected yet enormously productive and rewarding activity for human geographers. In Chapter 10, Pauline McGuirk and Phillip O'Neill explore the ways in which questionnaire surveys can be used in qualitative research in human geography. This chapter has also been revised substantially to take greater account of on-line techniques. It may be helpful to read this chapter in conjunction with the chapter on research design and rigour by Bradshaw and Stratford. In the final chapter of this section, Gordon Waitt has prepared a careful review of the conduct of discourse analysis. In this chapter, which was completely rewritten for the third edition, Gordon suggests how a Foucauldian approach to understanding 'texts' might be applied. Although it may be more complex than some of the other chapters in this book, I strongly encourage you to persist with it. The rewards will justify your efforts.

Two chapters deal directly with issues of observation and participation. In Chapter 12, Robin Kearns reviews the purposes and practice of observational techniques in human geography, giving new attention to photographic and video observation. Then, taking up some of the issues about participant status for researchers that Robin has introduced, Sara Kindon offers a fine chapter on participatory action research. There is strong complementarity between Sara's chapter and that of Richie Howitt and Stan Stevens on cross-cultural research, and it would be helpful to consider them in conjunction with each other.

Five chapters in a section on 'interpreting and communicating' follow. In the first, Meghan Cope sets out a succinct and very helpful discussion of coding in qualitative research. She introduces new material to this edition on coding and mapping. It is worth reading Meghan's chapter together with that by Gordon Waitt on discourse analysis. The following chapter (15), by Robin Peace and Dutch geographer Bettina van Hoven, discusses the relationships between computers, qualitative data, and geographic research. This chapter has been revised substantially for this edition to reflect the extended ways in which computers are being used to help make sense of qualitative data.

The book then turns from explicit discussions of interpretation to matters of communication, but with Juliana Mansvelt and Lawrence Berg, who explore this matter in detail in Chapter 17, I would again like to emphasize the dialogic, relational nature of the connection between communicating and research. In Chapter 16, Jan Monk and Dick Bedford draw from their vast range of research and research management experience to discuss compelling ways of proposing qualitative research projects to others. Juliana Mansvelt and Lawrence Berg provide the penultimate chapter of the book, discussing some of the practical and conceptual issues surrounding representation of qualitative research findings. As I note above, they make the point that writing not only reflects our findings but constitutes how and what we know about that work. Finally, Dydia DeLyser and Eric Pawson offer help to qualitative researchers beginning with the vital task of communicating results for public consumption.

This book has chapters of varying lengths, 'densities', and complexity. While some of the chapters are challenging to read, I would like to think that the challenges lie in comprehending the ideas presented, not in wading through obfuscatory text. In every case, authors have striven to present material that is as clear and well-illustrated as possible. As an editor with a career interest in effective communication, I have devoted considerable attention to the matter of clarity.

While the chapters of this volume offer a comprehensive overview of qualitative research methods in human geography, the book is not intended as a 'one-stop' resource or prescriptive outline (indeed, how could it be?) for qualitative researchers and students. Instead, it is meant as a starting point and framework. Accordingly, each chapter directs readers to a number of useful resources that may be consulted to follow up the material introduced here. There is, of course, a consolidated list of all references cited for readers who might wish to inquire even further. Chapters also include review questions intended for classroom discussion or as individual exercises. These questions might also serve as prompts for ideas for quite different exercises. Indeed, if you have discovered or created any useful exercises to illustrate some of the matters covered within these chapters, I would be delighted to hear about them.

The book features an extensive glossary. While the individual authors have tried to ensure that their chapters are written in a language that is accessible to undergraduate readers, there are—without doubt—terms that may be somewhat alien to many readers. The glossary should help resolve that sort of difficulty. Many of the terms in the glossary are drawn from the lists of key words associated with each chapter.

I owe thanks to many people who have been involved in the production of this book. Students in research methods classes at Flinders University encouraged me to put the book together and offered honest and helpful comments on many of its chapters. Each chapter was reviewed by at least two experts in the field who provided timely, critical, and comprehensive opinion. In alphabetical order, with their affiliation at the time of their review, these reviewers were: Stuart Aitken (San Diego State), Kay Anderson (Durham), Nicola Ansell (Brunel), Andrew Beer (Flinders), Alison Blunt (Queen Mary, London), Mark Brayshay (Plymouth), Michael Brown (Washington), Jacquie Burgess (University College, London), Jenny Cameron (Griffith), Garth Cant (Canterbury), Bev Clarke (Flinders), Mike Crang (Durham), Jon Goss (Hawai'i), Penny Gurstein (British Columbia), Ellen Hansen (Emporia State), Andy Herod (Georgia), Richie Howitt (Macquarie), Mark Israel (Flinders), Jane Jacobs (Melbourne), Lucy Jarosz (Washington), Ron Johnston (Bristol), Minelle Mahtani (New School), Murray McCaskill (Flinders), Pauline McGuirk (Newcastle), Eric Pawson (Canterbury), Meryl Pearce (Flinders), Chris Philo (Glasgow), Joe Powell (Monash), Lydia Mihelic Pulsipher (Tennessee), Lyn Richards (La Trobe), Noel Richards (Flinders), Mark Riley (Exeter), Mark Rosenberg (Queen's, Ontario), Pamela Shurmer-Smith (Portsmouth), Robert Summerby-Murray (Mount Allison), Lynn Staeheli (Colorado), Sarah Turner (McGill), and Eileen Willis (Flinders).

I must also express a special note of thanks to the contributors to this volume. As with previous editions, authors have to put up with repeated e-mailed requests and comments from me, often as long as the chapters themselves! I remember sending out those messages, uttering as I clicked the 'send' button a quiet prayer to this effect: 'please, please don't let them spit the dummy and pull out of this project!' I admire the authors' tolerance, persistence, and fortitude in the face of both my editing style and any personal and professional adversity they may have encountered while working on the project. I thank each of them for their efforts, their patience, and their continuing dedication to this collection.

Iain Hay
Adelaide

Part I

'Introducing' Qualitative Research in Human Geography

Qualitative Research and Its Place in Human Geography

Hilary P.M. Winchester and Matthew W. Rofe

CHAPTER OVERVIEW

This chapter aims first to provide an overview of contemporary qualitative research methods in human geography. The range of methods commonly used in human geography is considered and categorized, together with some of the ways in which those methods are used to provide explanation. Second, the chapter aims to review briefly the context from which qualitative research has developed in human geography. This is achieved by examining changing schools of thought within the discipline, recognizing that changes are messy, overlapping, and coexisting. Third, the chapter aims to link methodological debates to wider theoretical perspectives in geography.

INTRODUCTION

Contemporary human geographers study places, people, bodies, **discourses**, silenced voices, and fragmented **landscapes**. The research questions of today's human geographers require a multiplicity of conceptual approaches and methods of inquiry. Increasingly, the research methods used are qualitative ones intended to elucidate human environments, individual experiences, and social processes. This introductory chapter aims to set the scene of qualitative research in human geography and to highlight issues and techniques that are examined in more detail in later chapters. The chapter has three main and interlinked objectives.

First, it provides an overview of qualitative methods, their context, and the links between methodology and theory. The categories used and established here are relatively fluid; they are designed merely as a way of organizing this growing field and are not meant to be fixed or constraining. Indeed, some recent research, such as that on the body and embodied experiences in place, essentially defies categorization. A number of issues raised in this introduction are more fully developed later in the volume, particularly those that relate to ethical practices (Chapters 2 and 3), the positioning of the author relative to the audience (Chapters 2 and 17), as well as the broad issues of **credibility, dependability**, and **confirmability** (Chapters 4 and 5).

Second, the chapter outlines the arguments about, and differences (and some of the similarities) between, qualitative and **quantitative methods**. Further, it reviews the current debate on combining methods through **triangulation** and **mixed-method** approaches. The resurgence of qualitative techniques and the current qualitative/quantitative debate in human geography are set in the context of the discipline's development and evolution. Inevitably, a thumbnail sketch of the evolution of geographical thought in part of just one chapter will necessitate some broad generalizations. However, in this context it is important to recognize that controversy about the nature and validity of research methods has existed for decades. Methodological debates may entrench polarized positions—for example between quantitative and qualitative, **objective** and **subjective** (see, for example, Johnston 2000). The apparent polarity between those positions may in fact prove to be largely a false dichotomy. In the past, such debates in geography have raged around other extremes (Wrigley 1970). Geographers in the early decades of the twentieth century argued, for example, over the merits of determinism versus possibilism (the extent to which humans have control over or are controlled by their environment) (Wrigley 1970). While qualitative methods are well entrenched within human geography, the occasional airing of the qualitative versus quantitative debate may in one sense be likened to the resurrection of a dinosaur in the shape of a false dichotomy, while in another sense it is a sign of healthy debate and intellectual vigour within the discipline. Of greater import is the recent emergence of concern within human geography that qualitative methodologies have come to rely too heavily on a narrow range of specific methods and hence limit the very flexibility that qualitative approaches champion (see Davies and Dwyer 2007 and Thrift 2000). As we will discuss, this need not be the case, because a range of new, experimental techniques are emerging that continue to enrich the qualitative approach.

Finally, the chapter focuses particularly on links between theory and methodology and raises issues of ethics, authorship, and power. The contemporary use of qualitative methods in human geography is positioned within the theoretical debates and intellectual evolution of the discipline, while raising issues for consideration by reflexive and committed researchers. Intense arguments about methodology are often as much to do with researchers' beliefs and feelings about the structure of the world as about their regard for a particular research method, such as **participant observation** or in-depth **interviews**. While creationists and catastrophists clearly subscribe to a particular world view (belief in the biblical account of Earth's creation and catastrophic geomorphological events such as the Great Flood), other geographers may hold equally strong but perhaps less obvious views about the order, structure, measurement, and knowability of phenomena. In complex ways, **ontology** (beliefs about the world) and **epistemology** (ways of knowing the world) are linked to the methods we choose to use for research (Grbich 2007; Hesse-Biber and Leavy 2004; Sayer and Morgan 1985).

What Is Qualitative Research?

What Questions Does Qualitative Research Answer?

Qualitative research is used in many areas of human geography. In a broad sense, qualitative research is concerned with elucidating human environments and human experiences within a variety of conceptual frameworks. The term 'research' is used here to mean the whole process from defining a question to analysis and interpretation. 'Method' is used as a much more specific term for the investigative technique employed. A huge range of methods is used in many different situations. Some of the variety of methods that spring to mind range from interviews about disability rights in the Australian state of Victoria (Smith 2003), through participant observation with homeless children (Winchester and Costello 1995), to **deconstruction** of media events and textual material involved in the place-making of former industrial cities such as Newcastle (M℃Guirk and Rowe 2001) or Port Adelaide (Szili and Rofe 2007). Inevitably, it is difficult to summarize the questions addressed by such a variety of research. However, it is instructive to recall the answer to a similar question posed three decades ago in relation to statistical analysis in geography. In that text, Ron Johnston (1978, 1–5) argued that the gamut of statistical techniques answered two fundamental questions. Those questions were either about the relationships between phenomena and places or the differences between them. The elegant simplicity of Johnston's questions can be paralleled by a different but similarly broad pair of questions that qualitative research is trying to answer.

The two fundamental questions tackled by qualitative researchers are concerned either with **social structures** or with individual experiences. This dualism may be hard to disentangle in practice, but it is of fundamental importance in explanation. The behaviour and experiences of an individual may be determined not so much by their personal characteristics but by their position in the social structure, 'together with their associated resources, constraints or rules' (Sayer 1992, 93).

The first question may be phrased as:

> **Question One:** *What is the shape of societal structures, and by what processes are they constructed, maintained, legitimized, and resisted?*

The structures that geographers are analyzing may be social, cultural, economic, political, or environmental. Structures may be defined as internally related objects or practices. Andrew Sayer (1992, 92–5) gives the example of the landlord–tenant relationship in which structures exist in relation to private property and owner- ship, rent is paid between the two parties, and the structure may survive a continual turnover of individuals. Further, he emphasizes that tenants almost certainly exist within other structures; for example, they may be students affected by educational structures or migrants constrained by racist structures. The coexistence of rented housing, students, and minority groups produces a complex linkage and mutual reinforcement of structures within which individuals live out their lives. Qualita- tive geographers balance the fine line between the examination of structures and processes on the one hand and of individuals and their experiences on the other. Structures constrain individuals and enable certain behaviours, but in some cir- cumstances individuals also have the capacity to break rather than reproduce the mould. An overemphasis on structures and processes rather than on individuals could lead to a dehumanized human geography. On the other hand, individuals do not have all-powerful free will and ability, which would enable them to overcome the powerful structures embedded in society, such as capitalism, patriarchy, or racism.

Sayer (1992, 95) considers that the key question for qualitative researchers about structures may be phrased as: 'What is it about the structures which produce the effects at issue?' Geographers have studied structures qualitatively in a number of ways. A significant focus has been on the ways in which they are built, reproduced, and reified: for example, Kay Anderson (1993) has analyzed the documentary hist- ory that has led to stigmatization of the suburb of Redfern in Sydney as an Austral- ian 'ghetto'. More recently, Wendy Shaw (2000; 2007) has examined the physical and ideological context of Redfern's stigmatization as a manifestation of the dis- proportionate politics of race that disempowers indigenous Australians. In this example, the structures are essentially indistinguishable from the processes that build, reinforce, and contest them. However, at the level of analysis, both authors have chosen to emphasize one particular scale of social agency for the purpose of framing, understanding, and representing the processes evident in Redfern. Simi- larly, other authors have considered either the material or symbolic **representations** of structures: in their 1994 analysis of British merchant banking, McDowell and Court emphasize media representations of banking patriarchs and the importance of dress and body image for younger female and male bankers. A further aspect of the investigation of structures may be concerned with their oppressive or exclu- sionary nature: Gill Valentine's (1993) interviews with lesbian workers considered the ways in which workplace structures naturalize heterosexual norms and thereby contribute to the oppression and marginalization of workers who do not conform to these norms. More recently, her work on young people 'coming out' is placed within the structures of the family (Valentine, Skelton, and Butler 2003). Most of the qualitative geographical work on structures in fact emphasizes the processes and relations that sustain, modify, or oppose those structures rather than focusing specifically on their form and nature.

The second question is concerned with individual experiences of structures and places:

Question Two: *What are individuals' experiences of places and events?*

Individuals experience the same events and places differently. Giving voice to individuals allows viewpoints to be heard that otherwise might be silenced or excluded: Jenny Pickerill's (2009) examination of the complex and at times contested negotiations between indigenous Australians and environmental groups gives voice to Aboriginal perspectives on naturalized notions of land and country (see also Johnson and Murton 2007). For Pickerill (2009, 66) these negotiations 'provide clues as to how commonalities across difference . . . [can be] built'. The upsurge of academic research into indigenous knowledge and politics is a global phenomenon creating what Johnson et al. (2007, 118) refer to as 'anti-colonial geographies'. At the heart of this project is the opening up of geography to previously unheard voices typically positioned as curios on the margin. However, as Wilson and Peters (2005) reveal in their examination of the spatial identities of First Nations people in Canada, indigenous populations are increasingly urban. Indigenous peoples' existence within and identification with the urban 'core' of the nation-state disrupts naturalized understandings of indigeneity, thereby blurring the boundaries between the core and the margin. For Native Hawai'ian academic Renee Pualani Louis (2007, 137), decolonizing indigenous voices has heralded the emergence of indigenous-specific and/or -sensitive methodologies that can help 'invigorate and stimulate geographical theories and scholarship while strengthening Indigenous peoples' identities and supporting their efforts to achieve intellectual self-determination'. Louis's (2007, 130) published challenge, 'can you hear us now?', is a contemporary answer to earlier calls to give 'voice' to silenced groups (McDowell 1992a), which in itself reflects a wider political project answering David Smith's (1977, 370) challenge to develop a 'new paradigm for geography . . . [that being] a . . . geography about real people'.

Dissident or marginalized stories may be 'given voice' through diaries, oral histories, recordings of interviews and conversations, or the use of 'alternative' rather than 'mainstream' media. Participant observation by immersion in particular settings allows multiple viewpoints to be heard and acknowledged. A study of the post-school celebrations of Schoolies Week on the Gold Coast of Australia (Winchester, McGuirk, and Dunn 1999) gives voice not only to the partying Schoolies (i.e., high school students celebrating the end of the academic year) but also to agents of control (for example, police). Women's activist groups in urban and rural areas of Australia used public and private spaces differently and strategically to give voice and space to their concerns (Fincher and Panelli 2001). The experiences of individuals and the meanings of events and places cannot necessarily be generalized, but they do constitute part of a multifaceted and fluid reality. Qualitative geographical research tends to emphasize multiple meanings and interpretations rather than seeking to impose any one 'dominant' or 'correct' interpretation.

The experiences of a single individual have been used in a generalizable sense to illuminate structures and structural change (a matter that is taken up in detail in Chapter 5 of this volume). For example, Rimmer and Davenport (1998) use the travel diaries of Australian geographer Peter Scott as an example of the huge changes in mobility and technology that have characterized air transport since the 1950s. And an autobiographical account by Reginald Golledge (1997) tells his personal story to outline the difficulties experienced by geographers and other academics faced with major physical disabilities. The power of such works is demonstrated by the emergence of disability studies as a sub-field of human geography (see Imrie and Edwards 2007) and wider pedagogic research on the development of disability-sensitive learning environments within geography and beyond (see, for example, Golledge 2005 and Treby, Hewitt, and Shah 2006). In some cases, the boundary between structure and individual experience is blurred: in a study of the daily geographies of caregivers, Wiles (2003, 1307) found that 'the social and physical aspects of the many interconnected scales and places which caregivers negotiate on an everyday basis both shape and are shaped by caregiving.' In other words, the experience itself could not be analytically separated from the structures that form the context for that experience.

Types of Qualitative Research

It is clear, even from the brief preceding section, that qualitative research in geography is now used to address a huge range of issues, events, and places and that these studies utilize a variety of methods. Indeed, it can now be asserted that 'there is . . . a maturity about qualitative methods in geography' (Crang 2005b, 225). While this 'maturity' can be said to reflect a rigorous and well-laid epistemological foundation, Crang (2005, 225) cautions that with it has also come 'a certain conventionality of approaches'. While a number of new, so-called 'experimental' techniques are emerging, (such as **discontinuous writing**, **photo-elicitation**, and **go along interviews**), the weight of qualitative analysis rests on a limited range of conventional techniques. Reflecting this view, Davies and Dwyer (2007, 257) assert that oral techniques (e.g., interviews, focus groups), textual analysis, and associated observationally based ethnographies embody a 'suite' of methods that 'remains the backbone of qualitative research in human geography'. In this vein, this section discusses the three main types of qualitative research employed in human geography: the oral (primarily interview-based), the textual (creative, documentary, and landscape), and the observational. It also considers a number of experimental techniques that are currently being developed within human geography.

Clearly, the most popular and widely used methods are oral. Talking with people as research subjects encompasses a wide range of activities. The spoken testimony of people is used in ways that range from the highly individualistic (oral histories and autobiographies) to the highly generalized (the individual as one of a **random sample**). The latter type of survey technique borders on the quantitative, in which responses can be counted, cross-tabulated, and analyzed statistically (for example, Oakley 1981). The former approach, often achieved through **oral history** methods, lies at the more qualitative, individualistic end of the spectrum. Such methods are considered in more detail in Chapters 6 and 7. A middle ground is occupied by the increasingly popular technique of using **focus groups** (see Chapter 8). Jackson and Holbrook (1995) effectively used focus groups differentiated by age, gender, and social group to analyze the complex meanings of the 'everyday' activity of shopping. The range of ways in which **oral methods** are utilized in geography—allowing subjects to speak in their own voice—is outlined in Box 1.1. It should be noted that the research questions will to some extent shape the methods that will be used. In particular, the methods range from answering the research question about individual meanings and experiences at the biographical end of the spectrum to answering the research question about societal structures at the survey end. Surveys are undertaken to obtain information from and about individuals that is not available from other sources. While an interview is undertaken with an individual, a survey involves a more standardized interaction with a number of people. Oral surveys of personal information, attitudes, and behaviour usually (but not inevitably) utilize questionnaires. Questionnaires are more closely structured and ordered than interviews, and every **respondent** answers the same question in a standard format. They are discussed fully in Chapter 10.

Oral Qualitative Methods in Human Geography		Box 1.1
General Method	*Specific method*	*Research Questions*
Biography	Autobiography	Individual
	Biography	
	Oral history	↑
Interviews	Unstructured	
	Semi-structured	
	Structured	
	Focus groups—open-ended	
Surveys	Surveys—structured	↓
	Questionnaires—structured	General/Structural

The second major type of qualitative research is **textual analysis.** Such texts are wide-ranging but more diffuse in the human geography literature than the oral testimonies described above. Important groups of textual methods utilize creative, documentary, and landscape sources. Creative texts are likely to include poems, fiction, films, art, and music. Documentary sources may include maps, newspapers, planning documents, and even postage stamps! Rose (2003) uses family photographs in a study of domestic space, while Waitt and Head (2002) examine the role of postcards in Australian frontier myths. Landscape sources may be very specific, such as the micro landscapes of the retailing street (Bridge and Dowling 2001) or the 'cemeteries and columbaria, memorials and mausoleums' that Lily Kong (1999) uses in her analysis of deathscapes. Frequently, landscape sources are more general, such as the landscapes of suburbia as indicators of social status (Duncan 1992). The analysis of creative sources, including fictional literature, film, art, and music, has shown increasing complexity in recent years (for example, see Lukinbeal 2004 on cinematic geographies, Pinkerton and Dodds 2009 on radio and acoustic spaces, and Butler 2007 on the ethics of photographing torture and suffering; see also Chapter 11 of this volume). Geographers have searched such sources for underlying structures, looking at paintings, for example, to understand changing perceptions of landscape (Heathcote 1975, 214–17; Lowenthal and Prince 1965) or using film to examine both the impact of city restructuring and the ways in which it is represented (see Winchester and Dunn 1999 on *Falling Down*). In the past decade, a plethora of methodological monographs has been produced, exploring and explaining the epistemologies and practices of textual analysis. Within human geography, Rose's (2001) *Visual Methodologies*, Winchester, Kong, and Dunn's (2003) *Landscapes: Ways of Seeing the World*, and Wylie's (2007) *Landscape* are prime examples. Each of these texts reveals the fecund data available through deconstructive analysis.

Written texts have also been used as the source of underlying discourses that underpin and legitimate social structures. Analysis of media representations demonstrates their myth-making power, whether related to myths of the inner city (Burgess and Wood 1988) or to the imagery of national identity (Pickles 2002) and both urban and rural place-making (for urban examples, see Dunn, McGuirk, and Winchester 1995 and Schollmann, Perkins, and Moore 2000; for rural examples, see Rofe and Winchester 2007 and Winchester and Rofe 2005). Herman (1999) has argued that changing placenames in Hawai'i reveal a transformation from Hawai'ian political and cultural economy into Western capitalist forms, while Sparke (1998) shows convincingly some of the ways in which *The Historical Atlas of Canada* is enmeshed in the post-colonial politics of that country.

A significant and controversial source of textual analysis is the landscape itself. Landscape here is understood as the entwining of social meaning with, and its expression through, the physical environment. This entwining is crucial to the creation of *place* as a definable and knowable geographic location. Thus, landscape is a specific and highly contextual 'way of seeing' (Winchester et al. 2003). As a social construction, physical landscapes can be 'read' to reveal their social meanings and intentions. Reading a landscape is a challenging process, but it offers important insights into the history of places and struggles over them. Lewis (1979, 12) astutely argued 30 years ago that the 'human landscape is our unwitting autobiography . . . [where] our cultural warts and blemishes are . . . exhibited for anybody who wants to find them and knows how to look for them.' The argument that landscape may be read as text is epitomized in the work of Duncan and Duncan (1988) in which the residential landscape is decoded of its social nuances. A study of roadside memorials in the Australian state of New South Wales concluded that the roadside crosses and flowers were indicative not only of individuals' behaviour but also of a 'problematic masculinity' characterized by aggression, fast driving, and reckless behaviour (Hartig and Dunn 1998). Textual analyses of particular landscapes such as model housing estates use techniques derived from **semiotics** (the language of **signs**) to demonstrate literally the inbuilt naturalization of social roles according to gender and family status (Mee 1994). From a feminist perspective, Massey (1994) demonstrates how planning decision-making processes behind seemingly innocuous land-use decisions such as the provision of sporting facilities reveals 'systems of gender relations' (Massey 1994, 189) that are inequitable. Rofe (2007), after Massey (1994), has problematized the belief that the revitalization of derelict industrial areas heralds the emergence of a more progressive and equitable city, concluding that inequitable gender relations remain implicitly encoded within the post-industrial landscape. Schein (1997) interprets landscape architecture, insurance mapping, and other elements of a 'discourse materialised' to explore the ways they symbolize and constitute particular cultural ideals. Waitt and McGuirk (1996) examine both documentary and landscape texts to explore the selective representation of the heritage site of Millers Point in Sydney; the choice of particular buildings as 'heritage' both reflects and reproduces a white, male colonizer's view of Sydney's history while silencing other views and voices. Bishop (2002) uses the Alice Springs to Darwin railway as a corridor of 'difference, struggle and reconciliation' in the redefinition of national identity and its relationship with the land.

The third significant type of qualitative research in human geography consists of forms of participation within the event or environment that is being researched (see Chapters 12 and 13). The most common form of qualitative geographical research involving participation is participant observation. Within participant observation, there may be a wide variation in the role of the observer from passive to proactive (Hammersley and Atkinson 1983, 93). All forms of observation involve problems of how the author should be positioned in relation to the subject of the research (for example, Smith 1988). In particular, very active participation may clearly influence the event that is being researched, while researchers who are personally involved (for example, researching the community in which they grew up) may find it hard to wear their 'community' and 'researcher' hats at the same time. We have firsthand experience of these issues from our research into the scripting and performance of a hyper-masculine motorcycling identity (Rofe and Winchester 2003). Rofe was a long-standing member of the subculture under study. While his background afforded detailed insight into and access to a community notoriously suspicious of 'outsider' scrutiny, it had to be balanced against the potential for bias in the interpretation and reporting of results. To avoid these problems, we employed a creative tension within the research and writing process whereby understanding and explanation were not just determined between us and the research participants but also negotiated between ourselves as researchers. Clearly, issues surrounding the position of the researcher or his or her **reflexivity** are critical in participant observation. Despite such tensions, participant observation allows the researcher to be—at least in part—simultaneously 'outsider' and 'insider', although differences in social status and background are hard to overcome (Moss 1995). The positioning of the researcher in relation to the 'researched' raises some significant ethical issues, especially if the research is covert (Evans 1988, 207–8). It can, however, have some important advantages. For example, a student doing research at a fast-food outlet might be given paid work there (see Cook 1997 for an excellent account of this and other student projects). In-depth participant observation is essentially indistinguishable from ethnographic approaches, which often involve lengthy fieldwork. That fieldwork can enable meaningful relationships to develop with the research subjects and may facilitate deep understanding of the research context (Cooper 1994, 1995; Eyles 1988, 3). Cook's conclusion (1997, 127) that participant observation is the means or method by which ethnographic research is undertaken is useful in this context.

Beyond the backbone of the methods outlined above, a range of experimental techniques have emerged in recent years. These techniques stem from the backbone methods but, reflecting the evolving nature of inquiry in human geography, constitute new tools to address emerging questions. Rather than representing radically new methods, these experimental techniques embody adaptations of existing ones aimed at making them 'dance a little' (Latham 2003; 2000). Such techniques include discontinuous writing (Meth 2003), photo-elicitation (Harper 2002; 2003), **diary photographs** and **diary interviews** (Latham 2003), and go along interviews (Kusenbach 2002). Latham's (2003) use of diary photographs and diary interviews is most instructive in terms of the rich insights available into seemingly ordinary experiences and their geographies offered by methodological experimentation. Observing that '[e]veryday life and everyday culture are two of the great frontiers of contemporary human geography', Latham (2003, 1996) encouraged his respondents to keep personalized journals recording their social–geographic activities. The resulting journals represented deeply personalized and flexible 'time-space collages' that according to Latham (2003, 2009) 'can . . . enliven our sense of what human geography should look like'. Even more ephemeral aspects of human existence are pursued through other, newly emerging techniques such as observant listening and participant sensing (Wood and Smith 2004). Davies and Dwyer (2007) position the development of these techniques as stemming from human geographers, increasing interest in the geographies of performance and emotion. For Wood and Smith (2004, 534), the 'challenge of understanding emotions has . . . fuelled a search for new ways of knowing: ways that range beyond the visual and representational traditions which have for so long dominated social thought.' In these pursuits, new experimental techniques offer new insights into the lived reality of the social world, pioneering what Wood, Duffy, and Smith (2007) refer to as 'unspeakable geographies' (see also Davidson, Bondi, and Smith 2005). The unspeakable nature of such geographies relates to the inherent difficulties of capturing emotion, which by its nature is elusive, and the problems associated with its conversion into research data and representation as research findings. This echoes Law's (2004, 3) belief that human agency is 'emotive and embodied, rather than cognitive', and therefore fluid. Musing on the challenges facing qualitative researchers, Davies and Dwyer (2007, 258) wonder whether 'the world is so textured as to exceed our capacity to understand it', let alone represent it. In the face of this question, new experimental techniques such as those briefly identified above are essential.

The Contribution of Qualitative Techniques to Explanation in Geography

In this chapter, we have discussed two fundamental questions of geographic inquiry, those concerned with individuals and those concerned with social structures. We have also indicated three main groups of methods: the oral, the textual, and the participatory. There is no simple relationship between the method used and the research questions posed. It is tempting to say that oral methods may be directed predominantly towards elucidating the experiences of individuals and their meanings; however, this is overly simplistic. People's own words do tell us a great deal about their experiences and attitudes, but they may also reveal key underlying social structures. In work on lone fathers, Winchester (1999) found that the in-depth interviews illuminated underlying structures of patriarchy and masculinity in ways that were much more profound than anticipated. Depths of individual anger and despair reflected mismatches between those individuals' romanticized expectations of marriage and gendered behaviour and their actual experience of married life. In this sense, the oral method chosen elucidated both individual experiences and social structures in the holistic sense that would most frequently be associated with participant observation.

Similarly, it might appear that textual methods would most commonly be employed to throw light on the social processes that underpin, legitimate, and resist social structures. This generalization would probably be more widely accepted than any equation of oral methods with research questions that focus on the individual. Textual methods have indeed been used to analyze some of the many social processes studied by contemporary human geographers. Examples that spring to mind include the discursive construction of place (Dunn et al. 1995; Mee 1994; Szili and Rofe 2007), processes of social exclusion (Duncan and Duncan 1988), marginalization (Hay, Hughes, and Tutton 2004), and expressions of 'problematic' masculinity (Hartig and Dunn 1998; Rofe and Winchester 2003).

A fruitful area of study in human geography focuses on the body and on our embodied experiences. New Zealand geographer Robyn Longhurst's research into embodiment, its construction and intersections with issues of gender, space, and power, is an excellent example of work in this field. From her original studies into the experiences of pregnant women in shopping malls (see Longhurst 1995; 2001) to her more recent work on obesity (see Longhurst 2005), these works reveal how the embodied experiences of individuals (of feeling marginalized, of needing more toilets, of being uncomfortable in public places) are indicative of the way certain bodies may be socially constructed as **Other** (i.e., oppositional to or outside the mainstream). Once designated as Other, such bodies come to be confined to particular places and roles, medicalized, and marginalized. Resistance to such marginalization is manifested in a number of ways, documented notably in Longhurst's (2000) study of 'bikini babes', pregnant women who participated in and thereby destabilized a beauty contest for bikini-clad women. Despite such resistance, other studies affirm the power of social constructions of beauty and appropriate behaviours and how they shape and in some instances hinder the performative opportunities of individuals. Evans's (2006) recent work confirms this, revealing the complex entwining of societal pressures, personal body image, and declining sports participation rates among young women in the UK. The study of the body may also be as a text or as a landscape that may be marked or shaped in particular ways either as a form of identity (McDowell and Court [1994] identified bodily performances of male and female merchant bankers, while McDowell [2005] investigated differing performances of masculinity according to class) or as a form of resistance (Bell et al. [1994] commented on 'lipstick lesbians' as practising resistance to heterosexual norms). The study of the body as text, as performance, or as social construction illuminates some of the richness of methods that cannot easily be pigeon-holed into the types of qualitative method and types of geographical explanation identified for convenience earlier in the chapter.

THE RELATIONSHIP BETWEEN QUALITATIVE AND QUANTITATIVE GEOGRAPHY

During the late twentieth century, the pendulum of geographical methods within human geography swung firmly from quantitative to qualitative methods. With the two mythological camps generally characterized as being in opposition at best or at worst as conflicting methodologies, qualitative methods have been in the ascendant since the 1980s. This trend has been chronicled in and stimulated by recent books and collections on qualitative methods and mixed methods (Brannen 1992b; Denzin and Lincoln 2008; Eyles and Smith 1988; Flowerdew and Martin 1997; Grbich 2007; Hesse-Biber and Leavy 2004; Holland, Pawson, and Shatford 1991; Limb and Dwyer 2001; Lindsay 1997; Moss 2002). More recently, some form of equilibrium, in the form of multi-methodologies, has been achieved between these approaches. Rofe's (2000; 2004) work on gentrification in and the re-imaging of Newcastle, once Australia's industrial heartland, is indicative of this development. Drawing on UK-based work by Loretta Lees (1994), Rofe was able to quantify the extent and significance of inner-city change while also providing a richly contextualized qualitative account of people's lived experience of these processes. The result was a nuanced and holistic image of urban change in an Australian city undergoing an industrial to post-industrial transition.

Typically, the perceived gulf between qualitative and quantitative methods has been presented as a series of dualisms. Hammersley (1992) listed seven 'polar opposites' between qualitative and quantitative methods (Box 1.2). Similarly, Brannen

---

Dualisms Identified between Qualitative and Quantitative Methods	Box 1.2

Qualitative Methods	*Quantitative Methods*
Qualitative data	Quantitative data
Natural settings	Experimental settings
Search for meaning	Identification of behaviour
Rejection of natural science	Adoption of natural science
Inductive approaches	*Deductive approaches*
Identification of cultural patterns	Pursuance of scientific laws
Idealist perspective	Realist perspective

Source: After Mostyn 1985 and Hammersley 1992.

(1992a) characterized qualitative approaches as viewing the world through a wide lens and quantitative approaches as viewing it through a narrow lens. A dualistic view of methods is highly problematic, as Hammersley (1992, 51) recognized: it represents quantitative methods as focused, objective, generalizable, and, by implication, value-free. On the other hand, qualitative methods are often presented as soft and subjective, an anecdotal supplement, somehow inferior to 'real' science.

Such a view is misleading in that it represents quantitative methods as objective and value-free; increasingly, however, this assumption about the nature of science has been questioned (see Chapter 17). Our choice of what we study and how we study it reflects our values and beliefs. For example, much early feminist geography uncovered sexist assumptions in how geographers had typically studied and measured human behaviour (Monk and Hanson 1982). Measurements of migrants and shoppers that ignored 'half of the human population' were clearly shown to be non-objective and value-laden and in many cases strongly coloured by naturalized assumptions about gendered roles and behaviour. If one acknowledges the subjectivity and value-laden nature of all research methods, then the apparent gap between the two groups of methods is reduced dramatically. Geographers using qualitative methods often declare their personal subjectivity and possible sources of **bias**, summarizing their own background as researchers and their relationship to the research and to its intended audience. Further, the cultural 'turn' within human geography has made such openness more commonplace across the discipline, encompassing not only social and cultural geographers but those with more economic and political interests as well (for example, Coe, Kelly, and Yeung 2007 and Thrift 1996). Indeed, researchers who define their own position in relation to their research might be more objective than their colleagues who point to the supposed objectivity of quantitative methods and fail to reveal the many subjective influences that shape both the research question and the explanations that they put forward. However, Bourdieu (2003, 282) takes aim at an excess of reflexivity within the social sciences generally. Specifically, he ridicules the self-indulgent 'fashion' of 'observing oneself observing, observing the observer in his [*sic*] work of observing' (Bourdieu 2003, 282). Given the intellectual complexity and introspection for which Bourdieu is renowned, this is a most interesting comment. However, Bourdieu is emphasizing what is termed the crisis of representation—the conundrum of celebrating the subjective nature of the social world while striving to ground this very subjectivity in objectivity. As Crang (2005, 226) clarifies, 'Bourdieu frets that . . . reflexivity recreates the myth of the exceptional researcher set apart from their respondents not now by the clarity of their knowledge, but by their level of introspection, doubt and anxiety.' In short, equating objectivity with the quantitative and subjectivity with the qualitative is highly contested (Philip 1998). This contest is discussed further in Chapters 4 and 17.

As qualitative methods have again become prominent within the discipline, they have increasingly had to be justified in a scholarly environment that had come to value measurement and scientific observation more highly than individual experience or social process. Within a largely unfriendly hegemonic scientific framework, advocates of qualitative studies have generally drawn from three arguments. First, some studies of individual experiences, places, and events have been represented as essentially non-generalizable case studies that have meaning in their own right but are not necessarily either representative or replicable (for example, Donovan 1988). The second argument, appropriate to some large-scale studies, has been to suggest that these studies have generated sufficient data to allow general, and sometimes quantified, conclusions to be drawn from the research (for example, Oakley 1981; Wearing 1984). Third and more usually, however, qualitative methods have been justified as a complementary technique, as an adjunct or precursor to quantitative studies from which generalizations can be drawn and as explorations in greater depth as part of multiple methods or triangulation (Burgess 1982a). In making these arguments, qualitative geographers have often been on the defensive, aiming to present their studies as legitimate in their own right and as research that produces not just case studies or anecdotal evidence but that has added immensely to the geographical literature through powerful forms of geographical explanation, including analysis, theory building, and geographic histories. Mixed methods are increasingly being used in an effective and powerful way without the self-consciousness evident in some of the studies from the 1990s (Longhurst 2000; Nolan 2003).

Classically, qualitative and quantitative methods, such as interviews combined with questionnaires, are seen as providing both the individual and the general perspective on an issue (for example, England 1993), while similar arguments have been raised for mixed methods more broadly (McKendrick 1996; Philip 1998). This triangulation of methods and use of multiple methods are sometimes deemed as offering a cross-checking of results in that they approach a problem from different angles and use different techniques. Brannen (1992a, 13), however, has argued that data generated by different methods cannot simply be aggregated, because they can only be understood in relation to the purposes for which they were created. This question of purpose is intimately related to the theoretical perspectives from which the techniques derive and is considered further in Chapter 4.

THE HISTORY OF QUALITATIVE RESEARCH IN GEOGRAPHY

As a discipline, geography enjoys a long history. For much of that time, geographical work has been dominated by qualitative research of a scholarly and informed but unquantified nature, drawing assessments of evidence from both physical and human environments (see, for example, Powell 1988). It should be recognized that qualitative methods of many sorts were used widely throughout the twentieth century, particularly in the development and writing of sensitive and nuanced regional geography, such as that of Oskar Spate on the Indian subcontinent (Spate and Learmonth 1967), in the landscape school of both human and physical geographers with 'an eye for country' (see, for example, Sauer's 1925 and Zelinky's 1973 work on American landscapes and culture), but also in interviewing and field observation (Davis 1954; Wooldridge 1955). The postwar era of the 'quantitative revolution' may in hindsight be seen as an aberration rather than the revolutionary **paradigm** (mode of thought) that it was claimed to be at the time (Wrigley 1970).

The early history of geographical thought has been represented classically as a series of paradigm shifts, each triggered by dissatisfaction with the previous prevailing paradigm. A schematic representation of paradigm shifts in academic geography is presented in Box 1.3 (Holt-Jensen 1988; Johnston 1983; Wrigley 1970). For example, dissatisfaction with the crude environmental determinism of the early twentieth century prompted the study of unique places around the world. When this regional approach degenerated into stale layers of facts and geographers had totally marginalized themselves from the academy by their commitment to 'the region'—both as object of study and research method—then an alternative, more credible to the academic community, was sought. The strategic alliances of the discipline shifted away from history and geology to newer and more innovative disciplines such as psychology and economics. By the 1960s, the quest for academic credibility, combined with a technological and data revolution, propelled more scientific ways of thinking into the discipline. This scientific approach combined the use of quantitative method, model-making, and hypothesis-testing. The regional idiosyncrasies were condemned as 'old hat', and geographers turned themselves into spatial scientists. This very compressed 'history' has some validity for the earlier years of this century, although the notion of paradigm shifts has been challenged (for a concise review, see Gregory 1994).

The notion of paradigm shifts essentially becomes inapplicable in the confusing and exciting world of post-quantitative human geography (Billinge, Gregory, and Martin 1984). It is recognized that in recent human geography, there are coexistent, contradictory, and competing communities of scholars adhering to different views of the world, different schools of thought, and different approaches to research questions. Box 1.3 shows that the recent period is occupied by a number of competing viewpoints jostling for space and credibility. The reactions against normalizing spatial science have spawned a huge diversity of approaches; by the early 1980s, radical, feminist, and environmental geographers were reasserting the importance of the social, the agency of the individual, and the particularity of place. Qualitative research requiring qualitative methods reasserted its respectability.

Box 1.3

Paradigm Shifts and Research Methods in Geography from the Twentieth Century*

Time	Paradigm	Research Questions	Research Methods	Characteristics	Trends
Early 20th century	Exploration/ discovery	Discovery	Exploration	Colonial	
	Classical geography	General/theoretical/ contextual	Qualitative/quantitative	Environmental determinism	Decrease in spatial and time scale
	Regional geography	Unique/empirical	Qualitative/regional delineation and description	Region both method and object of study	
	Spatial science	General/theoretical and empirical	The 'quantitative revolution'	Scientific method	Increasing separation of physical from human geography
1980s +	Critical social science	Theoretical/structural/ individual	Qualitative	Pluralist	Increasing diversity in approaches
	Radical	Theoretical/structural			
	Feminist	Structural/individual			Rise of environmental studies
	Phenomenological	Unique/individual	Qualitative	Local	
	Postmodern	Theoretical			
	Post-colonial/ subaltern	Theoretical/empirical	Qualitative	Global and local	

* The shifts identified post-1980 are most relevant to human geography rather than to geography in general.

From the schema outlined in Box 1.3, a few major points can be drawn:

1. The period of spatial science is unique and aberrant in focusing on quantitative methods.
2. The paradigm shifts within geography have involved an increasing separation in the methods and philosophies of human geography from those of physical geography. The reactions against 'scientific' geography established since the 1970s have drawn human geography more and more into the realms of critical social science, while physical geography has remained essentially within the scientific paradigm.
3. The questions that geographers have asked have oscillated between elucidating general trends and patterns in one period to examination of the individual and unique in subsequent phases of geographic inquiry. The multitude of post-quantitative approaches allows both individual and structural research questions to be tackled.
4. In general, the scale of geographic inquiry has shifted from the global, to the regional, to the local. However, recent writings are concerned not only with the specifics of individual experiences and places but have re-engaged with both the theoretical and the global.

Qualitative methods are currently used by all the major groups within the critical social science approaches utilized in human geography and identified in the lower half of Box 1.3. Much of the drive for qualitative research has come initially from humanistic geography of the late 1970s, which focused geographers sharply on values, emotions, and intentions in the search to understand the meaning of human experience and human environments (for example, Ley 1974). Another significant influence in the reassertion of qualitative methods has been the work of feminist geographers establishing links between the personal and the political. A clear example of this might be Cupples and Harrison's (2001) work on media representations of sexual assault by establishment figures or Mackenzie's earlier (1989) work on women and environments in a postwar British city. This approach also predominates in recent studies that reconfigure the 'everyday' domestic spaces of women as entwined with very public circuits of community and even globalization (see, for example, Isabel Dyck 2005 on female caregivers). Similarly, and as discussed earlier, studies that might be grouped as **post-colonial** increasingly give voice to people defined as 'other', enabling multiple interpretations of events to be heard. The power of this scholarship has not just been an opening up of geographic inquiry but equally an embracing of what Joanne Sharp (2009) would refer to as the transformative power of emotional geographies. Much qualitative work within contemporary human geography cannot be clearly categorized within any of the schools of thought listed in the final section of Box 1.3 (critical social science) but is concerned with the broad questions of elucidating human environments and human experiences within a variety of conceptual frameworks.

CONTEMPORARY QUALITATIVE GEOGRAPHY — THEORY/METHOD LINKS

Contemporary human geography adopts a broad range of research methods. Although not tied specifically to particular theoretical and philosophical viewpoints, the methods discussed in the preceding sections of this chapter are often more frequently associated with one standpoint than with another. For example, feminist geography is often associated with qualitative methods through a naturalized association of the feminine with the 'softer' qualitative approaches. However, feminist questions can be stated within a variety of theoretical frameworks and may use a variety of methods (Lawson 1995).

Qualitative methods have been used more widely in human geography throughout the past century than is commonly believed. They have been used in conjunction with quantitative methods in a search for generality and have also been used to explain difficult cases or to add depth to statistical generalizations. Above all, they have traditionally been used as part of triangulation or multiple methods in a search for validity and corroborative evidence. However, qualitative methods have also been used in different conceptual frameworks to reveal what has previously been considered unknowable—feelings, emotions, attitudes, perceptions, and cognition. Overwhelmingly, qualitative methods have been used to verify, analyze, interpret, and understand human behaviour of all types.

Studies that utilize qualitative methods in their own right to express individual meanings are much more limited in number. Although humanistic geographers of the 1980s laid claim to this territory, the output of the humanistic school per se was both limited and short-lived. Even by 1981, Susan Smith was calling for 'rigour' in humanistic method in a way akin to the current calls for **validity** and **replicability** and (by implication) respectability (Baxter and Eyles 1997; Philip 1998).

Similarly, the aims of critical realism expressed by Sayer and Morgan (1985) gave pre-eminence to individuals' actions and their meanings, yet contributions in this mould to geography literature have been slim. The schools of thought that may have made the greatest contribution to answering qualitative research questions have been the feminist and the post-structural (including the postmodern and the post-colonial). Both these frameworks recognize that multiple and conflicting realities coexist. They deliberately give voice to those silenced or ignored by hegemonic (modern, colonial) views of histories and geographies. They embody and acknowledge previously anonymous individuals. Paradoxically, however, the voice of the oppressed not only speaks for itself: it is part of a wider whole. Reality is like an orchestra: post-structural approaches differentiate the instruments and their sounds and bring the oboe occasionally to centre-stage; usually dominated by the strings, the minor instruments too have a tune to play and a thread that forms a distinct but usually unheard part of the whole. It is the voices of the women and children, the colonized, the indigenous, the minorities that, when released from their silencing, enable a more holistic understanding of society to be articulated (for example, Lane et al.'s [2003] examination of Aboriginal and mining company conflicts over the Coronation Hill mine project in northern Australia and Hibbard, Lane, and Rasmussen's [2008] examination of the potential role of planning processes in empowering Aboriginal people).

Sayer and Morgan (1985) make the point that exactly the same research technique can be used in different ways for different purposes according to the theoretical stance of the researcher. Interviews, for example, may be used to gain access to information from gatekeepers about structures or to give voice to silenced minorities. John McKendrick (1996, Table 1) considers the relationship between methods and their applications in different research traditions. He contrasts the use of interviews in the humanistic tradition to 'explore the meaning of the migration of each individual migrant' with their use in a postmodernist framework whereby in-depth interviews with women may be used 'to "unpack" their rationalizations of their migrations'.

Qualitative methods raise an immense number of difficult issues that are considered in more detail in several of the chapters of this volume. Among these issues are concerns over authorship, audience, language, and power (discussed, for example, in Chapters 2, 3, 13, 17, and 18). Ethnographic research is often highly complex within which the individual subject and the audience for the research are often intermingled and mutually dependent. The position of the author as observer in relation to the object of research raises issues of power relations and control. The engagement of the researcher does not necessarily allow the voices of the researched to speak, since they are mediated through the researcher's experience and values. The language of research reporting may also exclude those researched, although language varies according to the audience towards which the research is directed. The key issue of the outsider gazing, perhaps voyeuristically, at those defined as Other is an intractable problem that needs to be recognized even if it cannot be solved. Even participant observation cannot surmount inbred and naturalized class differences, as demonstrated by Canadian geographer Pamela Moss (1995) in her immersion in manual labour in hotels. Moss never managed to bridge the cultural, linguistic, and social gap between herself as middle-class researcher and the housemaids who spent their lives in manual labour. The engagement with human research subjects raises significant questions about ethical research practice, which are only now being addressed seriously within the discipline.

SUMMARY

This chapter has outlined three major groups of qualitative methods currently used in human geography: oral, textual, and participatory. As noted earlier, these techniques may be considered the 'backbone' of qualitative human geography (Davies and Dwyer 2007). For Thrift (2000, 3), there has emerged a 'convention' within qualitative approaches to human geography that is based on this 'narrow' range of methodological skills and that may unwittingly register a narrow range of life, brought back from the field and represented in 'nice', even predictable ways. This is a stinging critique and one with which the discipline must engage. We assert that Thrift's critique is being responded to; new and experimental techniques and debate over their merits and limitations indicate a healthy and robust methodological climate within qualitative human geography.

Regardless of whether backbone or experimental techniques are employed, qualitative human geography methods are used to answer two broad research questions relating either to the experiences of individuals or to the social structures within which they operate.

Qualitative methods have often been categorized as oppositional to quantitative methods, yet in many respects this is a false dichotomy. The differences between qualitative and quantitative methods are related to the conceptual frameworks from which they have been derived. In elucidating human experiences, environments, and processes, qualitative methods attempt to gather, verify, interpret, and understand the general principles and structures that quantitative methods measure and record. Further, qualitative methods have very frequently been used in conjunction with other methods. The use of qualitative methods alone to explore human values, meanings, and experiences has been more limited. Currently, some of the most rewarding qualitative research in human geography is being carried out in feminist and post-colonial frameworks to enable silenced voices to be heard and to foster better comprehension of those naturalized discourses that exclude and marginalize certain groups.

KEY TERMS

bias	dependability
confirmability	diary interviews
credibility	diary photographs
deconstruction	discontinuous writing
deductive	discourse

epistemology	post-colonial
focus group	quantitative methods
go along interviews	random sample
inductive	reflexivity
interviews/interviewing	replicability
landscape	representation
mixed methods	respondent
objective	semiotics
ontology	sign
oral history	structures/social structures
oral methods	subjective
Other	text/textual
paradigm	textual analysis
participation/ participant observation	triangulation
	validity
photo-elicitation	

REVIEW QUESTIONS

1. What research questions may be answered by qualitative methods? Give examples.
2. What are the main types of qualitative research methods used in human geography?
3. How may different types of qualitative methods be linked to theoretical approaches within the discipline?
4. Outline what is meant by the quantitative/qualitative debate.

USEFUL RESOURCES

Davies, G. 2007. 'Qualitative methods: Are you enchanted or are you alienated?' *Progress in Human Geography* 31 (2): 257–66. This paper provides a current overview of methodological advances in three significant sub-fields of human geography: embodied and emotional geographies, geographies of nature, and performing places.

Laurier, E. 2004. 'Participant observation'. In N.J. Clifford and G. Valentine, eds, *Key Methods in Geography*. London: Sage. This chapter focuses on the third type of method discussed in this chapter (i.e., participatory). It contains many useful hints and stories that make it very useful for students thinking about using participatory research.

Marshall, D. 2004. 'Making sense of remembrance'. *Social and Cultural Geography* 5 (1): 37–54. This article analyzes landscape, described in this chapter as the second main type of qualitative method. It investigates war memorials in Britain, addressing issues of sight, sound, and touch as being significant aspects of landscapes of remembrance.

Nijoux, D., and J. Myers. 2006. 'Interviews, solicited diaries and photography: 'New' ways of accessing everyday experiences of place'. *Graduate Journal of Asia-Pacific Studies* 4 (1): 44–64. A highly readable article that provides a clear discussion of the methods employed.

Winchester, H.P.M., P.M. McGuirk, and K. Everett. 1999. 'Celebration and control: Schoolies Week on the Gold Coast Queensland'. In E. Teather, ed., *Embodied Geographies: Spaces, Bodies and Rites of Passage*. London: Routledge. This chapter uses oral qualitative evidence to examine the cultural phenomenon of a rite of passage. It provides an example of oral methods.

Winchester, H.P.M., and M.W. Rofe. 2005. 'Christmas in the "Valley of Praise": Intersections of the rural idyll, heritage and community in Lobethal, South Australia'. *Journal of Rural Studies* 21: 265–79. This article provides significant insight into the development and use of a qualitatively based multi-methodology. Specific methods successfully employed in this research include interviews, participant observation, textual analysis, and landscape deconstruction.

Power, Subjectivity, and Ethics in Qualitative Research

2

Robyn Dowling

CHAPTER OVERVIEW

This chapter aims at introducing issues that arise because qualitative research typically involves interpersonal relationships, interpretations, and experiences. I discuss three issues of which qualitative researchers need to be aware: (1) the formal ethical issues raised by qualitative research projects; (2) the power relations of qualitative research; and (3) objectivity, subjectivity, and intersubjectivity. Rather than advocating simple prescriptions for dealing with these issues, the chapter proposes that researchers be 'critically reflexive'.

INTRODUCTION: ON THE SOCIAL RELATIONS OF RESEARCH

Chapter 1 outlined the types of research questions asked by qualitative researchers—namely, our concerns with the shape of societal structures and people's experiences of places and events. This chapter takes you one step closer to conducting qualitative research. It discusses some of the implications of research as a social process. Collecting and interpreting social information involves personal interactions. Interviewing, for example, is essentially a conversation, albeit one contrived for research purposes. Interactions between two or more individuals always occur in a societal context. Societal norms, expectations of individuals, and structures of power influence the nature of those interactions. For instance, when you are conducting a focus group, you may find men talking more than women and people telling you what they think you want to hear. Societal structures and behaviours are not separate from research interactions. This places all social researchers in an interesting position. We may use a variety of different methods to understand society, but those methods cannot be separated from the structures of society. The converse is also true. The conduct of social research necessarily has an influence on society and the people in it. By asking questions or participating in an activity, we alter people's day-to-day lives. And communicating the results of research can potentially change social situations.

Both qualitative and quantitative researchers recognize this lack of separation between research, researcher, and society. What typically distinguishes qualitative researchers' approach to this issue is the emphasis they give to it. For those who use the qualitative methods discussed in this volume, the interrelations between society, the researcher, and the research project are of critical and abiding significance. They permeate all methods and phases of research. These relationships cannot be ignored and raise key issues that must be considered when designing and conducting research. This chapter outlines three issues that arise because of the social nature of research and suggests an approach to dealing with them.

The chapter begins with a discussion of research ethics and the philosophies and practicalities of institutional review of research ethics within universities. It introduces key criticisms of institutional review and moves beyond these guidelines to introduce the concept of critical reflexivity. The chapter then focuses on the ways power traverses the conduct of qualitative methods. Finally, the chapter considers the significance that a researcher's subjectivity and intersubjective relations have for the collection of qualitative data.

A word of caution to begin. In general, most of the chapters that follow offer practical guides to different qualitative methods. They explain how to be a participant observer, how to conduct a focus group, and so forth. This chapter is different in two important respects. First, it is not about any specific method. It is about issues common to each of the methods discussed in this volume. This chapter should be used in combination with those on particular methods to help you think about the specific challenges you are likely to encounter in your research. Second, this chapter does not—and cannot—offer hard and fast rules on conducting ethical research that is responsive to matters of power and **intersubjectivity**. The conduct of good, sensitive, and ethical research depends in large part on the ways you deal with your unique relationships with research participants, peers, and other organizations at particular times in particular places.

RESEARCH ETHICS AND INSTITUTIONAL REVIEW OF RESEARCH ETHICS

All research methods necessarily involve ethical considerations. Decisions about which research topics to pursue, appropriate and worthwhile methods of investigation, 'right' ways to relate to sponsors of and participants in research, and appropriate modes of writing and communication of results involve ethical questions. These questions include how researchers *ought* to behave, the role of research in the pursuit of social change, and whether and how research methods are 'just'. Research **ethics**, broadly defined as being about 'the conduct of researchers and their responsibilities and obligations to those involved in the research, including sponsors, the general public and most importantly, the subjects of the research' (O'Connell-Davidson and Layder 1994, 55), constitute an issue that must be dealt with in your research. Researchers' ethical practice is regulated in a number of ways. The Association of American Geographers (2005), for example, has published a statement on professional ethics for its members that is designed 'to encourage consideration of the relationship between professional practice and the well-being of the people, places and environments that make up our world'. The late twentieth century saw increasing formalization of ethical review (Israel and Hay 2006). Important here are statements on ethical research by government agencies charged with the oversight of research. In Canada, for example, the 'Tri-Council policy statement: Ethical conduct for research involving humans' (Tri-Council 2005) provides both general guidelines and prescriptive rules for ethical conduct. These statements and their attendant rules are then implemented by ethics committees (known, for example, as institutional review boards [IRBs] in the United States, research ethics boards [REBs] in Canada, and human research ethics committees [HRECs] in Australia), typically associated with organizations where a considerable amount of research is conducted, such as schools, hospitals, and indigenous organizations, as well as universities.

University ethics committees are of most relevance to student research, although in some research, the ethics committees of other institutions like schools may also need to be consulted. In most universities, post-graduate students or those in charge of courses are required to obtain the formal approval of a university ethics committee before beginning research that involves people. University ethics committees focus on the researcher's responsibilities to research subjects and apply guidelines about what researchers should and should not do. It is useful to consider such formal guidelines as a first step in thinking through the social context of your research. The committee may not evaluate your research design but will want to know the aims of the research and the methods you will use. It will be concerned primarily with your responsibilities to research participants with regard to matters of privacy, informed consent, and harm.

Privacy and Confidentiality

Qualitative methods often involve invading someone's privacy. You may be asking very personal questions or observing interactions in people's homes that are customarily considered private. Ethics committees are concerned that these private details about individuals not be released into the public domain. Accordingly, you may have to show that your original field notes, tapes, and transcripts will be stored in a safe place where access to them will be restricted. You may also need to ensure that your research does not enable others to identify your informants. There are various ways of ensuring the anonymity of informants, including using pseudonyms and masking other identifying characteristics (for example, occupation, location) in the written version of your research. You should note, however, that when dealing with significant public figures, it is sometimes not possible or desirable to ensure anonymity. For example, in O'Neill's (2001) research on the Australian-based transnational BHP (Broken Hill Proprietary Company Limited), he identified both the firm and the executives with whom he spoke.

Informed Consent

For most geographical research, participants must consent to being part of your research. In other words, they have to give you permission to involve them. However, this criterion is somewhat stricter than a simple 'yes, you can interview me.' It must be *informed* consent (see the helpful discussion of this in Chapter 3). Informants need to know exactly what it is that they are consenting to. You need to provide participants with a broad outline of what the research is about, the sorts of issues you will be exploring, and what you expect of them (for example, the amount of time required to complete an interview). Most ethics committees recognize that there are exceptions to informed consent. Simple observation of people in a place like a public shopping mall, for example, may not need the explicit consent of those individuals. Indeed, it may be physically impossible to secure the consent of everyone involved. Sometimes informed consent may be waived, although an ethics committee will typically ask you to justify that decision. There are some relatively rare instances when research may involve deception. Deception is when research participants either do not know that you are a researcher (for example, Routledge's 2002 work in India) or do not know the true nature of your research. Set against the principle of informed consent, deception is clearly an ethically difficult issue, and you should think carefully and seek advice before contemplating a research project that involves deception. Moss's (1995) discussion of why and how she used deception in her study of domestic workers may be useful.

Harm

Your research should not expose yourself or your informants to harm—physical or social. As social scientists, it is highly unlikely that you will be subjecting people to physical harm. You may, however, be bringing them into contact with 'psycho-social' harm. You may raise issues that may be upsetting or potentially psychologically damaging. This does not mean that your research cannot proceed. Rather, it means your research should cater to this possibility. For example, in work with gay men in western Sydney, Stephen Hodge (reported in Costello and Hodge 1998) had the contact details of a counselling service ready in case participants became upset during interviews. You should also avoid putting yourself at risk during the research. For example, a young woman planning a participant observation project on single women's safety at night on public transport could meet a very cautious response from an ethics committee or research supervisor because of the potential danger to her while conducting that work.

Criticisms of, and Moving Beyond, Ethical Guidelines

Although important, ethical rules and ethics committees are not unproblematic for qualitative researchers in human geography. At a general level, the universalist ethical stance and biomedical model of research embedded in IRBs involve rigid codes that cannot always deal with 'the variability and unpredictability of geographic research' (Hay 1998, 65). More specific criticisms of how the issues of informed consent, privacy, confidentiality, and harm are framed within IRBs, and their potential incompatibilities with qualitative research, now abound. They include the individualized nature of informed consent when a community's or collective consent may be more appropriate (Butz 2007) and the ways the unpredictable nature of qualitative research make gaining informed consent difficult. Whether guarantees to privacy and anonymity can and should be given if illegal behaviours are involved is also at issue. Finally, in relation to harm, it is not always possible to predict the impact of research on participants, especially over a longer term (Bailey 2001).

Critical Reflexivity

Despite these problems, ignoring institutional ethical review is neither desirable nor possible (Martin 2007). Instead, ways of practically negotiating with these codes are preferable. For Hay and Israel (2006, 142), this involves a commitment to theoretically informed, self-critical ethical conduct, revolving around awareness of how to identify and resolve ethical dilemmas when they arise. Implicit in practical engagements with ethical codes is the suggestion that as geographers engaged in research, we must constantly consider the ethical implications of our activities. Because research is a dynamic and ongoing social process that constantly throws up new relations and issues that require constant attention, self-critical awareness of ethical research conduct must pervade our research. Our engagement with ethical behaviour does not end when we submit our research proposal to an ethics committee.

As the foregoing discussion of ethical research conduct might imply to you, human geographers have come to appreciate more fully than ever before the social nature and constitution of our research. Indeed, we now recognize and acknowledge this location through the concept of **critical reflexivity**. Reflexivity, as defined by Kim England (1994), is a process of constant, self-conscious scrutiny of the self as researcher and of the research process. In other words, being reflexive means analyzing your own situation as if it were something you were studying. What is happening? What social relations are being enacted? Are they influencing the data?

Critical reflexivity is difficult but rewarding. It is rewarding in that, as some of the examples used in this chapter indicate, it can initiate new research directions. Critical reflexivity is, however, difficult in two respects. First, many geographers do not write about the research process in their published work. Linda McDowell (1998) comments, for example, that many of the details of how her merchant banking research proceeded and, as a result, her reflections on the process do not appear in the book based on that research. However, discussions of reflexivity are becoming more common, and a guide to some of them is provided in the Useful Resources section at the end of the chapter. Second, reflexivity is difficult because we are not accustomed to examining our engagement with our work with the same intensity as we regard our research subjects. You may be helped in this matter by keeping a **research diary**, as outlined in Box 2.1.

The Research Diary as a Tool for the Reflexive Researcher | **Box 2.1**

Your efforts to be reflexive will be enhanced if you keep a research diary. The contents of a research diary are slightly different from those of a fieldwork diary. While a **fieldwork diary**, or field notes, contains your qualitative data—including observations, conversations, and maps—a research diary is a place for recording your reflexive observations. It contains your thoughts and ideas about the research process, its social context, and your role in it. You could start your research diary by writing answers to the questions posed in the checklist set out in Box 2.3 at the end of the chapter.

The rest of the chapter focuses on two of the important issues about which you need to be reflexive: power and subjectivity. For readers who are especially interested, Chapter 17 expands on the discussion of reflexivity.

POWER RELATIONS AND QUALITATIVE RESEARCH[1]

One important outcome of the social character of qualitative research is that research is also interwoven with relations of power. Power intersects research in a number of ways.

It can enter your research through the stories, or interpretations, you create from the information you gather. Power is involved here because knowledge is both directly and indirectly powerful. Knowledge is directly powerful through its input into policy. Some studies are specific analyses of policy issues, and their results have a direct impact on people's lives. Knowledge is also indirectly powerful. The stories you tell about your participants' actions, words, and understandings of the world have the potential to change the way those people are thought about. Power relations also exist beyond the relationship between researcher and participant. Qualitative research often investigates issues that are the subject of intense societal and political scrutiny, like indigenous disadvantage or childhood poverty.

Power is also involved in earlier parts of the research process. In undertaking qualitative research, you are attempting to understand—participating in, and sometimes creating—situations in which people (yourself included) are differently situated in relation to social structures. Both you and your informants occupy different 'speaking positions'. Not only do you and your informants have different intentions and social roles, but you also have different capacities to change situations and other people (see Liamputtong 2007).

Social researchers typically enter one of three different sorts of power relations in parts of their work. **Reciprocal relationships** are those in which the researcher and the researched are in comparable social positions and have relatively equal benefits and costs from participating in the research. You may, for example, be conducting focus groups with your fellow students on how they are adjusting to university life. Although not absent, power differences in this relationship are minimal compared with two other sorts of relationships. The other two types of research relationships are forms of **asymmetrical relationship**, characterized by significant differences in the social positions of researcher and those being researched. On the one hand, those being studied may be in positions of influence in comparison to the researcher. Interviews with senior executives of large corporations can fall into this category because of such executives' relative access to cultural and financial resources (see McDowell 1992a). This kind of asymmetrical research relationship was first termed '**studying up**' by anthropologist Laura Nader (1969). On the other hand, the researcher may be in a position of greater power than the research participant. This is termed a **potentially exploitative relationship.** Kate Swanson's research with young people who beg on the streets of Ecuador is an example of a potentially exploitative relationship with ethical and power dimensions that have been considered thoughtfully (Swanson 2007; 2008).

Power cannot be eliminated from your research, since it exists in all social relations. Human geographers typically have one of two responses to issues of power. The first, responding directly to potentially exploitative relationships, is to involve participants in the design and conduct of the research. In their research on young people's experience of crime and victimization, Pain and Francis (2003) explicitly sought young people's perceptions of what required investigation and then gave participants a number of opportunities to verify or refute the researchers' interpretations. Pain and Francis also conducted a number of meetings and workshops to disseminate their findings. Through these two strategies, they hoped to effect some social change and alter potentially exploitative relationships.

Such participatory forms of research are not always the most appropriate for every research project or for student researchers. A second response that recognizes and negotiates relations of power is critical reflexivity. For example, when collecting data, our responsibility to the research participants is such that we should not take advantage of someone's less powerful position to gather information. In the case of homeless youth, for example, you would not make the possibility of gaining access to shelter dependent on that person's participation in the study. But you cannot eliminate the power dimension from your research, since it exists in all situations. The best strategy is to be aware of, understand, and respond to it in a critically reflexive manner. Critical reflexivity does not necessarily mean altering your research design, but it does imply that you reflect constantly on the research process and modify it where appropriate. When you are formulating your topic, think about the various ethical and power relations that may be enacted during your research (see, for example, Box 2.2). Are you happy with the situation? Would you like to do anything differently? Could you justify your actions to others? You should also think about how you communicate the results. Have you reflected as faithfully as possible what you have been told or observed without reproducing stereotypical representations? Are you presenting what you heard and saw or what you expected to hear and see? Remember, the stories you tell may change the worlds in which you and your research participants live (for more detail on this matter, see Chapter 17).

Increasingly, research in human geography is connected to organizations outside the university. Government, industry, or community groups may directly commission human geographers to carry out research on a particular topic, or student research projects may be carried out in consultation with such groups. These situations introduce new power dynamics to the research process. There may be attempts to control methodological protocols, research findings, and their dissemination, or you may find that your desire to please others involved in the research raises some moral dilemmas. If such issues arise, awareness of how other researchers have navigated such relationships is a necessary first step (see Hallowell, Lawton, and Gregory 2005). Israel and Hay (2006, 112) usefully employ the notion of research integrity as an additional prompt for all researchers. Research integrity encompasses intellectual honesty, accuracy, fairness, and collegiality. Its first two elements, honesty and accuracy, can guide researchers' interactions with other organizations involved in research projects. Sponsor dissatisfaction with research findings, for example, can be discussed by reiterating the rigour of the research methodology and the researcher's confidence in the robustness of the analysis.

Researchers also have relationships with peers. Undergraduate students may conduct a research project as a group activity, graduate students may be working on one element of a larger research project directed by their supervisors, or you may be the geographer on a multidisciplinary research team. The second two aspects of research integrity—fairness and collegiality—come into play here. Nurturing relationships through open and transparent communication is one method of successfully negotiating such situations, although again, the issue of power relations (e.g., between student and supervisor, contract researcher and employer) remains.

Sexism in Research   Box 2.2

Sexism can present problems in many different sorts of research projects. Eichler (1988) identifies four primary problems of sexism in research:

- *androcentricity/gynocentricity*: A view of the world from male/female perspectives respectively. For instance, concepts of 'group warfare' developed through reference to men's experiences only.

- *overgeneralization*: A study is only about one sex but presents itself as applicable to both sexes. For example, a study that is exclusively concerned with men's location decisions might be misleadingly entitled 'Residential location decisions in Vancouver'.

- *gender insensitivity*: Ignoring gender as an influential factor in either the research process or interpretation. For instance, a study of the geographical effects of a free trade agreement that fails to consider any gender-specific effects

- *double standards:* Identical behaviours or situations are evaluated, treated, or measured by different means or criteria (for example, drawing different conclusions about men and women on the basis of identical answers to a survey or aptitude test).

Eichler also identifies three 'derived' forms of sexism:

- *sex appropriateness:* The notion that some characteristics and behaviours are accepted as being more appropriate for one sex than for the other (for example, designing a research project on parental perceptions of children's play space and interviewing women only because you assume that they will know more about the issue).

- *familism:* Using family as the unit of analysis when the individual might be more appropriate or vice versa (for example, working at the family scale in evaluating the social costs and rewards of in-home care for the

elderly rather than exploring the different implications it might have for males and females within the family).

- *sexual dichotomism:* Postulating absolute differences between women and men (for example, women are sometimes considered 'naturally' more timid than men).

Objectivity, Subjectivity, and Intersubjectivity in Qualitative Data Collection

Objectivity has traditionally been emphasized in geographic discussions of quantitative research methods. Objectivity has two components. The first relates to the personal involvement between the researcher and other participants in the study. The introduction to this chapter suggested that it is impossible to achieve this sort of objectivity because of the social nature of all research. Objectivity's second component refers to the researcher's independence from the object of research. This implies that there can be no interactive relationship between the researcher and the process of data collection and interpretation. Clearly, however, dispassionate interpretation is difficult if not impossible, because we all bring personal histories and perspectives to research. **Subjectivity** involves the insertion of personal opinions and characteristics into research practice. Qualitative research gives emphasis to subjectivity because the methods involve social interactions. As will become evident in later chapters, you need to draw on your personal resources to establish rapport and communicate with informants. Discourse analysis also involves your subjectivity in that your everyday understanding of the world helps you to decipher texts. If subjectivity is important, then so too is intersubjectivity. This refers to the meanings and interpretations of the world created, confirmed, or disconfirmed as a result of interactions (language and action) with other people within specific contexts. Collecting and interpreting qualitative information relies upon a dialogue between you and your informants. In these dialogues, your personal characteristics and social position—elements of your subjectivity—cannot be fully controlled or changed, because such dialogues do not occur in a social vacuum. The ways you are perceived by your informants, the ways you perceive them, and the ways you interact are at least partially determined by societal norms.

Critical reflexivity is the most appropriate strategy for dealing with issues of subjectivity and intersubjectivity. Although you cannot be entirely independent from the object of research, trying to become aware of the nature of your involvement and the influence of social relations is a useful beginning that can help you to identify the implications of subjectivity and intersubjectivity in your research.

Geographers' work on gender provides some good examples of the role of inter-subjectivity in research. Gender is important, because we often ascribe character-istics to people on the basis of gender. Furthermore, personal interactions vary with the gender of participants; we tend to react differently to men and women. Therefore, gender is a factor that can influence data collection. For instance, Andy Herod (1993) found that male union officials restricted the sort of information he was given during interviews because he was a man. Specifically, his informants downplayed the role of women in the union's struggles. Herod attributes this not only to the perspectives of the union officials but also to the social (masculine) context of the interviews. The union officials assumed, Herod thinks, either that he was not interested in gender or that it was inappropriate to raise the issue in the context of a male-to-male conversation. By contrast, Hilary Winchester (1996) suggests that being a woman interviewing men aided her research on lone fathers. She found herself adopting a typically feminine role of facilitating conversation with men, which helped considerably in gathering the men's stories.

Your ability to interpret certain situations also depends on your own charac-teristics. Important here is a debate about the relative merits of being an **insider** or an **outsider**. An insider is someone who is similar to their informants in many respects, while an outsider differs substantially from their informants. Coming from an 'out' lesbian interviewing other lesbians, Valentine's research on sexual-ity could be considered insider research, whereas she was an 'outsider' in her inter-views with parents and children about childhood (see Valentine 2002). One pos-ition in the debate is that for you as an insider, both the information you collect and your interpretations of it are more valid than those of an outsider. People are more likely to talk to you freely, and you are more likely to understand what they are saying, because you share their outlook on the world. If you are not a member of the same social group as your informants, then establishing rapport may be more difficult (see Chapters 3 and 13). And since you do not share their perspec-tive on the world and their experiences, then your interpretations may be less reliable. But being an outsider can also bring benefits to the research. It may mean that people make more of an effort to clearly articulate events, circumstances, and feelings to the researcher. Interviewing trade union officials in eastern Europe, Herod (1999) found it helped to be from a different country because that guaran-teed a warm reception and insightful two-way exchanges of information.

A final perspective to consider is that you are never simply either an insider or an outsider. We have overlapping racial, socio-economic, gender, ethnic, and other characteristics. If we have multiple social qualities and roles, as do our informants, then there are many points of similarity and dissimilarity between ourselves and research participants. Indeed, becoming aware of some of these commonalities and discrepancies can be one of the pleasures and surprises of qualitative research. As a British researcher, Mullings (1999) was far removed

from the social worlds of the Jamaican factory workers and employers she interviewed. To her surprise, however, her position as a person of African descent led to shared experiences with local interviewees and subsequently rich interviews.

Intersubjectivity also means that neither you, your participants, nor the nature of your interactions will remain unchanged during the research project. Your general outlook and opinions may indeed change as result of your research. If you are a participant observer, for example, you will be immersing yourself in a situation that will invariably affect you. Robin Kearns (1997) discusses how both he and his research project changed in response to his interactions as an observer. Kearns began his project on Maori health in rural New Zealand in the waiting room of a medical clinic. He did not intend to interact with patients in the waiting room. His attempts to be unobtrusive failed because of the inquisitiveness of patients who talked to him incessantly. These interactions had two immediate effects. First, they forced him to adopt a different stance in his research: he had to be more open about his presence and his research. Second, he modified his method by actively involving the local community in the research design and the findings (Kearns discusses this more fully in Chapter 12).

Critical reflexivity can give rise to new and exciting directions in one's research. In a very different context, Gilbert (1994) began a research project believing that the women she was speaking with were not feminists. But responses such as 'she had devoted her life to God because she was not going to let any man stop her from surviving' (1993, 92) helped to challenge her own preconceived ideas about feminism and shed new light on the empirical material.

SUMMARY AND PROMPTS FOR CRITICAL REFLEXIVITY

Using the qualitative methods described in subsequent chapters will involve you in various social relations and responsibilities. I have advocated critical reflexivity—self-conscious scrutiny of yourself and the social nature of the research. Critical reflexivity means acknowledging rather than denying your own social position and asking how your research interactions and the information you collect are socially conditioned. How, in other words, are your social role and the nature of your research interactions inhibiting or enhancing the information you are gathering? This is not an easy task, since it is not always possible to anticipate or assess accurately the ways in which our personal characteristics affect the information we accumulate.

I shall not end this chapter with answers. Instead, I offer a preliminary set of prompts that might help you reflect critically on issues of ethics, power, and intersubjectivity in different types and phases of research (see Box 2.3).

How to Be Critically Reflexive in Research

Box 2.3

Before beginning:
- What are some of the power dynamics of the general social situation I am exploring, and what sort of power dynamics do I expect between myself and my informants?
- In what ways am I an insider and/or outsider in respect to this research topic? What problems might my position cause? Will any of them be insurmountable?
- What ethical issues might impinge upon my research (for example, privacy, informed consent, harm, coercion, deception)?

After data collection:
- Did my perspective and opinions change during the research?
- How, if at all, were my interactions with participants informed/constrained by gender or any other social relations?
- How was I perceived by my informants?

Remember to take notes throughout data collection and keep them in a research diary.

During writing and interpretation:
- Am I reproducing racist and/or sexist stereotypes? Why and how?
- What social and conceptual assumptions underlie my interpretations?

KEY TERMS

asymmetrical power relation	objectivity
critical reflexivity	outsider
ethics	potentially exploitative power relation
fieldwork diary	reciprocal power relation
informed consent	research diary
insider	studying up
intersubjectivity	subjectivity

REVIEW QUESTIONS

1. What is critical reflexivity? Read a study in human geography in your field of interest that uses qualitative methods. Is the researcher reflexive about the research process? If so, how? If not, what sort of questions would you like to ask the researcher about the research process?
2. What forms of power relations may be part of a qualitative research project?
3. Read a piece of qualitative research in human geography. Can you identify any or all of Eichler's (1988) forms of sexism in the way the research has been carried out or in the interpretation?
4. How are qualitative methods intersubjective? How might social relations like gender influence the collection of data?
5. Outline and explain the issues of concern to university ethics committees. Find out how your university's committee deals with these issues.

USEFUL RESOURCES

Mark Israel and Iain Hay's highly readable 2006 book *Research Ethics for Social Scientists: Between Ethical Conduct and Regulatory Compliance* (London: Sage) more thoroughly canvasses many of the issues addressed in this chapter.

Kate Swanson (2008) offers a novel account of the practice of critical reflexivity that revolves around the place of her pet dog in her research in her paper 'Witches, children and Kiva-the-research-dog: Striking problems encountered in the field' (*Area* 40 [1]: 55–64).

Gillian Rose provides a sophisticated discussion of critical reflexivity and some of its problems in her 1997 article 'Situating knowledges: Positionality, reflexivities and other tactics' (*Progress in Human Geography* 21: 305–20).

Note ──

1. See Chapter 3 for further discussion of power-related issues.

Cross-Cultural Research: Ethics, Methods, and Relationships

3

Richie Howitt and Stan Stevens

CHAPTER OVERVIEW:
AN EXPLANATION OF OUR WRITING METHOD

Working across the differences that constitute 'cultures' is a common challenge for geographical researchers. Ideas about research are profoundly shaped by cultural contexts. As geographers, we are aware of and engaged by the complex and dynamic relationships between culture and geography—our thinking is shaped by where we conduct research and the people we work with. In discussing how we might write about cross-cultural research as part of a wider discussion of qualitative methods, we initially thought that each of us might write sections of the chapter, with a jointly developed introduction and concluding discussion. This approach, we thought, would enable us each to have our own voice (and responsibility for positions that not all might fully share!). But the wonders of e-mail allowed us to explore some alternatives. We decided that it might be better to think of our writing in terms of a conversation across cultures and geographies, with each of us framing key questions arising from our readings of each other's work and the wider literature on cross-cultural research and our frustrations with existing practices and their impacts on people we work with. In the end, we have written the chapter as a conversational text. In many ways, the text parallels what we seek to emphasize about cross-cultural field-based research— it involves respectful listening, difficult and challenging engagements, careful attention to nuances in the lives of 'others', and a critical, long-term consideration of the implications of methods in the construction of meaning.

MODES OF CROSS-CULTURAL ENGAGEMENT: COLONIAL, POST-COLONIAL, DECOLONIZING, AND INCLUSIONARY RESEARCH

Richie: Can we start our conversation by talking about the intercultural spaces in which we might think about 'cross-cultural' research? Conventionally, field-work is seen as 'the heart of geography' (Stevens 2001, 66), but there has been a lot of discussion about just what we mean by 'the field' in 'fieldwork' (see, for example, special issues of *Professional Geographer* [1994] and the *Geographical Review* [2001], including Blake 2001; DeLyser and Starr 2001; Gade 2001; Katz 1994; Kobayashi 1994; Nast 1994; Zelinsky 2001) and about the multiple versions of 'out-there-ness' and 'in-here-ness' that are constructed around ideas of what constitutes the 'cross-cultural' in 'cross-cultural research'.

As I see it, most human geographic research is cross-cultural, because we are drawn into thinking about other people's constructions of place, other people's ways of reading their cultural landscapes—even when they are the landscapes that we live in ourselves in our everyday lives! I remember as an undergraduate being struck by the idea of a 'geographical expedition' into Detroit (for example, Horvath 1971; Pawson and Teather 2002). That was juxtaposed, for me at least, with the whole 'boy's own adventure' representation of fieldwork and the ethical issues of interpreting someone else's culture for one's own reasons.[1] The critique of that androcentric-style in geography was simultaneously about its sexism *and* its ethnocentricism.

Stan: I agree with you, Richie. A lot of cultural geographical research is cross-cultural. Geographers often face similar ethical and methodological issues whether working close to home or on the other side of the planet. I've chosen to spend a quarter of my time over nearly 30 years, living and working in the Himalayas with the Khumbu, Pharak, and Katuthanga Sharwa peoples of the Chomolungma region (better known to the rest of the world in British colonial nomenclature as the Sherpas of the Mount Everest region) (Stevens 1993; 1997; 2001; 2004; Stevens and Sherpa 1993). But one need not go to the other side of the planet to engage with 'others', given the often complex dimensions of diversity created by societal and group constructions of regional, ethnic, linguistic, class, racial, gender, sexual, religious, ideological, and other difference. I suspect that the challenges and moral considerations of cross-cultural communication, learning, and activism might not be much different if I were to undertake research with my neighbours in the rural Massachusetts community I've lived in for the past few years or in the neighbourhoods of nearby cities.[2] Accordingly, I think that there may well be broader applicability and utility to some of the lessons I've learned from Sharwas, my Berkeley advisor Bernard Nietschmann (1973; 1979; 1997; 2001; Maya People of Southern Belize, Toledo Maya Cultural Council, and

Toledo Alcades Association 1997) and his work with the Miskito and the Maya, and post-colonial studies that challenge 'conventional'—in the sense of common, long-established, and unexamined—views of fieldwork.[3] These lessons revolve around the importance of rejecting the attitudes, assumptions, purposes, and methodologies of what post-colonial theorists refer to as colonial research in favour of those of 'decolonizing', 'post-colonial' research.

Richie: Stan, these are issues that fieldworkers have grappled with for a long time. Even in 'colonial' modes, people such as Clifford Geertz, Charles Rowley, and Nugget Coombs made valuable and very critical contributions.[4] Among geographers, we might think of the work of people such as Fay Gale, Janice Monk, Elspeth Young, Keith Buchanan, Oskar Spate, and Jim Blaut. You also identify 'inclusionary' research as relevant to your work. How do you differentiate these different approaches to research? Do you see them as adopting different positionalities vis-à-vis difference, or do they adopt different methods, conceptual emphases, or perhaps purposes for the research that leads to your particular wording?

Stan: You've written insightfully about these distinctions yourself, Richie (Howitt 2002a; 2002b), which are critical to rethinking research ethics, methods, and relationships. I see colonial and post-colonial research as differing fundamentally from all of the standpoints you've just identified. Colonial research reflects and reinforces domination and exploitation through the attitudes and differential power embodied in its research relationships with 'others', its dismissal of their rights and knowledge, its intrusive and non-participatory methodologies, and often also its goals and its use of research findings. Post-colonial research, to me, is a reaction to and rejection of colonial research and is intended to contribute to the self-determination and welfare of 'others' through methodologies and the use of research findings that value their rights, knowledge, perspectives, concerns, and desires and are based on open and more egalitarian relationships. **Decolonizing research** goes further still in attempting to use the research process and research findings to break down the cross-cultural discourses, asymmetrical power relationships, representations, and political, economic, and social structures through which colonialism and neo-colonialism are constructed and maintained. I owe the term **inclusionary research** to your use of it for a particularly revolutionary kind of decolonizing research aimed at helping to empower subordinated, marginalized, and oppressed others and providing training and tools they can use to 'overturn their world' (Howitt 2002a).

Colonial research has unfortunately long been the dominant mode of cross-cultural research in geography and anthropology and continues to be widespread today despite feminist and post-colonialist critiques. Much research continues to be imposed by outsiders for their own purposes and benefits on indigenous peoples, on many of the non-indigenous peoples of Asia, Africa, and Latin America, and on women and ethnic and cultural minorities in many societies throughout the world.[5] So much research has been conducted from such ethnocentric perspectives for morally suspect purposes and through ethically dubious ways that as Linda Tuhiwai Smith (1999, 1) notes from Maori and other indigenous peoples' perspectives in her important book, *Decolonising Methodologies: Research and Indigenous Peoples*, 'the word itself, "research" is probably one of the dirtiest words in the indigenous world's vocabulary.' Vine Deloria Jr made the same observation and indictment, declaring that 'Indians have been cursed above all other people in history. Indians have anthropologists' (1988 [1969], 78).

Richie: The Australian (Wadi Wadi) poet Barbara Nicholson makes a similar point in her 2000 poem 'Something there is . . .' (see Box 3.1).

Something There Is...   Box 3.1

by Barbara Nicholson

. . . that doesn't like an anthropologist.
You go to a university
and get a bit of paper
that says you are qualified.
Does it also say that you
have unlimited rights
to invade my space?
It seems that you believe your bit of paper
is both passport and visa to my place,
that henceforth you have the right
to scrutinise the bits and pieces
of me.
You have measured my head,
indeed, you preserved it in brine
so that future clones of your kind
can also measure and calculate my cognition.
You've counted my teeth and compared them with
the beasts of the forest,
you've delved into my uterus,
had a morbid fascination with my sacred practices of incision and concision,
with the secret expressions of my rites of passage.
On your bit of paper you record how I dress,
earn my money,
what I eat and drink,
with whom I mix and with whom I don't,

where I go and don't go,
what I spend my money on,
the physical, mental and moral state of my being,
my marriage habits,
my birthing rituals,
my funerary rites,
the position I hold
in my society.
You analyse to a fine point
my art and music,
dance and composition,
horticulture and agriculture,
pharmacology and technology.
Nothing escapes your keen eye
and your pen records it
so that other aspirants to your elevated state
may draw on your findings and further explore
the intricacies
of me . . .
and perpetuate the invasion.

Oh yes . . . something there is.
If I were to go to university
and get a bit of paper that says,
'Wadi Wadi woman, you are an anthropologist',
will that give me the right to invade your space,
to visit you
in your three-bedroom brick veneer,
note how many rice bubbles
go into your breakfast bowl,
what colour is on the roll
in your bathroom,
and see if the bathroom is clean?
Will I have the right to sit
on the end of your bed
and count every thrust
as you make love?
You will not complain
when I calculate your expenditure
on alcohol and yarndi,[6]
or count the cost when you visit McDonalds?
Remember, because I am an anthropologist,
my bit of paper gives me the right!
From now on I have carte blanche
to all the above

in your society,
and I can invade your space,
and I can record my findings
so that for generations to come
my kin can pursue
a relentless investigation
into the fabric of your existence,
into the bits and pieces
of you
so that you will be better able to fit into.

And resulting from my research
into the common cold and its effects
on you, a representative sample
of a cross-section of the population
of Double Bay, Sydney, 1994,
an avalanche of vultures from the media,
the government,
and the tourism industry
will descend on you
in ever increasing hordes
to see for themselves
if what I said
could really be true.
They will take over your lounge-room
and lay down laws for you to live by
—all for your own good of course;
they will point out to you
the necessity of changing your way of life
the prescriptive patterns of social behaviour
devised by them on your behalf.
I will be to you,
in the guise of humane academic inquiry,
as you have been to me, invader!
something there is . . .

From *Urban Songlines*

Used with kind permission of the poet from K. Reed-Gilbert, ed., *The Strength Of Us as Women: Black Women Speak*, pp. 27–30 (Ginninderra Press, 2000). For other work by Barbara Nicholson, see the journals *XText* (1996) and *Law, Text and Culture* (2003).

Stan: Exactly! I hadn't come across Barbara Nicholson's poem, but it should be required reading for prospective cross-cultural researchers. These kinds of **subaltern** critiques of colonial research condemn not only the ways in which research has objectified 'others', violated their privacy and their humanity, and promoted colonizing agendas but also the ways in which Western science and scholarship have (mis)represented non-Western, indigenous, and subaltern peoples and groups. Linda Tuhiwai Smith accordingly identifies research as 'a significant site of struggle between the interests and ways of knowing of the West and the interests and ways of resisting of the Other' (Smith 1999, 2).[7]

Although colonial research has typically claimed positivistic objectivity, validity, and **reliability**, from the perspectives of the colonized and researched it has instead been seen as subjective, superficial, and misinformed. Smith (1999) argues that this reflects subjectivity, positionality, situated and partial knowledges, and ethnocentrism. She draws on Said's (1978) seminal post-colonial critique of the West's construction of the Orient, Foucault's (1972) concept of Western knowledge as a cultural archive in which many different cultures' knowledges, histories, and artefacts have been collected, categorized, and represented, and feminist critiques of positivism that highlight the partial and situated character of knowledge (Haraway 1991; Reinharz 1992; Stanley and Wise 1993). Colonial research also reflects and embodies unequal power relationships (and associated discourses and methodologies) and the ramifications they have for cross-cultural relationships and interaction. Imposed, exploitative research denies respect for alternative ways of knowing. It undermines trust and sabotages communication and collaborative exploration. By thus shaping the interactions and relationships on which cross-cultural fieldwork relies, this colonial model of unreflective fieldwork produces distorted and ethnocentric findings and analysis, fostering **Orientalism** in its depictions of 'others'. When it ventures into the realm of activism in the name of development or conservation, such research tends to support paternalism and imposed, external intervention, often with equally regrettable results. Post-colonial research rejects the assumptions and methods of colonial research. Research is instead envisioned as a means of contesting colonialism and neo-colonialism through fostering self-determination and cultural affirmation (Smith 1999). In my view, post-colonial research aims at being emancipatory not simply through being more culturally sensitive or seeking local research approval but through respect for the legitimacy of others' knowledge and their ways of knowing and being and through activism in support of their pursuit and exercise of self-determination. This requires acknowledging and repudiating the dynamics of power that shape colonial research interactions with subordinated and marginalized peoples and groups, attempting to overcome whatever ethnocentrism and paternalism we bring to the research and whatever suspicion we are greeted with, persuading people that we are worthy of being taught and capable of learning, and being willing to put aside preconceptions (and academic and activist preoccupations), to listen, and to be of service to local concerns and projects.

Having drawn such strong contrasts between colonial and post-colonial research, I would like to be clear that I am not suggesting that there has been a simple historical progression from colonial to post-colonial research, nor even that there is always a clear dichotomy between the two. Nor would I advocate that post-colonial researchers casually dismiss and ignore research findings from less critically-informed research or assume that all work prior to the development of post-colonial research methodologies was uniformly exploitative, misinformed, and paternalistic. We should recognize that much work today, even work cast within post-colonial frameworks, may remain colonial to some degree. And certainly there is much to learn from the research findings and experience of colonial research (as well as to react to and to revise), despite the ethical issues that it raises and the difficulties of evaluating the degree to which its research findings have been distorted by the goals, attitudes, relationships, and methodologies that shape such fieldwork. There is also much to respect and learn from in the work of earlier researchers whose fieldwork in some ways broke with the prevalent colonial research climate of their time.[8]

Richie: Deborah Rose refers to the need to develop critical perspectives on our own work and to guard against deep colonizing (1999, 177). She suggests that even the educated and sensitive 'Western' self risks enclosing itself within a 'hall of mirrors' in which it mistakes the endless reflections of its own common sense for a universal truth—and truthfulness. And we often see a romantic idealizing of the 'indigenous Other' as a source of some alternative universal truth. How do you see the task of doing research using methods and approaches that foster decolonization of peoples' lives and domains—and even of ourselves and our discipline?

Stan: That raises important issues of attitudes, intentions, and relationships. I'd like to first suggest a few practical strategies for such a decolonizing approach and then discuss the post-colonial perspectives and attitudes towards difference that they reflect and embody. I think we need to work to avoid 'deep colonizing' and foster decolonizing in multiple, interactive ways. One vital part of this is 'critical self-reflexivity' (England 1994; Rose 1997; see also Chapter 2 of this volume). Another is fostering relationships with the people we work with that make it possible for them to voice their concerns and other feedback about us and the research project in an open and honest way. This is important on an everyday level, but it can also include seeking formal or informal local authorization for research, as you've written about, Richie (Howitt 2002a) or placing the project under community supervision (Smith 1999). **Local research authorization** or **community supervision** both involve reaching an understanding on appropriate research goals, methodologies (including the **cultural protocols** within which it will be conducted), and how knowledge will be shared and used. If there is no formal process for this, as has been the case in the Sharwa regions I've worked in, one can try to achieve something similar through working with a set of local residents or—still better—with local mentors and co-researchers. As I see it, the greatest potential for fieldwork to be decolonizing for all involved is to give up 'control' of the research and develop either truly collaborative research or to contribute to local or indigenous research in which we offer our skills as colleagues, consultants, or allies (see Chapter 13 for further discussion; see also Herlihy 2003; Herlihy and Knapp 2003; Maya People of Southern Belize, Toledo Maya Cultural Council, and Toledo Alcades Association 1997; Park 1993; Nietschmann 1995; D.A. Smith 2003).

Decolonizing research thus requires a fundamental shift in research approach and conduct that begins with a perception and response to difference and 'others' that is the antithesis of the perspective and practice of colonial research. Colonial research is grounded in the binary, polarizing perception of other peoples as the 'Other', a position of distance in opposition to the 'Self', 'Us', (and often also to the 'West') that is fraught with ethnocentrism and hierarchy and undergirds cross-cultural and interpersonal dynamics of dominance and exploitation.[9] To me, post-colonial research is instead grounded in the perception of other peoples as 'others' who are different but not intrinsically different or alien, who differ culturally but not in essential humanity and value. It is attracted to difference rather than wary of it; it seeks to interact rather than to remain distant; it coexists with, respects, and honours difference rather than dominating or exploiting it. This difference of attitude and intention makes for very different conceptions of the purposes of research. Colonial research promotes, deliberately or inadvertently, a colonizing (and now globalizing) agenda that includes political, economic, and cultural imperialism and neo-colonialism; national, transnational, and global exploitation and commodification of 'resources' (natural and human); Western or other colonial discourses about 'civilization', 'development', or 'conservation'; Western conceptions of 'science'; and promotion of a Western over-preoccupation with 'self' and self-gratification (not excluding such selfish motivations as the pursuit of academic status or the satisfying of intellectual curiosity). With post-colonial approaches, the purposes of research instead become cross-cultural understanding, the celebration of diversity, and especially empowerment or emancipation—a decolonizing project based on rejecting the ethnocentrism and exploitation of colonialism through cross-cultural respect and through support for self-determination.

These differences in attitudes and intentions in turn tend to be manifest in research topics and conceptual frameworks. I would like to think that post-colonial researchers approach research with a high level of idealism, that they may tend to be less concerned with research topics for their self-serving value in enhancing their careers or bringing financial gain or with scholarly or scientific intellectual pursuits as ends in themselves.[10] Post-colonial researchers are often strongly drawn to research topics that enable them to engage in advocacy and activism and to theory and concepts that enable them to study and affect exploitative or oppressive political, social, and economic relationships, processes, and contexts.

This has attracted post-colonial researchers particularly to **postmodernism** (including post-structuralism as well as post-colonialism itself), feminist theory, neo-Marxism, and the perspectives and language of independence struggles, 'subaltern' social justice movements, and indigenous rights campaigns.

From these perspectives, values, and conceptual frameworks, post-colonial research has developed methodologies that differ in significant ways from those of colonial research—a topic that will be the focus of much of our remaining discussion in this chapter.

Richie: That makes me think of David Harvey's call as president of the American Association of Geographers for an **applied people's geography** (1984, 9) in which the skills of geographical analysis might be harnessed to the service of marginalized 'others'. That idea always sat comfortably in my mind with the educational philosophies of Paulo Freire (for example, 1972b; 1976) and more recent discussion of using Goulet's ideas of **border pedagogy** in geographical education and research settings (for example, Cook 2000; Howitt 2001). But universities are themselves a major element in the construction of power and privilege. Academic institutions harness well-intentioned efforts to their own purposes—not all of which are consistent with the liberal values of education and research for the common good. In some cases, much narrower institutional concerns and specific ideological concerns become dominant. Yet universities also continue to provide opportunities for discursive, and even practical, dissent. Indeed, allowing a small space for difference while reinforcing the status quo of privilege is a defining characteristic of the liberal academy. The task of harnessing these opportunities in pursuing the core values of social justice, economic equity, ecological sustainability, and the acceptance of cultural diversity has been a central element in my projects for inclusionary geographies—they straddle intercultural spaces both discursively and practically (Howitt 2001).

Stan: I would certainly agree with that. I'd add one more point here relative to the discussion of 'deep colonizing', though—that prospective researchers should be aware that despite good intentions, even what might be intended as post-colonial, applied people's geography can still be exploitative, paternalistic, and ethnocentric in practice. This is one reason that some indigenous advocates of post-colonial research have strong reservations about research by outsiders, particularly outsiders who are not under indigenous guidance (Smith 1999). It is not enough, for example, to *intend* to carry out advocacy and activism-directed research if the means we use reflect our own ethnocentric baggage or if our preconceived notions of what needs to be studied and what needs to be done are considered patronizing, irrelevant, or threatening to the people we wish to work with and for. I am reminded here, Richie, of your call (Howitt 2002a) for 'inclusionary research', your accounts of your early experience in Australia, and my own early experiences with the Sharwas of the Chomolungma region. I arrived in Khumbu in 1982 with the idea that I could best 'help out' by documenting and exposing the adverse impacts of international tourism. That project was welcomed by local Sharwa leaders, and from the outset of the fieldwork in 1982–3 Sharwas actively participated in and assisted the research. Yet when I returned in 1984 and was taken under the guidance of two Sharwa mentors, I was gently taught that Sharwas had their own perspectives about what kinds of research they needed, their own critique of previous research (much of which had been 'colonial' in character), and their own ideas about the kinds of contributions that I could make. I came as a result to work closely with them and with other Sharwa co-researchers on a series of projects that Sharwas thought important, including documenting oral traditions and oral histories at the request of elders who were concerned about their possible loss, reporting the perspectives of Sharwas who wanted to correct outsiders' misimpressions and 'orientalizing' of their culture and homeland, and working with Sharwas to document 'counter-knowledge' and 'counter-history' that they can use in their efforts to redress the Nepal government's seizure of their territory and authority over their land use and management (Stevens 1993; 1997; 2004).

GETTING STARTED: RESEARCH LEGITIMACY AND LOCAL AUTHORIZATION

Richie: Yes, I remember heading off to western Cape York in late 1978 as an enthusiastic young student researcher intent on 'studying land rights around the Weipa bauxite mine'. I called in on a regional Aboriginal organization in Cairns on my way to Weipa, expecting to be welcomed with open arms. But I was taken aside and told in no uncertain terms that the last thing local Aboriginal people wanted or needed was a study that told the mining companies about their land rights. 'If you want to help, why don't you do something useful?' I was asked. 'Why don't you do a study of the companies that we can use?'

This question reoriented not just my first foray into geographical research but my relationship with the discipline. I read Laura Nader on 'studying up' (1972) and enthused about 'applied people's geography' (Harvey 1984). I shifted my audience from a scholarly focus to a community focus and began to ask questions that other social scientists hadn't really asked before. When I started working on research projects with Aboriginal people, no formal process of ethics review was required—although Mary Hall, my honours supervisor, required detailed research protocols and critical engagement with participating or affected community groups about ethical concerns. In many ways, she anticipated the shift away from an assumption that academic researchers have a 'right to research' and the increased accountability for formal ethical oversight of research. The legal and institutional requirements for ethical oversight and the support for ethical engagement from formal institutional quarters has been quite an important shift in the past 10 years. In that time, most institutions have increasingly put procedures in place that require researchers to establish strong ethical foundations for cross-cultural work, particularly with vulnerable groups, rather than allowing them to assume a right to research non-Western 'others'. For example, recent guidelines for researchers working with Aboriginal and Torres Strait Islander groups in Australia advocate 'building of robust relationships' as the basis for ethical engagement, with no implicit checklist to ensure conformity (**NHMRC** 2003, 3).[11]

Stan: I've found university human subjects protocols a useful starting point for pursuing decolonizing research. For cross-cultural work generally, and not only with indigenous peoples, however, I think formal or informal community-based research agreements are critical to creating a space in which non-local researchers are held accountable to local, cultural research protocols. In breadth and specificity, these protocols may go far beyond the usual university research protocols with 'informed consent', confidentiality, and avoiding harm to those with whom we work, identifying appropriate research goals and questions, appropriate ways to seek knowledge (culturally specific, appropriate methodologies), and appropriate ways for research findings and knowledge to be shared. There is usually no place for this in the other kinds of 'research permission' that may be required for foreign fieldwork. The process of obtaining research permission from the government of Nepal and from the Department of National Parks and Wildlife Conservation, for example, in part involves crafting research proposals and logistical arrangements to meet the self-interest of various government agencies and institutions and requires no consultation with the communities where research would be carried out, no consideration of culturally appropriate research methods, and no concern with the research being co-ordinated with local residents or with knowledge or other research products being shared with them. However, regardless of whether or not university, governmental, or local research agreements specify cultural protocols, I would suggest that as researchers we have an ethical responsibility to learn about and respect them. How well we work within them, moreover, will often greatly shape the reception we receive, the kinds and quality of information and insight we obtain, and indeed whether we can carry out research at all.

Richie: Yes. That ethical responsibility can be focused in the process of complying with formal institutional ethics requirements very constructively, I've found. For example, in supervising graduate students, it's helpful to take the student through my university's application form for ethics clearance and ensure that

they have considered the ethical and cultural dimensions of various methodological choices as well as the implications of philosophical pluralism and multiple viewpoints in the real-world context of their research interests (for example,
Howitt 2001; Howitt and Suchet-Pearson 2003).[12]

DOING THE WORK

The Scale Politics of Cross-Cultural Research Projects

Richie: One of the things that I've been drawn to over the years is the ways in which
a politics of scale is implicated in the construction of cross-cultural research (for
example, Howitt 1992; 1997; 2001; 2002c; 2003). In the indigenous studies domain, there is an assumption—often rooted in the ways that power relations are
constructed between indigenous and non-indigenous domains in Australia—that
the 'correct' entry point for cross-cultural research of any sort is through a 'local'
or 'regional' Aboriginal organization or through a 'community'. Indeed, many
ethical review procedures are predicated on the need for researchers working with
indigenous Australians to demonstrate their accountability to 'local' or 'community' interests. This implies, however, a relatively naive conceptualization of scale.
In contrast, people working in overseas locations often depend on national agency
approval as an entry point to their research topic and are forced to conceptualize
their study to conform to the window of opportunity offered by such links.

Working on the social impacts of transnational mining company strategies
or major development projects in mining, tourism, conservation, and transport
over recent years, however, made it impossible for me to restrict my vision to the
naively 'local'. Multi-locational fieldwork in locations linked by various aspects
of the mining production cycle (e.g., common ownership, downstream process
integration, competition, government policies) meant that the local quickly
emerged as a set of particular kinds of relationships that linked to a much wider
set of scale relationships rather than as a singularity focused on a bounded location. And any local was always and inescapably contextualized for me by a range
of critically important power relations that were constructed at several scales: in
a nation that denied its indigenous populations a right to self-government (or
self-determination), in industries characterized by corporate strategies of integration, cartelization, and financial innovation.

This raises a number of methodological questions, both in relation to how we
might conceptualize the places in which we situate ourselves as cross-cultural researchers and in relation to how we conceptualize the range of topics suitable for
research in cross-cultural settings. Indeed, in my work, it has even raised serious
questions about just what we might mean by 'research' and how we might be held

ethically responsible for our work as 'researchers'. How does one engage research participants in 'informed consent' for their involvement in, for example, a PhD study, when not a single person from the language group has completed high school and no one has been to a university? What sort of sense might they make of the question of informed consent? And even if one concedes that some sort of consent can be constructed in such circumstances, how is one really held accountable for one's immediate or subsequent actions in relation to the people involved, their representations of their lives and cultures, and one's interpretation of them for other audiences?

Stan, you've raised some important questions that also seem to have implications in terms of scale and the scale politics of research in 'foreign' settings, whether they are an unfamiliar part of one's hometown or a location on the other side of the world. Your identification of 'research affiliations' as an issue is, to my mind, immediately drawing in some critically important scale issues, and I've just had a discussion with some young colleagues about the ways in which development studies and indigenous studies enter the field via quite different (scaled) windows of 'government' (or NGO—non-government organization) and 'community' respectively. You also identify many of the key issues of collaboration and accountability.

Stan: The scale issues you've raised, Richie, are quite familiar. They are likely to face researchers carrying out political ecological work with indigenous peoples anywhere. This often involves research with—and about—multiple actors and processes in order to analyze patterns of regional land use and environmental change. Typically, this requires not only fieldwork with indigenous communities but also with government agencies and often with regional, national, and transnational NGOs and national entrepreneurs or transnational corporations (Blaikie and Brookfield 1987; Bryant and Bailey 1997; Zimmerer and Bassett 2003). When my Sharwa mentors, co-researchers, and I began to examine forest issues in Khumbu in 1984 and in the Pharak and Katuthanga regions in 1994, for example, they made it clear from the outset that it would be necessary to look at forest use and management by a range of people (Sharwa villagers, non-local Sharwa timber merchants, Nepali foresters and national park staff, and international tourists) and to assess this within the complex politics of contestation over control of territory and the management of forest commons that was being waged within and between villages and between villages and the central government's district forest office and department of national parks. In such situations, the research 'site' becomes not only indigenous settlements but also offices and archives in regional centres, national capitals, and abroad. This also often required conducting more than one kind of cross-cultural research, because in Nepal, as in many countries where indigenous peoples live in states controlled by other peoples, trying to understand local economic and environmental change involved work with non-indigenous government officials, NGO staff, and entrepreneurs (Stevens 1993; 1997; 2004; Stevens and Sherpa 1993).

Cross-cultural research often not only requires work at multiple scales but also requires negotiating relationships at those scales. This takes place at the national level (and with central government officials stationed at the local level), moreover, within a highly-charged context, because in many domestic and foreign fieldwork situations research authorization processes are often controlled by state agencies that seek to influence the direction of research, benefit from it, and prescribe procedures to hold researchers accountable to its dictates. Often, research funding agencies and universities expect researchers to adhere to these research authorization requirements. In practice, however, there is often a great range of variation in the degree to which researchers are controlled, how much cooperation and co-ordination is expected, and how well potentially awkward expectations can be defused. Researchers who aim to carry out decolonizing work may be inclined to focus on the ethics of the relationships they develop with indigenous communities and may well find government requirements inimical. Negotiating these ethical and practical dilemmas may be critical to our work and can derail it before we can begin. Such situations can be particularly intimidating when one lacks familiarity with the internal politics, institutional culture, and bureaucrats involved. It can help to seek advice from other researchers with recent experience.

Richie: In these settings, it is important to contextualize one's work—to recognize that the state and its agencies have ambiguous relationships with many indigenous and tribal minorities and their territories, that there is a scale politics in the development discourse itself, and that research is easily drawn into that scale politics in ways that are not obvious to new researchers in a region. There is a substantial literature on the ways in which nation-states, even mature democratic nation-states, respond to developmentalist imperatives to develop remote corners of 'their' territories and the people whose homelands they have been since time immemorial (see, for example, Berger 1991; Clarke 2001; Howitt 2001; Wolf 1982; also, more generally, Tully 1995). Yet there are also elements of accountability, ethics, and responsibility to grapple with in working with state agencies, transnational corporations, and environmental NGOs. In many ways, there is a very different sort of responsibility and accountability in being an informed critic in such complex situations rather than an ignorant critic!

Social/Cultural Transmission and Creation of Knowledge

Richie: We face a tension in how we conceptualize and bring the creation of knowledge to life and negotiate its various purposes. In some situations, research orientation is towards demonstration of a theory or salvaging a 'lost' or fragile way of life (for example, Singer and Woodhead 1988). But in many situations, the sorts of research local people themselves prioritize is concerned with exploration of the possibilities of the intercultural domain or explanation of 'the other side' of the cross-cultural relationships they experience. For many of my research students, it is their capacity to explain how the institutions, values, and practices of non-Aboriginal society work that is their greatest value for Aboriginal people—not their expertise in cross-cultural matters. Indeed, it is often our lack of facility in listening to and learning from local research participants and collaborators that limits our capacity to frame knowledge and understanding of our own culture and its operations in ways that are accessible and meaningful to local people. It is in listening to and engaging with interpreters about the concepts underpinning the knowledge we create that we make some of the most important realizations about our work. For example, in work on Aboriginal contributions to the central Australian economy in the late 1980s (Howitt, Crough, and Pritchard 1990), we found that Warlpiri interpreters were translating the concept 'wealth' with the Warlpiri term for 'money'. Our analysis was seeking to explore the contributions of non-monetary wealth to Aboriginal futures—including culture, children, health, and so on—and we needed to tackle the terms in which these ideas were being conveyed in workshop discussions with Warlpiri speakers. More recently, in South Australian work (Agius, Howitt, and Jarvis 2003; Agius et al. 2004), Antikarinja interpreters and speakers spent half a day discussing the conceptual differences between ideas of 'agreement' and 'negotiation' after they realized we were using different words that were being translated into the same Antikarinja term. This discussion of the different implications of a term, one signifying a process and another addressing an outcome, was a careful exploration of how meaning is constructed and the implications for meaning-making in cross-cultural settings. It is through such conversations with the most direct users of our research that I often find my most inspirational collaborators and peers.

Although producing materials for these groups is important, there are other times when producing more conventionally academic publications is important. Ensuring the credibility of our work through peer review procedures, for example, can ensure that work that local or indigenous collaborators are relying on in their own efforts to change their circumstances has credibility with governments and others. And we have a responsibility to contribute ideas into those academic discourses that reflect the learning made possible by the work we do in our intercultural engagements. Those more academic papers can provide transformational opportunities well beyond the confines we might have previously imagined for our modest endeavours.

Collaborative and participatory research[13]

Stan: Yes, we've been talking all along about methodology in the sense of approaches to knowing rather than as only a set of research techniques. As a reaction against colonial research, there has been increased interest in research that is more culturally sensitive and emancipatory in its orientation and often focused on **collaborative research**, locally guided research, and indigenous research. In some indigenous societies, as Smith (1999) describes for the Maori, histories of colonial research have led some people to demand that outsiders cease research on indigenous peoples and issues, and in such cases some have called instead (as she so powerfully does) for indigenous research and research methodologies. I feel that both research by 'insiders' and cross-cultural research by 'outsiders' are important, and on the basis of my own experience, I have come to believe that 'others' can value decolonizing relationships and friendships with outside re-searchers, consider outsiders' cross-cultural perspectives and insights into their societies and situations useful, seek to mentor and work with outside research-ers out of interest in the research and belief in its significance, and view outside researchers as useful advocates and allies. While Smith (1999) is sceptical of re-search by non-indigenous researchers, she and other Maori researchers and crit-ics of colonial research indicate some avenues to cross-cultural research that are both ethical and concerned with outcomes that benefit indigenous peoples and that would be equally valuable in other cross-cultural research. In her view, such efforts begin with critical awareness of how research is shaped by relationships, power, and ethics. Researchers also can make an effort to work in more culturally sensitive ways, prepare for research by learning the local language, interact with 'others' on their terms in their own social/political community venues, become informed about local concerns, seek local support and consent for research, and honour local cultural research protocols and negotiated research agreements. And they can change the nature of their research by making local participation integral to it. This can be done in several ways, as Graham Smith (in Linda Tuhi-wai Smith 1999) has suggested for research with indigenous peoples. Research-ers can work under the guidance of local mentors. They can become adopted as members of the community, with all the lifelong obligations and responsibilities this entails. They can establish a 'power-sharing' approach in which they seek community support for research. And they can adopt an 'empowering outcomes' approach in which research becomes a vehicle through which 'others' can obtain information they seek and which they can use to their benefit. To these Linda Tuhiwai Smith adds a more deliberately 'bicultural or partnership' approach in which local and outside researchers design and carry out a project together, a process that requires negotiation and agreement on many important aspects of research methodology, design, and use.

In exploring the idea of 'bicultural or partnership research', I feel that it is worth drawing a distinction between the attitudes and relationships embodied by participatory research in general and what I would call truly collaborative research. In participatory research, an effort is made to create a space for more involvement by 'others' as an integral part of the research approach. Participation itself does not, of course, necessarily represent a break from colonial research, since it can amount to nothing more than enlisting local cooperation in a research project that continues to be driven by outside researchers' definitions of its purposes, methods, and use. Participatory research can also be carried out with an agenda of activism and empowerment by addressing locally relevant issues, supporting self-determination, human rights, and indigenous rights, providing new knowledge that is of use to subordinated peoples, groups, and individuals, and transferring research skills that they can employ in the future on their own behalf.[14] Even in this case, however, there may still be attitudes of intellectual arrogance, paternalism, and evangelism (Bishop 1994; Smith 1999). Who, it might be asked, is being given the opportunity to participate in whose project?[15]

Richie: Stan, this is important in so many settings, where young, well-intentioned researchers feel that their university credentials give them an expertise in situations they know relatively little about. The risk is in allowing this assumption to frame the collaboration. In native title research in Australia, for example, we often find Aboriginal groups having to tell an expert who they are, how their society works, and what their historical and geographical setting is, because only an accredited 'expert' can give evidence on these topics to a court. The claimants themselves are not trusted to represent their societies adequately (or perhaps truthfully), and the intervention of (well-paid) experts is mandatory—an extraordinarily arrogant misreading of the nature and source of such knowledge. Although it is the evidentiary system that drives this situation, many of the researchers accept the misreading without challenging it.

Stan: Collaborative research—truly collaborative research—reflects a sharper break from imposed, colonial research based on different attitudes towards 'others' and different relationships among researchers. Collaborators work as equals on a mutual project. This decolonization of the relationships within the research team can generate an interactive, cross-cultural synthesis of knowledge and skills through which research conducted with community and other local authorization and within local cultural protocols can be used to address community concerns. Local and non-local researchers conceive and design the research together, including making the key decisions on defining research goals and questions, where and how to seek funding, affiliation, and authorization, who should be on the research team, what methodology should be used, how cultural research protocols should be honoured, how the day-to-day conduct of fieldwork should be handled, what kinds of analyses should be attempted, and how research findings should be shared and used. This requires non-indigenous researchers to give up 'control' over a project and for all involved to contribute their time and efforts in order to work together towards shared goals. This is not easy to do and can only really take place if all researchers can move beyond often quite strong assumptions and behaviours conditioned by their statuses in their own societies and by asymmetrical cross-cultural power relationships. On the basis of my own experience of working within collaborative research relationships with several different Sharwa co-researchers, I would advise not to underestimate the time, care, emotional commitment, self-reflection, learning, and stress it can entail on everyone's part.[16]

Collaborative research approaches and locally-guided ones are based on conducting research that works within what are considered culturally appropriate ways of seeking and transmitting knowledge. There are several important dimensions to this. One of the most basic, and often also the most difficult for researchers to accept, is that in many societies, knowledge and information are not necessarily shared openly within communities, much less with outsiders. There may thus be particular types and levels of knowledge that are considered appropriate to share with or withhold from outside researchers (Stevens 2001).[17] Knowledge may also be considered ethnic-, gender-, age-, class-, religion-, or subculture-specific. Researchers should also be aware that there may be types of knowledge that may be shared with outside researchers with the understanding that they are not to be otherwise shared with outsiders. Outside researchers should accept that honouring these concerns may greatly affect the design and conduct of research and the publishing of research findings.

Cultural protocols about acquiring knowledge also may include an understanding that knowledge must be earned by working within culturally-sanctioned methodologies. This may apply only to certain kinds of knowledge and may not preclude the use of other culturally-sensitive and authorized research methods to learn about local conditions and practices, methods that may produce new knowledge, insight, and community empowerment. But the importance of working within existing local systems of transmission of knowledge should not be ignored. This may mean, for example, that one must be found worthy of being mentored and then undertake a possibly long period of instruction. Researchers whose projects do not have the documentation or translation of a particular type of specialist's knowledge as a goal may be inclined to decline involvement in such a relationship (and knowledge specialists may have their own reasons for declining to take one on as a student). But it may often be the case that a great deal of information that is important to the project can only be obtained through such culturally-sanctioned means. Researchers must then weigh whether to make the commitment to the relationship, the process, and the project. Those who persevere may find great personal and research rewards from a kind of cross-cultural communication and learning that ordinary interviewing (much less group meeting-based rapid research) cannot approach.

Cultural protocols also may have considerable ramifications for the use of particular research techniques. The use of questionnaires or formal, structured interviews, for example, may be considered intrusive, rigid, and exploitative by some peoples. Informal, semi-structured interviews, on the other hand, may not conflict with local etiquette about social interaction and communication because they can be interactive discussions or conversations in which there can be reciprocal exchanges of information. Interviewees may indeed value such interviews as social occasions that provide an opportunity to get to know the researchers, inquire about research findings, and learn about the outside world. Researchers may find that elders and knowledge specialists do not wish to be interviewed, or may think that requests for assistance are appropriate only after a process of acquaintance and relationship, or may initially speak only in generalities. Yet in other cases, if properly approached, they may be willing to act as mentors. People may be uncomfortable with group discussions or consider them essential. Mapping may be highly suspect if one is not well known and respected, while in other contexts it may be invited and welcomed. Cultural protocols and negotiated local research agreements may also influence research design in terms of whom it is appropriate to interview, what topics are considered suitable, and when interviews are and are not appropriate (such as not being appropriate during festivals or times when people are busy with subsistence activities). Local etiquette may also influence conversation in very specific ways, such as when there are taboos against using the names of the dead.

Another aspect of honouring local cultural protocols that can significantly affect research arises in cases in which societal discourses and practices promote relationships and interactions of inequality and domination between women and men or between people of different ethnicity, race, class, religion, age, or other socially-defined categories and groups. Such cultural protocols can pose a formidable dilemma for outside researchers who wish to respect local concerns but object to the character of the interactions and relationships created by societal discourses and asymmetries of power and do not wish to engage in or to tacitly accept them. Local societal attitudes towards difference may also complicate the interactions and relationships of team members and affect the degree to which their research is fully collaborative. And cultural protocols that discourage or constrain interaction between women and men or between people of different ethnicities, classes, castes, or other social groups can make research across socially-defined boundaries of difference difficult or impossible for both local and outsider researchers.[18]

Constructing Legitimacy

Stan: The reception that we and our research projects receive depends in part on local perception of the projects' legitimacy in terms of whether or not they address local concerns, needs, and interests, partly on how the projects are conducted and whether they meet cultural protocols, and partly on local perception of our character and those of our co-researchers. The importance of the perception of our character should not be underestimated. Often, it will be based primarily on our personal qualities, the evaluation of which can vary considerably among cultures but may well include whether we are considered to have a good heart or spirit, whether we can be trusted, how we treat and interact with people, how well we listen, and what skills and resources we bring to the community (see Smith 1999). While our academic achievements and status may matter little, we should be aware that how we are perceived can be very much affected by the company we keep or are perceived to keep. Research teams will often also be judged by the character, reputation, and actions of all of their members, including local ones (see, for example, Berreman 1972). And we may need to work hard to counter suspicions that we are agents of the central government, a transnational corporation, or a locally unpopular NGO and to reassure people that research findings will not find their way into such hands. However, affiliation with local, regional, national, or transnational NGOs or with particular government programs can also enhance one's legitimacy when they are well regarded locally, and this can sometimes both smooth the process of negotiating for research authorization from government agencies and serve as an easy entree into communities. The problem is that it can be difficult to know how agencies, organizations, and programs are locally perceived until one has spent time in an area and people trust one well enough to be candid. This is one of many reasons why it is often ideal to carry out an extended reconnaissance of a potential research site before establishing affiliations and research authorizations. In any event, we need to be clear with community members about our relationships (if any) with government agencies and NGOs and to be prepared to rethink those relationships based on community concerns.

Richie: Of course, this means that many students have to rely on the credibility and experience of a supervisor. That can open doors, but in my experience it is ultimately the students' own credibility and qualities that carry them through into worthwhile working relationships with people. Because those relationships are so dependent on personal integrity rather than on status, it becomes important for students and new researchers to think about what this integrity might mean in the setting they hope to work in. But it also means that the time frames involved in negotiating these matters can be inconsistent with the bureaucratic expectations around a student's candidature.

Stan: Legitimacy is also created through social relationships. Working in places and within communities fosters the development of relationships, and these relationships bring with them expectations of reciprocity and diverse responsibilities.

Meeting these obligations over time can greatly enhance our acceptance in a community and attitudes towards our work. And over time, we also establish networks of friends and allies who will vouch for our character and intentions.

MAKING SENSE, REACHING CONCLUSIONS

Richie: One of the really important issues that we've hardly touched so far is the question of audiences in cross-cultural research. Working in an academic environment, we are inevitably influenced by the publication culture of our institutions. Our own legitimacy as intellectuals is constructed in our 'public' work. But in many cross-cultural settings, the primary audience for our work is neither academic nor reached by academic publications. 'Publishing' to community-based audiences is often the most important element of cross-cultural research. It is the point at which we become accountable to our participants and collaborators for the ideas and knowledge our work produces. And it often involves exceptionally rigorous scrutiny, with multilingual discussion, careful (re)contextualization of ideas and information, checking of facts and interpretations, challenges, and debate. While this is some of the most demanding peer reviewing one experiences as an academic, it is not acknowledged as such by the institutional academic community. Many of the publication formats it involves (for example, community newsletters, comic books, radio broadcasts, long debates in community settings, joint submissions to inquiry processes, manifestos, and community statements) will not count for credit when it comes to thesis examinations, academic tenure, and promotion. Yet if we are in the business of producing ideas that change the world, this can be our most effective and influential work, and it is powerfully tested in communities' efforts to actually change their own circumstances!

Of course, we cannot escape the requirement to publish in more academic and professional settings, but that also raises important questions about how one represents the intercultural domain for other audiences. It is easy to be cast in a role as an 'expert' on another culture—however limited our grasp of or engagement with that culture in all its complexity. In some cases, a range of outputs or different types and formats for publications can reframe quite sophisticated academic outputs for multiple audiences (Coombs et al. [1989] and Rose [1996b] offer impressive examples).

Stan: Yes. There is a responsibility for researchers to share findings with the people they work with. This is more than the courtesy of ensuring that copies of our subsequent academic publications are widely available in the research area. We need to seek out culturally appropriate and effective ways of sharing knowledge. Academic publication is often not a very effective way of doing this in any cross-cultural situation, and non-academic writing specifically produced for local audiences may have a limited impact in societies where literacy is rare. In societies where knowledge is shared face-to-face and orally, it may be effective to discuss the research findings in group meetings, community meetings, or workshops. It can also be very effective and rewarding to discuss findings directly with individuals, although this takes considerable time to do widely. These methods also have the major benefit that they are interactive and provide opportunities for feedback that may correct misimpressions and overgeneralizations, provide alternative information and analyses, clarify concerns with the communication of some information or particular portrayals to the outside world, and provide opportunities to discuss how findings from the research can be used by individuals and communities in their lives and actions. Co-researchers also play a major role in disseminating research findings, since they become in-place, living repositories of that knowledge and the skills and experiences through which it was created and they can use them directly in their own lives and to inform community discussions and action. This can be much more powerful than anything an outside researcher would be able to contribute.

Finally, for me, another important aspect of making sense of the research is coming to terms with how that experience reshapes our lives and the lives of those with whom we work. Cross-cultural research can lead to long-term relationships between researchers and 'others', and with them come obligations and responsibilities that can far transcend the sphere of the research itself and the relatively brief time that most researchers devote to living in places and carrying out fieldwork. The closeness that we develop with co-researchers, mentors, and friends can lead to continuing, lifelong affection and interaction, and in these relationships there can be strong cultural expectations of generosity and mutual aid that extend to each other's relatives, children, and associates. For some researchers (as Barney Nietschmann and I both found), field experiences and relationships can create a strong sense of commitment to a people, particular communities, and a region that leads us to return repeatedly to the place, the people, and further rounds of research and activism.

Richie: Yes, what might be conceptualized academically as 'cross-cultural' is simultaneously interpersonal, and it has profound implications for just what sort of human beings we imagine ourselves to be or to be capable of becoming! To some extent, one is drawn into more activist and advocate roles than many of one's colleagues and even into framing our academic roles as 'teacher' and 'researcher' somewhat differently. I have long considered the links between my own intellectual nourishment from research and teaching and the construction of critical engagements with students and the wider societies of which I am part to be a central element of my responsibility as a public intellectual. That sort of engagement is not limited to the classroom. It happens in a wide range of places where one tries to make new sorts of sense that might disrupt the certainties of colonial (and deep colonizing) practices. For many researchers, finding the balance between scholarship and activism is far from easy.

Postscript, January 2009

Stan: Richie, resuming our dialogue for the revised edition of the book, there are a few points I would like to underscore in support of post-colonial, decolonizing, and inclusionary research. First, indigenous peoples and other marginalized and oppressed peoples, communities, and groups continue to struggle against colonial research in many parts of the world. Colonial research continues despite increased attention in some academic fields and in some countries to research ethics, including codes of conduct and research protocol requirements implemented by universities and funders (Israel and Hay 2006). Some peoples and communities are responding by developing their own research authorization and conduct protocols and requirements. The Inuit, for example, have developed such procedures for research in Nunavut territory, Canada (the Nunavut Research Institute's website, http://www.nri.nu.ca, includes research authorization forms and a number of other useful materials) and have produced a booklet, *Negotiating Research Relationships with Inuit Communities: A Guide for Researchers* (Nunavut Research Institute and Inuit Tapiriit Kanatami n.d.), that is useful for researchers and communities more generally and includes a sample negotiated research relationship agreement. Yet many peoples and communities lack the political self-governance and self-determination to be able to legally require that researchers obtain their prior, free, informed consent for research on their lives, lands, and waters, much less to ensure that research is collaborative, respects local cultural protocols, and is decolonizing. The Sharwas of the Mount Everest region are one of the many peoples who find themselves in this circumstance. Under an onslaught of recent, largely unwelcome research conducted without their consent, collaboration, or respect, Sharwa leaders have attempted to confront and reform (with modest success) egregious research programs and recently have been developing an informal, voluntary research authorization and conduct protocol. Sharwa leaders hope that most researchers will welcome working together with them once they are aware that procedures are in place.

Richie: That's a familiar story—but it's often the case that the demands and expectations of researchers exceed the capacities of and opportunities for local groups to intervene informally. Strategies such as the Nunavut website and disciplinary mentoring can also go a long way in addressing the colonizing consequences of otherwise well-intended research. But opportunistic and exploitative research is much harder to address. The persistence of colonial imperatives of possession, erasure, and denial is reinforced by many of the elements of exploitation that are glossed as 'globalization' and seems to fly in the face of the powerful idea of the 'stickiness' of place, which makes those relationships of governance, accountability, and ethics at the local scale that we emphasized previously so important. It reminds me of just how important Nader's idea of 'studying up' is in cross-cultural research (Nader 1974). We often need to train the focus of our research on the structures of privilege, power, and marginalization to really get at the questions of what local groups can or can't do. In many cases, it is the lack of capacity among the government agencies and corporate interests to actually work on the ground with people in their community groups and organizations that frustrates people so profoundly—and that can be investigated by social researchers. In many of the bureaucratic systems that dominate local peoples' lives, there is a genuinely wicked complexity that develops from the fragmented, conflicting, and competitive systems of agency 'silos', which communicate with each other only by imposing contradictory requirements on indigenous or local groups unlucky enough to receive funding support from multiple agencies.

Stan: In seconding your call for 'studying up', Richie, I would add that research on the varied intercultural relationships and dynamics between marginalized 'others' and national and international NGOs can also be a contribution. Much of what you've observed about the relationships between communities and corporate interests or government agencies can also apply to interactions with outside NGOs. Indigenous peoples and other communities and groups may find 'scaling up' through alliances with national and international NGOs a useful strategy in their struggles for recognition, respect, and self-determination. But in other cases, NGO interventions can create significant problems and indeed perpetuate or accentuate subjugation, dispossession, and assimilation, and affected communities may value research that illuminates these dynamics and contributes to transforming existing relationships and interactions. Researchers should not assume that affiliation with NGOs is a straightforward way of facilitating local authorization, legitimacy, and collaboration.

Before we conclude our dialogue, I also want to highlight that although indigenous peoples are not well represented in academe in many countries, some indigenous geographers are actively engaging with issues of research methods, including both **indigenous methodologies** and the methodological and ethical issues raised in cross-cultural, collaborative research. I would like to call attention to Jay Johnson, Renee Pualani Louis, and Albertus Pramono's (2005) discussions and Renee Pualani Louis's paper (2007). These issues are also being much discussed within professional associations, including the Indigenous Peoples Specialty Group of the Association of American Geographers, the Canadian

Association of Geographers Native Canadians Study Group, the Indigenous Issues Study Group of the Institute of Australian Geographers, and the Indigenous Peoples' Knowledges and Rights Commission of the International Geographical Union. I hope that these voices will be heard and listened to more widely in geography.

Richie: The paper by Renee Pualani Louis (2007) is a terrific piece. I think that the development of indigenous methodologies is moving rapidly at the moment. We are seeing the development of collaborative networks, research management frameworks, and ethical oversight and accountability arrangements that will require a dedicated chapter on this topic in the next edition of this book!

KEY TERMS

applied people's geography	inclusionary research
border pedagogy	indigenous methodologies
collaborative research	legitimacy
colonial research	local research authorization
community supervision	Orientalism
co-researcher	post-colonial research
cultural protocols	post-modernism
decolonizing research	reliability
deep colonizing	subaltern

REVIEW QUESTIONS

1. What are some key differences between colonial and post-colonial research in assumptions, attitudes, relationships, and methodologies?
2. How can researchers go about attempting to carry out decolonizing research and inclusionary research?
3. How can research be made more truly collaborative in all aspects of a project, and how might this affect the time, energy, and outputs involved and affect a student's progress towards her or his degree?
4. Why might university research protocols often need to be supplemented by local cultural protocols?
5. How might issues about research with indigenous peoples raised in this chapter be relevant to cross-cultural research in your own situation?

USEFUL RESOURCES

Ivanitz, M. 1999. 'Culture, ethics and participatory methodology in cross-cultural research'. *Australian Aboriginal Studies* 2: 46–58.

Louis, R.P. 2007. 'Can you hear us now? Voices from the margin: Using indigenous methodologies in geographic research'. *Geographical Research* 45 (2): 130–9.

Maya People of Southern Belize, Toledo Maya Cultural Council, and Toledo Alcades Association. 1997. *Maya Atlas: The Struggle to Preserve Maya Land in Southern Belize*. Berkeley, CA: North Atlantic Books.

Mullings, B. 1999. 'Insider or outsider, both or neither: Some dilemmas of interviewing in a cross-cultural setting'. *Geoforum* 30: 337–50.

Nunavut Research Institute and Inuit Tapiriit Kanatami. n.d. 'Negotiating research relationships with Inuit communities: A guide for researchers'. http://www.nri.nu.ca/pdf/06-068%20ITK%20NRR%20booklet.pdf.

Smith, L.T. 1999. *Decolonising Methodologies: Research and Indigenous Peoples*. Denedin: University of Otago Press; London: Zed Books.

Notes

1. For me (Richie Howitt), it was my awareness of the US government sponsorship of 'research' on social movements in Latin America and hill tribes in Indo-China in the 1960s and the disputes it produced within anthropology (e.g., Horowitz 1967; also Gough 1968; Jorgensen 1971). This helped push me into engaging with intercultural research ethics early in my strange career! It also drew me into a critical consideration of our own discipline's implication in colonizing efforts (e.g., Howitt and Jackson 1998).

2. One example of a similar approach close to home is Herman and Mattingly's (1999) effort to negotiate 'reciprocal research relations', cultivate 'the authority of self-repre-sentation', and implement 'ethical responsibility' through community action as well as research methodology in their work in the most culturally diverse neighbourhood in San Diego, California. Also see Chapter 13, Katz 1994, Kobayashi 1994, and Nast 1994.

3. Post-colonial can signify both 'after' colonialism and 'rejecting', 'against', or 'anti' colonialism. I use the term in the sense of 'rejecting' colonialism. Even in cross-cultural settings that are not appropriately characterized in these terms, rethinking research relationships in terms of participatory action frameworks (see Chapter 13) can redefine research outcomes towards mutual respect and benefit.

4. In Australian public policy, the work of Charles Rowley (e.g., 1970; 1971a; 1971b) and Nugget Coombs (e.g., 1978, but see also Rowse 2000) reflected a deeply reflective and non-paternalistic approach within a predominantly 'colonial' period. Reynolds (1998) refers to such dissent as a 'whispering in our hearts' and notes that 'In each generation people have expressed their concern about the ethics of colonisation, the incidence of racial violence, the taking of the land and the suffering, deprivation and poverty of Aboriginal society in the wake of settlement' (1998, xiv). Geertz (e.g., 1973; 1980; 1984) similarly throws a different light on 'colonial' efforts to understand cultural difference.

5. Many of the criticisms of colonial research can also be applied to much cross-cultural research with subordinated and marginalized peoples and groups in non-colonial contexts (although some post-colonial scholars would prefer to define the term more

narrowly), as in Nietschmann's (2001, 183) remark that 'Who studies and who gets studied reflects power, economics, status, class, color, and identity.'

6. Yarndi: marijuana.

7. **Stan:** I prefer to use 'others' in preference to 'the Other' to acknowledge cultural diversity, although I remain uncomfortable with the distance and attitudes implied in any form of the word and prefer less polarized language.

Richie: Yes, I find that too, but there is something important in acknowledging that the language of power reflects something significant about the relationships it represents. As the philosopher Levinas suggests, that awkward singularity of a generalized 'Other' annihilates something very significant in a relationship 'whose terms do not form a totality' (Levinas 1969, 39). In the case of the 'self–other relation', there is no larger concept higher in some implied hierarchy that encompasses these two terms. So for Levinas, aggregation of the self and the other does not produce a new, larger singularity because to do so would be to deny the ethical (or unethical) power relations that create differences. While the more generalized singularity 'human' might encompass such differences, the risk is that in seeking to challenge the realities of the relationships of power, we use language that obscures it. It's a good reminder of the need to problematize language and to reflect on and challenge many aspects of the hidden constructs of injustice in the work we do.

8. In the case of Sharwa studies, this has included research by pioneering anthropologists Christoph von Furer-Haimendorf (1964; 1975; 1984), Sherry Ortner (1978; 1989; 1999), and James Fisher (1990).

9. Said (1978) notes that such views of 'others' go back to the ancient Greek delineation of themselves and 'barbarian' others, although the ancient Chinese developed a similar perception early on and it might be argued that such ethnocentrism is very old, very widespread, and common outside of imperial and colonizing situations as well as in them.

10. In practice, many of us find we must negotiate multiple personal motives in our work, and I would suggest that while our ideals and commitments may often call for a measure of self-sacrifice in our work, it is also legitimate to be concerned with such matters as completing one's degree or attaining tenure.

11. These new guidelines seek to establish grounds for developing ethical relationships between indigenous communities and research groups rather than offering a checklist of how to make a project application look as though it conformed to institutional ethics requirements. This approach shifts the orientation of ethics oversight away from formal legal concerns about risks to the institution and constructs a framework for higher levels of accountability to those participating in or affected by the research and will require some degree of rethinking in institutional compliance structures—as does the recognition of native title.

12. There is material related to project-based agreements for research ethics available on Richie Howitt's website, http://www.es.mq.edu.au/~rhowitt.

13. Readers are encouraged to consult Chapter 13 for a complementary discussion of participatory approaches to research.

14. The term 'participatory research' is often used to refer to exactly this kind of collaborative, empowering research (see Park 1993).

15. This also brings up the issue of local 'research assistants', the term that it has long been conventional in geography and anthropology to use to refer to local members of a research team. In recent years, I have come to feel that this practice needs to be examined critically because of the way it can narrow local project members' participation in research by fostering hierarchical relationships among team members and restricting—and in some cases not properly acknowledging—local project members' contributions and roles.

16. On collaborative research, see Park (1993), Herlihy and Knapp (2003), and the approach that Barney Nietschmann and Berkeley geography students working for GeoMap developed with the Maya of Belize (Maya People of Southern Belize, Toledo Maya Cultural Council, and Toledo Alcades Association 1997). This 'community-based cartography' is both more collaborative and more empowering than most participatory mapping (see Wainwright 2008, however, for a critique of some aspects of the Maya Atlas project).

17. This is somewhat different from the issue of local perceptions of what levels of knowledge outside researchers are capable of understanding. Local residents may often over-generalize and simplify in response to what they perceive as an outside researcher's rather basic level of understanding of their homeland and ways of life. This is a major problem for short-term research and for researchers who do not realize that there are multiple levels of explanation and understanding.

18. For reflections on the issues raised by these situations, see the 2001 special issue of the *Geographical Review* on 'Doing fieldwork', the 1994 special issue of the *Professional Geographer* on 'Women in the field', and several other collections of essays by feminist geographers working within and outside of their own societies and communities (Jones III, Nast, and Roberts 1997; Moss 2002; Wolf 1996).

Qualitative Research Design and Rigour

4

Matt Bradshaw and Elaine Stratford

CHAPTER OVERVIEW

Careful design and rigour are crucial to the dependability of any research. Research that is poorly conceived results in research that is poorly executed and in findings that do not stand up to scrutiny. Thoughtful planning of research and the use of procedures to ensure that research is rigorous should therefore be central concerns for qualitative researchers. The research questions we ask, the cases and participants we involve in our studies, and the ways we ensure the rigour of our work all need to be considered in any dependable research.

INTRODUCTION

In this chapter, we focus on some matters of design and **rigour** that qualitative researchers need to consider throughout the life of a project to ensure that the work satisfies its aims and critical audiences. We outline various principles of qualitative research design as well as some specific means by which rigour can be achieved in our work.

The chapter is organized into three main sections. First, we discuss influences on us as researchers as well as the effects we have on the conduct of research. This discussion makes a link between the interpretive communities in which we work and the sorts of issues that are raised when we begin a research project. Second, we elaborate on how to select suitable cases and participants for study. In qualitative research, the number of people we interview, communities we observe, or texts we read is an important consideration but secondary to the *quality* of who or what we involve in our research and secondary also to *how* we conduct that research. Third, we outline some ways of ensuring rigour in qualitative research to produce work that is dependable.

Careful research design is an important part of ensuring rigour in qualitative research, and while texts and topics on research methods and design often imply that studies should be conducted in a particular way (Gould 1988; Mason 2004), no single correct approach to research design can be prescribed. For certain kinds of work, the order and arrangement of stages may be different, stages can overlap, other stages might well be included, and the combination of qualitative and quantitative research is also possible and not uncommon. Nevertheless, by the end of the chapter we will have moved through a number of stages of qualitative research design and summarized this process in three diagrams. We believe that you will find this movement through stages helpful in approaching your own qualitative research work.

Asking Research Questions

Each of us needs to acknowledge that our fellow geographers and other colleagues are *already* involved in our studies (Box 4.1). None of us ever formulates research questions or undertakes research in a vacuum. We are all members of **interpretive communities** that involve established disciplines with relatively defined and stable areas of interest, theory, and research methods and techniques (Butler 1997; Fish 1980). Interpretive communities influence our choice of topic and our approach to and conduct of study; this is because of what Livingstone (2005, 395) describes as 'the inescapably collective character of interpretation . . . [which, to] the extent that interpretive communities occupy material or metaphorical spaces, they fall within the arc of the cultural geography of reading' and, we would add, other geographical

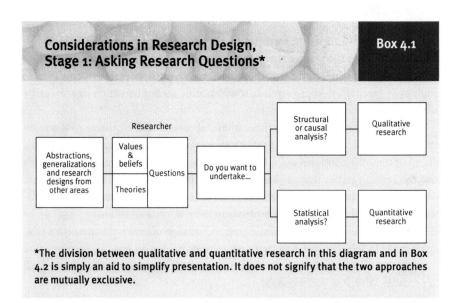

Considerations in Research Design, Stage 1: Asking Research Questions*   **Box 4.1**

*The division between qualitative and quantitative research in this diagram and in Box 4.2 is simply an aid to simplify presentation. It does not signify that the two approaches are mutually exclusive.

actions and analyses. We also fold our own values and beliefs into research, and they can influence both what we study and how we interpret our research (see Jacobs 1999; Flowerdew and Martin 2005; and Chapter 17 of this volume for more detail).

From Asking Research Questions to Conducting Research

Research aims affect research design. For example, asking 'how many skateboarders frequent a particular public place compared with other types of users?' will involve a research design different from one that aims to answer the question 'why do skateboarders use this place, and how do they interact with other types of users?' The first question focuses on quantification and statistical analysis, and the second is more concerned with the qualitative investigation of underlying causes of skateboarders' behaviour and how such behaviour is constitutive of various structural relations.

In considering the conduct of research, we also need to ask what to do with the information collected. Answering this question will influence the kind of research we do. Before making this decision, we need to be aware of some of the differences between qualitative (intensive) and quantitative (extensive) research. As Sayer (1992) has suggested, each method helps us to answer different research questions, employs different research methods, has different limitations, and ensures rigour differently. **Extensive research** is characterized by identifying regularities, patterns, and distinguishing features of a population, often through a sample that has been selected using a random procedure to maximize the possibility of generalizing to a larger population from which it is drawn. Extensive methods are designed to establish statistical relations of similarity and difference among members of a population, but they can lack explanatory power. For instance, we may be able to determine that N number of respondents in a sample think P, but we may not readily be able to determine why they hold this opinion. **Intensive research** requires that we ask how processes work in a particular case (Platt 1988). We need to establish what actors do in a case, why they behave as they do, and what produces change both in actors and in the contexts in which they are located. (This distinction is discussed further in Chapter 5.)

For example, the issue of multiple-use conflict in a public place can be investigated using quantitative methods. More than a decade ago, Wood and Williamson (1996) decided to distribute a standardized questionnaire to a random sample of the users of Franklin Square, an inner-city square in the city of Hobart, Tasmania, that had been partly claimed through day-to-day use by skateboarders. The data from their study were aggregated, and statements were made about the degree to which these data were likely to reflect the opinions of all the square's users about the presence of skateboarders. This extensive approach produced useful information suggesting the existence of common characteristics and patterns; for instance, skateboarders used certain parts of the square, while other users avoided them. But such findings did not account for the shifting quality of various people's different experiences of Franklin Square and of each other or the reasons behind their opinions. Also selecting Franklin Square as a case by which to examine multiple-use conflict in public places, Stratford (1998; 2002) and Stratford and Harwood (2001) used intensive methods such as in-depth interviews and observation to understand various responses to skateboarding in the square and around Tasmania more generally. Both types of approach have merit. However, if we are mainly interested in working through the elements of structure and process that arise from analyzing responses rather than in generating data that make statistical analysis possible, then we are pointed in the direction of intensive research.

In summary, in opting for a qualitative research design, we are influenced by the theories we are concerned with, by studies undertaken by other researchers in our interpretive communities that we have found interesting, and by the research questions we wish to ask—all of which are interrelated.

Selecting Cases and Participants

Definitions of the terms **case** and **participant** will differ among interpretive communities. However, it is our view that cases are examples of more general processes or structures that can be theorized. (See chapter 5 for a full discussion of case studies.) Researchers should be able to ask 'that categorical question of any study: "What is this case a case of?"' (Flyvbjerg 1998, 8). In our view, they should also resist any anxiety about questions related to the validity of case-based research. As Flyvbjerg (2006, 219) has noted, 'a scientific discipline without a large number of thoroughly executed case studies is a discipline without systematic production of exemplars, and a discipline without exemplars is an ineffective one.'

In the example cited above, Franklin Square is a *case* of multiple-use conflict in a public place, but it resonates outwards to embrace a number of general social and spatial processes involving, for example, the privilege of consumerism, the ways in which citizenship has come to 'attach' to acts of consumption, and the relations of government to capital (Stratford 2002). *Participants* make up some of the elements of the case in question—for example, skateboarders, the elderly, the business community, and city council. In some theoretical dispositions, participants can also include non-human elements, or actants, whose effects on the case are profound (see, for example, Callon 1986; Murdoch 2006).

Selecting Cases

Sometimes we find a case, and sometimes a case finds us. In both instances, selection combines purpose and serendipity (Box 4.2). On the one hand, we may read about multiple-use conflict in public places in other cities and want to see whether

Box 4.2

Considerations in Research Design, Stage 2: Selecting Cases and Participants

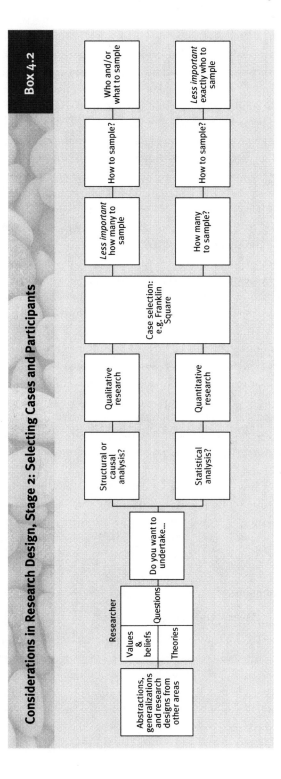

causal explanations advanced there have merit in—or inform our understanding of—situations with which we are familiar. In this instance, the general or theoretical interest 'drives' the research, and we must narrow the field, selecting cases and participants for research. On the other hand, a local government parks manager might draw attention to conflict among various groups in one site in the city and want an investigation of the options available to manage this conflict. In this situation, the case has 'found' the researcher—and theories about multiple-use conflict in public spaces are subsequently woven into it. It is worth noting that if, for example, a community development officer rather than a parks manager had contacted researchers about the same issue, the researchers might well have been presented with a different brief that would in some ways make for a 'different' case.

Regardless of how a case is selected, it is usually advisable to work at sites or on cases that are both practical and appropriate. In our example of skateboarders' use of public places, ambiguous sites—such as shopping malls, which are generally private places behind public facades—add a layer of complexity to investigation and often require researchers to secure a range of permissions to gain access to property, customers, and employees. One final issue needs to be considered in case selection. On the one hand, we might choose to work with so-called typical cases on the grounds that they will provide useful insights into causal processes in other contexts. Alternatively, we might deliberately seek out **disconfirming cases.** Such cases might include individuals or observations that challenge a researcher's interpretations or do not confirm ways in which others portray an issue. For example, we may have studied media reports suggesting that there is unmitigated conflict between youths and the elderly in a public square. However, interviews with elderly pensioners might lead us to understand that some aged people regularly frequent the square when youths are present because they enjoy the company of young people. Such disconfirming cases can be important in the research process, making us think through the way that different institutions and the practices they use (such as the media and their tendency to sensationalize events) create stereotypes. Such cases also require us to ask how various actors are represented (and for what reasons) and how they represent themselves.

Selecting Participants

Exploratory and/or background work (for example, reading, observation, viewing television documentaries, conducting preliminary interviews) will often give researchers the capacity to begin to comprehend the perspectives of participants that we think we want to interact with. Understanding their perspectives in complex cultural situations usually requires some form of in-depth interviewing (see Chapter 6 for details) or observational method (see Chapter 12 for details) that, though time-consuming, often results in a deep and detailed appreciation of the complicated issues involved (Geertz 1973; Herod 1993).

Generally speaking, the more focused our research interest becomes and the better our background information and understanding, the more certain we are about who we wish to involve in our research and why. Nevertheless, this certainty needs to be underpinned by a rigorous process of justification. As Mason (2004, 129) points out,

> Your answers to questions about which people to sample should therefore be driven by an interpretive logic which questions and evaluates different ways of classifying people in the light of the particular concerns of your study. Underlying all of this must be a concern to identify who it is that has, does or is the experiences, perspectives, behaviours, practices, identities, personalities, and so on, that your research questions will require you to investigate.

In this respect, it is conceivable that conducting in-depth interviews with a small number of the 'right' people will provide significant insights into a research issue.

The Nitty-Gritty of Participant Selection

Michael Patton's (2002) work on **purposive sampling** is among the more useful summaries of the topic available to researchers. Patton refers to various forms of purposive sampling, including the following seven commonly employed strategies. *Extreme* or **deviant case sampling** is designed to help researchers learn from highly unusual cases of the issue of interest, such as outstanding successes or notable failures, top-of-the-class students or dropouts, exotic events, or crises. **Typical case sampling** illustrates or highlights what is considered 'typical', 'normal', or 'average'. **Maximum variation sampling** documents unique or diverse variations that have emerged through adaptation to different conditions and identifies important common patterns that cut across variations (Williams and Round 2007). **Snowball** (or *chain*) **sampling** identifies cases of interest reported by people who know other people involved in similar cases (Kirby and Hay 1997; Stratford 2008). **Criterion sampling** involves selecting all cases that meet some criterion, such as involvement in natural resource management governance in Australia (Lockwood et al. 2007). **Opportunistic sampling** requires that the researcher be flexible and follow new leads during fieldwork, taking advantage of the unexpected (Clough et al. 2004). **Convenience sampling** involves selecting cases or participants on the basis of access (for example, interviewing passers-by on the street). While this final strategy saves time, money, and effort, it often produces the lowest level of dependability and can yield information-poor cases. Much purposive sampling combines a number of these strategies.

How Many Participants?

In both qualitative and quantitative research, it is usual that only a subgroup of people or phenomena associated with a case is actually studied. The size of this group is more relevant in quantitative research because representativeness is important. In qualitative research, however, the sample is not intended to be representative, since the 'emphasis is usually upon an analysis of meanings in specific contexts' (Robinson 1998, 409). (This is discussed in more detail in Chapter 5.)

The following analogy between a case and an island may help to explain the distinction. Suppose you are looking at a special kind of aerial photograph of an island, so detailed that you can see all of its inhabitants.

> Clearly, if the population of the island were ten thousand instead of ten, enumeration would count for a great deal. . . . But this is because of the investigator's limitations: [she or he] cannot really get to know ten thousand people and the various ways in which each interacts with others. The use of formalist techniques is a second-best approach to this problem because the ideal technique is no longer feasible. Even on this big island, the old technique will count for a great deal, but that is not the main point. The point is that counting and model building and statistical estimation are not the primary methods of scientific research in dealing with human interaction: they are rather crude second-best substitutes for the primary technique, storytelling. [Ward 1972, 185]

Numbers *do* tell us things about the island, and if what interests us happens to be the frequency and geographic distribution of the island's population, then we need no more than the photograph. If we are interested in a particular 'story', such as might revolve around an aspect of the cultural geography of the island—for example, multiple-use conflict in public places—then we will need more than the photograph to go on.

One way to conduct our investigation is to talk with the island's inhabitants. We could also engage in participant observation, or we could consult relevant texts such as submissions to government, letters to the editor of the island's newspaper, or television news stories that might give us an insight into multiple-use conflict in the island's public places. As researchers, however, we are usually resource-limited, both in terms of funding and time, and we must make decisions about what/whom to include and what/whom to exclude from our study. It is clear, however, that we still face the issue of how many people to talk with, or how many texts to read, and so forth. While it may seem disconcertingly imprecise, Patton's (2002) brutally simple advice remains accurate: there are few if any rules in qualitative inquiry related to sample size, and it depends on what is needed in the way of knowledge, on the purpose of the research, on its significance and for whom, and on logistics and resources. The richness of information, its validity and meaning, is more dependent on the abilities of the researcher than on size of sample. In the final analysis, then, it is you as the researcher who must be able to justify matters of case and participant selection to yourself, your supervisor, your interpretive community, and the readers and users of your work.

ENSURING RIGOUR

It is no frivolous matter to share, interpret, and represent others' experiences. We need to take seriously 'the privilege and responsibility of interpretation' (Stake 1995, 12). This responsibility to informants and colleagues means that it must be possible for our research to be evaluated. It is important that others using our research have good reason to believe that it has been conducted dependably. (An extensive literature on these matters, undertaken over many years, includes Anfara, Brown, and Mangione 2002; Baxter and Eyles 1997; Bogdan and Biklen 1992; Ceglowski 1997; Denzin 1978; Dey 1993; Flick 1992; Geertz 1973; Jick 1979; Johnson 1997; Keen and Packwood 1995; Kirk and Miller 1986; Lincoln and Guba 2000; Manning 1997; Mays and Pope 1997.)

Ensuring rigour in qualitative research (Box 4.3) means establishing the **trustworthiness** of our work (Bailey, White, and Pain 1999a, 1999b; Baxter and Eyles 1999a, 1999b; Golafshani 2003). Research can be construed as a kind of **hermeneutic circle**, starting from our interpretive community and involving our research **participant community** and ourselves, before returning to our interpretive community for assessment (Burawoy et al. 1991; Jacobs 1999; Reason and Rowan 1981). This circle is a key part of ensuring rigour in qualitative research; our participant and interpretive communities check our work for credibility and good practice. In other words, trust in our work is not assumed but has to be earned.

There are two steps in particular that need to be followed to ensure and defend the rigour of our research for our interpretive communities. First, strategies for ensuring trustworthiness need to be formulated in the early stages of research design and applied at various stages in the research process (Baxter and Eyles 1999a, 1997; Lincoln and Guba 1985). These strategies should include *appropriate* checking procedures in which our work is opened to the scrutiny of interpretive and participant communities (Mason 2004, but see also Bradshaw 2001 on some of the possible perils involved in these procedures). Second, we need to document each stage of our research carefully so that we can report our work to our interpretive community for checking; 'we should focus on producing analyses that are as open to scrutiny as possible' (Fielding 1999, 526).

Rigour must be considered from the outset of our research, underpinning the early stages of research design. It is important to incorporate appropriate *checking* procedures into our research process. These procedures are outlined in foundational qualitative research works by Denzin (1978) and Baxter and Eyles (1997) as the four major types of **triangulation**: multiple sources, methods, investigators, and theories. For example, as we move through various research stages, we might check: (a) our sources against others (re-search); (b) our process and interpretations with our supervisors and colleagues; and (c) our text with our research participant community to enhance the credibility of our research (although this check can be problematic if that community has considerable power, such as might be the case with a multinational corporation if its managers refuse permission for us to publish work related to findings derived from the corporation).

As indicated in the research stages in Box 4.3—which often overlap as they become a whole research composition—we also need to document our work fully: how we came to be interested in the research, why we chose to do it and for what purpose. We may declare our own philosophical, theoretical, and political dispositions, and we will almost certainly review literature dealing with both the general area of our research and the research methods we intend to use. This elaboration of context permits us to establish the plausibility of our research by demonstrating that we embarked on our work adequately informed by relevant literature and for intellectually and ethically justifiable reasons. We will most likely have checked the plausibility of our research with supervisors and/or colleagues before embarking on detailed research design. At the final stage of reporting research, we can also attempt to acknowledge limits to the **transferability** of our research due to particularities of the research topic, the research methods used, and the researcher (see Chapter 5). In this way, we confirm that the methods we use and the interpretations we invoke influence our research outcome. Thus, it is vital that we document all stages of our research process. Such documentation allows members of our interpretive and participant communities to check all of these stages and confirm that our work can be considered dependable.

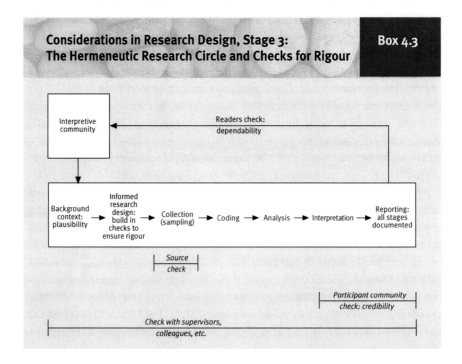

Considerations in Research Design, Stage 3:
The Hermeneutic Research Circle and Checks for Rigour   **Box 4.3**

FINAL COMMENTS

We began this chapter by suggesting that consideration of research design and rigour is essential to the conduct of dependable qualitative inquiry. We have addressed issues of case selection and participant selection and outlined some reasons to be concerned with rigour, as well as some means by which rigour might be achieved.

Most research is undertaken to be shared with others. We therefore need to ensure that our research can stand up to the critical scrutiny of our interpretive and participant communities. The work presented in this chapter provides some of the conceptual and practical tools by which this outcome of sharing plausible, credible, and dependable work can be achieved.

KEY TERMS

case

convenience sampling

criterion sampling

deviant case sampling

disconfirming case

extensive research

hermeneutic circle

intensive research

interpretive community

maximum variation sampling

opportunistic sampling

participant

participant community

purposive sampling

rigour

sampling

snowball sampling

transferability

triangulation

trustworthiness

typical case sampling

REVIEW QUESTIONS

1. Why is rigour important in qualitative research? Is it an ethical consideration?
2. What is an 'interpretive community'?
3. What is meant by the phrase 'participant community'?
4. What are some ways we might check our research to establish its dependability to members of our interpretive community?

Useful Resources

Anfara, V.A., K.M. Brown, and T.L. Mangione. 2002. 'Qualitative analysis on stage: Making the research process more public'. *Educational Researcher* 31 (7): 28–38.

Baxter, J., and J. Eyles. 1997. 'Evaluating qualitative research in social geography: Establishing "rigour" in interview analysis'. *Transactions of the Institute of British Geographers* 22 (4): 505–25.

Bradshaw, M. 2001. 'Contracts and member checks in qualitative research in human geography: Reason for caution?' *Area* 33 (2): 202–11.

Mason, J. 2004. *Qualitative Researching*. 2nd edn. London: Sage.

Patton, M.Q. 2002. *Qualitative Evaluation and Research Methods*. 3rd edn. Beverly Hills, CA: Sage.

Platt, J. 1988. 'What can case studies do?' *Studies in Qualitative Methodology* 1: 1–23.

Sayer, A. 1992. *Method in Social Science: A Realist Approach*. 2nd edn. London: Routledge.

Case Studies in Qualitative Research

5

Jamie Baxter

CHAPTER OVERVIEW

This chapter defines the case study as a broad methodology or approach to research design rather than as a method. Much of the chapter clarifies precisely what a case study is, describes the different types of case studies, and addresses some misplaced depictions (N=1) and criticisms (lack of generalizability) of case study research. Although most case study research is cross-sectional, conducted typically on one case at one point in time, this chapter reviews two major variants of multiple case studies that might appeal to geographers in particular: the within-case *temporal* comparison and the *spatial (place-oriented)* cross-case comparison. Much of the 'how-to' of case study methods and fieldwork is covered in companion chapters within Part 2 of this volume. This chapter instead focuses more on broader research design issues specific to case studies.

WHAT IS A CASE STUDY?

Gerring (2004, 342) provides a very concise and useful base definition of the **case study** as 'an intensive study of a single unit for the purpose of understanding a larger class of (similar) units'. However, we must be careful not to conflate sample size with the quality of case study research—a point that is addressed near the end of this chapter. Further, Gerring's definition may be amended to consider the notion of multi-case studies associated with temporal and spatial comparisons. Thus, case study research involves the study of a single instance or small number of instances of a phenomenon in order to explore in-depth nuances of the phenomenon and the contextual influences on and explanations of that phenomenon. Some examples of phenomena researched as case studies might include an event (e.g., a protest rally, a disaster), a process (e.g., immigration, discrimination, risk amplification, deforestation), or a particular place (e.g., a neighbourhood with a high crime rate, a community hosting a hazardous waste facility). Case studies are often used to better understand and sometimes directly resolve concrete problems (e.g., why is uptake of immunization so low in town X?). From an academic point of view, though, case studies are well suited to corroborating existing explanatory concepts ('theory'), falsifying existing explanatory concepts, or developing new explanatory concepts. Perhaps most important, the case study provides detailed analysis of *why* theoretical concepts or explanations do or do not inhere in the context of the case.

Thus, a case study is perhaps most appropriately categorized as an approach to research design or **methodology** (a theory of what can be researched, how it can be researched, and to what advantage) rather than as a **method** (a mechanism to collect 'data'). It is more an approach or methodology than a method because there are important philosophical assumptions about the nature of research that support the value of case research. The primary guiding philosophical assumption is that in-depth understanding about one manifestation of a phenomenon (a case) is valuable on its own without specific regard to how the phenomenon is manifest in cases that are not studied. This depth of understanding may concern solving practical/concrete problems associated with the case or broadening academic understanding (theory) about the phenomenon in general, or a case study may do both of these things. Other philosophical assumptions are described below as clarifications particularly concerning N=1 and **transferability/generalizability.**

It is worth examining where case studies fit relative to other approaches to social research. Case studies are often considered equivalent to field research, participant observation, **ethnographic research,** or even qualitative research. Although case study research certainly intersects with all of these, it deserves more attention than most social science textbooks provide. That is, case studies are typically mentioned within chapters that mainly concern these other aspects of qualitative research (this text and chapter notwithstanding). Further, the terms 'qualitative' and 'case study' are not entirely interchangeable, largely because quantitative researchers also conduct case studies and many use a combination of methods and methodologies within case study research. In fact, the relationship between qualitative, quantitative, and case study research looks more like the Venn diagram in Figure 5.1 whereby much of case study work is indeed qualitative but some is quantitative or a mixture of the two. Thus, researchers need to specify upfront precisely what type of case study is being conducted (quantitative, qualitative, or mixed) so that the quality of the work is fairly judged according to the methodology used (see Chapter 4). However, in keeping with the theme of this text, the focus in this chapter is on the qualitative case study and on qualitative aspects of mixed-method case studies.

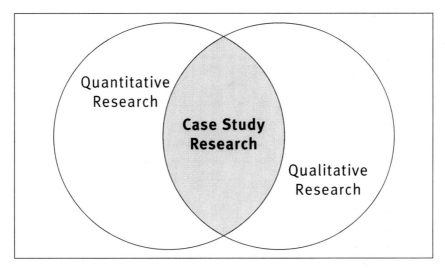

Figure 5.1: Intersecting Domains of Inquiry

Although there is much overlap, case study research is *not* synonymous exclusively with qualitative research. Rather, case study is a methodology or approach to research that can be either predominantly qualitative or predominantly quantitative or a mixture of both of these approaches.

THE HISTORICAL DEVELOPMENT OF THE CASE STUDY

The popularity of case study research has perhaps never been higher. Although the case study has a long history in the social sciences, its popularity faded during the quantitative revolution of the post–World War II era. However, its status has dramatically resurged during the past few decades (Platt 1992).

Most writers trace the origin of modern-day case study research to the **Chicago School of Sociology**. The so-called Chicago School emerged in the 1920s and 1930s during a time of rapid industrialization in North America. Industrialization in turn spawned the need to better understand the workings of rapidly growing cities. Major writers from the Chicago School include William Thomas (1863–1947), Robert Park (1864–1944), Ernest Burgess (1886–1966), and Louis Wirth (1897–1952), and their works have been heavily influential on a diverse array of qualitative researchers in areas such as urban and cultural sociology and geography. Much of the work was and continues to be ethnographic, combining quantitative survey work and qualitative participant observation, semi-structured interviews, and unstructured interviews (Platt 1992). For example, Thomas and Znaniecki (1918) produced a massive four-volume account of Polish immigration to America as a case study of the 'immigration experience'. Such work and much to follow was heavy on detail and conceptual development that included empathetic accounts of human experience. Thus, the Chicago School approach was appealing to humanistic geographers interested in the manifestation of various phenomena in 'places' imbued with contextualized meaning rather than conceptualized simply as 'locations' (Tuan 1977). However, the density and sheer length of the early Chicago School ethnographies made them largely inaccessible except through second-hand interpretation—in geography textbooks, for example. Indeed, the same issue remains a challenge for qualitative case study researchers today who want to provide rich detail but must produce work brief enough to appeal to a wide audience.

During the post–World War II period, quantitative aspects of case study work began to gain more prominence among scholars than its qualitative aspects. Nevertheless, the qualitative/ethnographic methodologies used in case studies continued to be honed throughout this era through such well-known breakthrough case studies as Whyte's (1943) *Street Corner Society* (Boston) and Liebow's (1967) *Tally's Corner* (Washington, DC). Both studies concerned men living in poor neighbourhoods, and the authors produced meticulous accounts of structural social issues about such phenomena as unemployment and gender relations in predominantly Italian and Black working-class neighbourhoods respectively. The fieldwork was measured in months and years rather than weeks. These works are lauded as much more concise and accessible (though still book-length) accounts than the early Chicago School works while retaining the conceptual/theoretical richness of the Chicago School studies. Further, they helped to develop both participant observation and less structured interview methods more fully.[1] Whyte and Liebow drew out concepts grounded in the case context in concrete terms (e.g. 'manliness') rather than academically abstract concepts. Although well-grounded in the case, such concepts did and still do resonate in other contexts that are similar to the ones in the Whyte and Liebow studies. What is important here is that good case studies are so richly described (theorized) that one generally finds it quite easy to make such comparisons with contexts outside the case. Indeed, this detail or richness is one of the best strategies for creating credible and trustworthy (rigorous) qualitative case study work (see Chapter 4).

It is important to note that one thread of the case study's history is the ongoing use of multiple methods, including the combination of qualitative and quantitative methods in the same case studies. Platt (1992) provides a detailed account of the history of the rise, fall, and rebirth of the case study, which is beyond the scope of this chapter. Yet in her account, she carefully shows how the Chicago School studies ushered in a long era during which researchers highlighted and, more detrimentally, exaggerated *differences* between quantitative ('objective') research and qualitative ('subjective') research, a tendency that still exists today. In the early years of the Chicago School, however, researchers were more inclined to highlight that quantitative and qualitative approaches were both powerful on their own but also complemented each other (Platt 1992). That perspective on combining methods has now resurged (Hesse-Biber and Leavy 2004), which means that there is a refreshingly diverse set of research design possibilities from which the case study practitioner can choose. Before we turn to some common ways that geographers design case studies, it is worth clarifying the importance of depth and context as opposed to sample size.

N=1 AND THE IMPORTANCE OF DEPTH AND CONTEXT

The term 'N=1' is often used to succinctly describe case study research, but as explained in Chapter 4, there is far more to case studies than the number of units studied. In fact, case study researchers tend not to think in terms of sample size. Thus, it is important to clarify what N=1 means in the context of case study research and to highlight that depth of understanding and contextualized understanding are far more important.

The use of the term 'N=1' to describe case studies can be confusing, because it grafts quantitative/statistical terminology onto non-statistical research. In statistics, 'N' refers to the population (the group about whom conclusions are drawn), while 'n' refers to a sample, or subset, of that population. Strictly speaking then, n=1 is more appropriate statistical terminology—one case from the population of N. Moreover, there is also selection within the case itself that must be accounted for—the sub-units within the case. That is, case study researchers often study large numbers of 'things' within their case. Yet qualitative researchers tend to think of their cases as wholes that cannot be fully understood when lumped together with a large number of other cases. That is, the context of the case is important, since it more often than not substantially influences the phenomenon in question (e.g., a change in national employment policy can affect local crime rates). That is why multiple sub-units of 'things' (e.g., people, newspapers, policies) are often studied—to get at these contextual influences. Thus, rather than study a few of each sub-unit across a wide array of communities (e.g., 10 people, one newspaper, and two policies each from 10 or 100 communities), the qualitative case study researcher instead prefers to study one carefully selected community intensively and holistically to understand how the various things studied interact with one another in, for example, one place. There is no statistical notation to adequately account for the importance of context, and any use of 'N' and 'n' does not do justice to the value of case study research.

Further, it is important to underscore that case study research is intensive rather than extensive research (see Chapter 4). Social scientists use the terms **idiographic** research and **nomothetic** research to describe this difference. Idiographic research is depth-oriented, since it tends to focus on the particular to understand a phenomenon in more detail. Nomothetic is breadth-oriented, since it focuses less on the details and more on investigating a limited number of things across several units (cases) simultaneously. Nomothetic researchers typically use probabilistic sampling to select those units. Although qualitative case study researchers do not typically choose sub-units based on probabilistic sampling, few case study researchers concede that their findings are exclusive *only* to the case. Indeed, qualitative case study researchers working in an idiographic frame expect and look for what might be common *between* cases. A case is viewed as neither entirely unique nor entirely representative of a phenomenon (i.e., a 'population' of cases, according to statistical terminology). This point is taken up again in the discussion of generalizability and transferability later in the chapter.

To explain the points about sub-unit selection and how to study context more fully, Figure 5.2 offers an example of how a case study can be based on multiples

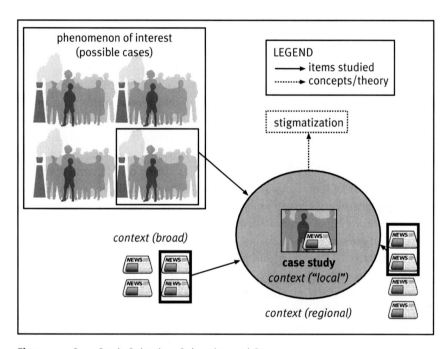

Figure 5.2: Case Study Selection, Sub-units, and Context

A case study is one manifestation of a broader phenomenon. Researchers carefully select the case to understand the practical/concrete aspects of the case itself but also to better understand the broader phenomenon. In this example, one community living with a hazardous facility (n=1) is selected from all possible communities with hazardous facilities (N). Further, multiple people will be studied within the case community, along with multiple newspapers both inside (local) and outside the community (regional and extra-regional). This helps to understand an array of contextual influences on the phenomenon (e.g., low risk perception) within the case to develop contextualized theory (e.g., stigmatization).

of different sub-units. Suppose a researcher is interested in the phenomenon of community hazard risk perception—specifically, why groups of residents vehemently oppose potentially hazardous or polluting facilities (e.g., industrial, waste, nuclear power) being located near their community.[2] This is a problem of practical importance, since at the very least, failed facility siting efforts cost millions of dollars, not to mention the potential negative impacts of such facilities on local residents. The researcher could look for a community that faces having a facility located in their midst. However, for the purpose of identifying a single valuable case relevant to the problem at hand, the researcher might instead flip the problem on its head and ask the related question 'why are some communities *unconcerned* and not opposed to these facilities?' Thus, the researcher needs to identify a town where residents seem to be supportive of a local hazardous waste treatment facility—a case study. What the researcher might actually do is use multiple methods whereby she might interview several different types of residents (councillors, facility workers, opposition group members) to understand their risk perceptions and conduct a discourse analysis on various types of facility-related media coverage to understand the influences on (a lack of) risk perception. These are some of the many design and selection considerations that are documented elsewhere in this book (e.g., Chapter 4). The intensive, holistic aspect of the study comes in part from trying to understand how the various contexts—which are conceptualized in Figure 5.2 as 'local', 'regional', and 'broad'—are involved with facility hazard risk perception. In this example, the analysis of local, regional, and national media coverage helps the researcher to better understand these contextual influences. The analysis could include other influences, such as recent national economic changes and policies at various scales. In the interviews, residents would be asked how they view these external influences, such as the reporting by various types of media. We will revisit this example later in the chapter.

| Studying One Person: A Case Study of 'Praxis' | Box 5.1 |

Sometimes cases are not identified as places or institutions with multiple people as 'sub-units' within to be studied. Instead, they may concern multiple manifestations of a phenomenon with one person only. Yet findings from such work can still be transferable beyond the single person. For example, Wakefield (2007) studies the phenomenon **praxis** in critical geography. Praxis concerns the way that researchers may use research itself to make positive change outside academia. The 'case' in this study is Wakefield's own experiences as an activist researcher in the 'food movement' operating in such contexts as her own household and Toronto food movement organizations, as well as within broader environmental justice organizations and 'globalized, corporate industrial agriculture' (Wakefield 2007, 332). That is, she studies how she uses research in all of these different contexts—how research praxis is manifest in each of these contexts. Wakefield draws on her personal experiences to expand on the underappreciated concept of 'activism at home'. This concept moves beyond academic teaching and writing, which she argues are the dominant forms of academic praxis. She highlights the value of taking *personal* action in the household and community, partly to set an example for others but also to better understand the challenges of alternative food consumption. Thus, most of her research is reflexive in the sense that she systematically studies her own experiences and interactions with others in relation to alternative types and sources of food. She has made a contribution to the general theory of academic praxis in terms of expanding the theoretical concept 'activism at home'. Further, one can easily imagine that activism at home resonates beyond the case of food movements and Wakefield alone. That is, it is relatively easy to see how the concept might apply to other cases, such as waste generation or fuel consumption.

TYPES OF CASE STUDIES

Theory Testing and Theory Generating Cases

As outlined above, case studies play two key but not necessarily mutually exclusive roles: to test theory and to generate or expand theory. The former may involve the search for negative or falsifying cases, while the latter concerns cases that are more typical.

For some, what distinguishes case studies from other approaches, such as grounded theory and ethnography, is that in case studies, theoretical propositions should be stated prior to entering the field (Yin 2003). Yet others tend to view qualitative case studies as primarily **theory generating** endeavours such that ethnography and grounded theory can be easily incorporated within a case study design. In practical terms, both positions are mainly a matter of degree, since most practitioners of grounded theory and ethnography do not commence field research without adequate knowledge of *some* theory. Nevertheless, if one decides to follow Yin's recommendation that formal propositions need to be stated upfront, two cautions are necessary.

First, qualitative researchers presume that propositions are contingent or context-dependent such that concepts describing relationships are only 'true' *under certain conditions* (Sayer 2000). That said, concepts are still 'true when . . .', and accounting for the context or contingencies within which a 'truth' happens certainly falls within the realm of what most qualitative researchers do in practice. Using our example of the low risk–perceiving, hazardous facility–hosting community depicted in Figure 5.2, the researcher might focus on the concept 'stigma' prior to field work. She might propose that such communities feel that the outside (non-local) media stigmatize the community, especially *when* accidents happen at the facilities and *when* these accidents are widely reported and *when* that reporting does not simultaneously report the social good the community is doing by hosting a facility that solves a widespread social problem (i.e., disposal of hazardous waste). Further, this stigma may minimize or redirect the concern or anger that residents once directed at the facility onto the media instead. The effects of stigma are not always 'true', but they seem to be under these specific conditions (and perhaps others).

The second caution is that there is a potential logical flaw in *ever* stating propositions upfront at the beginning of a study. That is, formal propositions typically require well-developed theory as their basis. Yet qualitative case studies are often used to delve into under-explored and thus under-theorized phenomena. Moreover, researchers tend to borrow from related areas of inquiry. Often, it is not necessary to 're-invent the (conceptual/theoretical) wheel'. In the case of our Figure 5.2 example, there might already be a well-developed theory on the negative effects of stigma on hazardous facility–hosting communities, but it might not go as far as suggesting that stigma can redirect or minimize concern about the facility itself. The concept of stigma, which may itself have been developed from case study work, is thus borrowed from the literature and elaborated upon.

This raises the question of how theory is actually *generated* using case studies. This question is covered in greater detail in Chapter 14 on coding data, but it is worth pointing out that qualitative research in practice is rarely a purely deductive or purely inductive endeavour. Rather, it tends to be more cyclical in the sense that theory stated initially either formally as hypotheses or loosely as budding ideas is explored (deductively) by studying the real world of the case and then that information is used to generate new concepts (theory) to explain what is observed (inductively) (Figure 5.3). These refined or new concepts are then further scrutinized by ongoing analysis of the real world of the case. All the while, the good researcher will remain aware of existing academic literature that might contribute to the explanations. Grounded theory—described in Chapter 14—is one approach to moving through this deductive/inductive cycle to form in-depth understanding of the entire case.

Case Studies Across Time and Space

Case studies need not be one-off, single studies of one case at one particular point in time. However, multiple case studies are generally not approached with the purpose of establishing *statistical* generalizability (see the discussion below). Instead, it is better to view multiple cases in one of two ways. In the first instance,

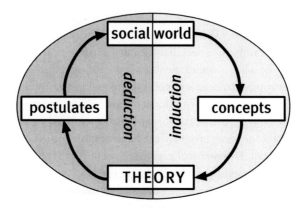

Figure 5.3: Cyclical Modes of Exploration in Case Studies

Often, qualitative case study researchers tend to emphasize the inductive mode of inquiry, moving from empirical observation to concepts/theory. In some cases, the main purpose may instead be to test an existing postulate (hypothesis). However, both inevitably involve 'multiple loops' of reasoning as the researcher tentatively develops concepts (induction) and then compares them to the details of the social world that comprise the case (deduction)

multiple case studies provide a broader basis for exploring theoretical concepts and explanations of phenomena. In the second instance, longer-term study of the case may be useful to **corroborate** and further explore theory as it relates to the case at hand. Both enhance the credibility and trustworthiness of the concepts and explanations by exposing them to different scenarios between cases in the first instance and within the case as it evolves over time in the second. Neither one is replication in the statistical or experimental sense of the term, but they may instead be seen as ways of both deepening and expanding the theoretical concepts (see the examples of 'stigma' and 'activism at home' above). Thus, for multiple cases, little attention needs to be paid to the sequence of the case studies—they can be studied in parallel. However, for longitudinal cases, more attention must be paid to the temporal sequencing of the research.

Time: Cross-sectional and Longitudinal Case Studies
The most common form of social research is **cross-sectional**—that is, it is conducted at one point in time (Bryman 2006). The definition of 'one point in time' can be somewhat fuzzy for the qualitative researcher, however, since he or she often spends extended time in the field collecting and analyzing data. Operationally then, a study may be considered cross-sectional if fieldwork is conducted in one block of time regardless of how long it takes. It would only be considered **longitudinal** if there was a revisit, with the researcher returning to the case after an intervening time period during which no appreciable research was done. Similarly, for studies

that do not involve fieldwork or face-to-face interaction (e.g., discourse analysis of media), identifying collections of data at specific time periods and analyzing them for changes from one period to the next would make them longitudinal.

A key advantage of longitudinal research on the *same* case is that it makes it possible for the researcher to address what may be considered the enduring versus the ephemeral by exploring the robustness of the original concepts and explanations (theory). When done as a follow-up study on the same case, longitudinal research amounts to a form of corroboration to determine whether the original explanations have endured over time. Nevertheless, case study researchers should not be so naive as to assume that social settings are invariant over time (see Chapter 1). For example, in our study of risk perception in the hazardous facility–hosting community case depicted in Figure 5.2, if the researcher were to revisit the community two years later and find that stigmatization and its effects were the same or stronger, then there would be good grounds for arguing that the original concepts and theory are credible, are trustworthy, and endure. If the follow-up research is done with the same participants as in the previous study, it serves as a tactic for guarding against such threats to rigour as **member checking** (Lincoln and Guba 2002) and **participant checking** (see Chapter 4).

However, a more challenging scenario is when the phenomena and the concepts that explain them seem to make only *some* sense still, but less so, or perhaps in a changed form. This condition may signal that the concepts or even the phenomenon itself are ephemeral and may have been relevant for only a brief period. For example, residents might talk less about the 'stigmatizing effect of non-local media' during the follow-up. This might threaten the original theory about stigmatization, leaving the researcher to explore the reasons for the apparent change of view. Such an exploration is potentially immensely valuable, because new insights may be gained. Nevertheless, if the changes regarding stigma were due for example, to the success of the community's legal action against a non-local newspaper, the theory would remain credible. That is, the community might have moved on, knowing that the 'stigmatizer' had been appropriately dealt with. If residents were *more* concerned about the facility (e.g., because of a recent accident) and tended to sympathize more with non-local media reporting on the community, it would pose a serious threat to the initial conceptualization. Recall that the original concept of the effect of stigma was that harsh criticism of the facility by outside media stigmatized the community and residents reacted by outwardly expressing little concern about the facility. That is, low concern about facility risks was a reaction to stigma that served to defend the community and enhance community pride. An apparent rise in concern, together with more sympathy towards outside media, go against the original thesis and would require either (1) the reconceptualization of 'stigma' or (2) abandoning it as an explanation of low concern altogether.

It is obviously very problematic to assume that social phenomena are static over time. The internal dynamics of a community typically change. For example, new people come into positions of power. Although the research context of the case may change, this does not *necessarily* invalidate (threaten the credibility or trustworthiness) of the original theory in relation to the overall phenomenon. Researchers need to be careful about identifying what aspects of their theory and concepts seem less or more relevant over time and, most important, *why*. Thus, longitudinal case studies are very good for tracking how the phenomenon changes over time in the case in question. Ultimately, this should lead to the development of well-rounded explanations. In this regard, the timing of follow-up research should be theoretically informed. For example, in our community with the hazardous waste facility, if stigmatization of the community by the media is important, then future research might solicit residents' opinions after the publication of key articles and reports that were highly negative—or highly positive— towards the local waste facility and/or the community.

Unlike the prospective longitudinal case studies described above, retrospective studies go back in time. They constitute the other major class of longitudinal study and may at first appear problematic for the qualitative case study researcher intent on primary data collection. For example, if semi-structured interviews require participants to 'think back' to the way things were and how they 'felt then', recall bias becomes a serious and unpredictable problem. Nevertheless, in a design that depends on the triangulation of methods and data including such things as diaries, letters, policy documents, and meeting minutes, it may be a far less serious threat.

Space: Comparative Analysis

Not all comparative analyses are spatial, but most conducted by geographers are spatial in one way or another. For example, there is a long tradition in human geography that emphasizes how phenomena may present very differently from one case to the next because of the place itself—i.e., the meanings derived from the interactions among residents. When conducted at the same point in time, case studies of multiple instances of a phenomenon are commonly known as **comparative analysis, comparative case study**, or **parallel case study**. For example, a comparative analysis might involve a parallel study of three of the communities shown in Figure 5.2. Comparative studies tend to share many of the same advantages as longitudinal case studies in that there are opportunities to generate and modify concepts and theory so that they explain commonalities across cases *despite* contingencies or context. Although research phenomena (e.g., risk perception) need not be place-specific, it is often useful for geographers to focus on place as an initial basis for comparison. It is also important to clarify that we are talking about field comparisons of two or more instances of a phenomenon rather than merely the comparison of a case study with existing, published case studies. Although situating one's research within the literature is essential for good academic work, the focus here is on multiple cases built into a study at the design stage. For example, in a special issue on comparative studies of transnationalism (e.g., remittances of money or visits back to the country of origin), Dunn (2008) suggests that much can be gained by comparing the impact of transnationalism on the same cultural groups in different places or on different cultural groups in the same place, as well as the impact of transnationalism on the places themselves. The emphasis is not on the number of cases per se but on understanding how the phenomena are manifest in different contexts.

Castree (2005) underscores the value of comparative analysis by making a call for more explicitly comparative case study work in geography concerning research on 'nature's neoliberalisation' (2005, 541). He complains that a growing number of such case studies focus too much on the particular and not enough on what is common *across* case studies. He wants to explore some general impacts (manifestations) of neo-liberalization on nature, irrespective of context. This is an ongoing tension in qualitative research generally—balancing the particular with the more abstract when developing explanations. We see this tension in what Castree himself writes: elsewhere in the same paper, he writes that he is also frustrated by *too much* abstraction. Despite his call for some theory that cuts across case studies, he complains that existing concepts quoted in the literature are 'clearly so abstract that [they] fail to tell us how and with what effects otherwise different neoliberalisations work, (2005, 543). These over-general concepts include privatization and deregulation. Castree's frustration is directed towards the literature and not towards any particular **cross-case comparison**. Indeed, a formal cross-case comparison of neo-liberalization's influence on nature in a single study is one solution to Castree's concern, as long as the concepts are sufficiently specific (but of course not *too* specific). These ideas are taken up further in the following discussion on generalizability.

ARE CASE STUDIES GENERALIZABLE?

The short answer is yes. Quite probably, the most common criticism of case study research is its supposed lack of generalizability (Campbell and Stanley 1966; see also Flyvbjerg 2006). Yet such concern may be exaggerated. That is, generalizability should not be a problem if case study research is designed appropriately and the analysis is attentive to the tension between concrete and abstract concepts.

Generalizability (or **external validity**) is a term used by quantitative social scientists, but many qualitative researchers prefer the term **transferability** (Lincoln and Guba 2002). Generalizability or transferability concerns the degree to which findings apply to other cases of the phenomenon in question. It may be interpreted as 'the more cases the theory applies to, the better'. However, this is only one way to look at generalizability, and qualitative case study researchers tend not to emphasize the more-cases-is-better approach. Instead, qualitative researchers are more concerned that explanations of the phenomenon as manifest in the case are credible. This distinction is partially explained as the difference between statistical generalization and **analytical (theoretical) generalization**. In the former case, generalization is achieved through large probability samples, while in the latter, transferability is accomplished by (1) carefully selecting cases and (2) creating useful theory that is neither too abstract nor too case-specific (Flyvbjerg 2006; Yin 2003).

Flyvbjerg (2006) provides some good examples of how generalizable, or transferable, theoretical concepts and explanations can be generated from a single case study (e.g., a single experiment). Perhaps the most famous is Galileo's debunking of Aristotle's theory of gravity. Aristotle claimed that heavier objects will accelerate faster than lighter ones. However, with a single experiment (case study) involving only a few different types of balls rolled down inclined planes, Galileo showed that balls of different weights accelerated at the same rate. This case study disproved the theory that weight is a determinant of gravitational acceleration, which paved the way for others like Sir Isaac Newton to suggest that the determinant was something else (i.e., the relationship between mass and distance). Thus, the choice of case (in the Galileo example, the choice of a ball and ramp experiment) can be critical. Further, it is important to recognize that falsifying existing theory simultaneously opens up avenues for new theory. Similarly, Popper (1959) argued that it takes only one black swan to falsify the theory that all swans are white. Keep in mind that although these examples serve to show both the value of single cases and the strategy of **falsification**, social scientists rarely expect any concept or theory to apply in all cases—i.e., to constitute a theoretical law. What is important is to describe *why* a theory does or does not apply in a particular case.

There are numerous examples of case studies in geography that falsify, particularly in areas like political economy and political ecology, which use case studies to show how social and environmental problems are connected to flaws in the global socio-economic system. For example, Weis (2000, 300) provides an explanation of unsustainable deforestation in the blue mountains of Jamaica that shifts blame for the problem away from the 'poor stewardship' of the peasants towards the 'grossly inequitable land regime' within a globalized food system. Thus, Weis's empathetic account of these Jamaican farmers is a form of falsification as it attempts to supplant a long-standing and traditional explanation of deforestation with a more compelling alternative one.

Finding cases that falsify existing theory is but one of many examples of the role case studies can play in terms of addressing generalizability or transferability.

For most qualitative researchers, the development of a coherent theory itself is of primary concern, not necessarily whether the findings challenge hegemonic wisdom or whether findings adhere in *all* or even *most* cases at the time they are studied (George and Bennett 2005). In practical terms, it is too great a burden for one study not only to understand the context, contingencies, and details of the case but at the same time to know all contingencies and contexts for all other cases of the phenomenon in question—whether risk perception, academic praxis, transnationalism, neo-liberalization, or deforestation.

Strictly speaking, the only way to truly know whether a theory applies in other cases is to study those other cases—but perhaps not in as much detail as the cases that were used to develop the theory in the first place. The true value of a qualitative case study, then, may not be known until several years after the research or policy community has had a chance to digest the concepts and theory. The theory may only *eventually* prove useful for explaining similar phenomena in different cases. Indeed, this is one of the main reasons for the enduring fame of Whyte's (1943) and Liebow's (1967) studies. Although they focused on Italian and Black men in Boston and Washington respectively, the studies' insights apply in numerous cities among a wide array of ethnic and minority groups. Their concepts have resonated in several contexts over space and time. They seemed to strike the right balance between describing concepts that explain the concrete details of the case but are sufficiently abstract to apply to similar phenomena (i.e., living in a poor ethnic neighbourhood in a large urban centre). The take-home message is to generate theoretical concepts and explanations that potentially resonate in other (as yet unstudied) contexts.

Conclusion

Case studies have a long and rich history in the social sciences and geography. Case study methodology is a powerful means by which to both (1) understand the concrete and practical aspects of a phenomenon or place and (2) develop theory. That is, case studies may be used to understand and solve practical problems relating to the case alone, and they may be used to test, falsify, expand, or generate explanatory theoretical concepts. Case studies are valuable because when done well, they produce deep, concrete explanations of social phenomenon that are attentive to a variety of contextual influences at various scales. Most of what qualitative researchers do involves a case study methodology. This research often uses various combinations of qualitative (and quantitative) *methods* to support data collection (e.g., interviews, focus groups, participant observation) as well as a variety of analytical *strategies* (e.g., grounded theory, discourse analysis). These methods and analytical strategies may be used in longitudinal analyses of the same case, cross-case comparisons of different cases, or one-time analysis of a single case. It is important to recognize that although a case study may only involve a sample of one, a carefully chosen and well-studied case can be used to produce very robust, credible, and trustworthy theoretical explanations. These explanations are generalizable, or transferable, in the analytical sense rather than in the statistical sense. Good theoretical explanations are those that are well rooted in the concrete aspects of the case yet sufficiently abstract that others in similar situations can see how they might apply to their own context.

KEY TERMS

analytical generalization

case study

Chicago School of Sociology

comparative analysis

comparative case study

corroboration

cross-case comparison

cross-sectional case study

ethnographic research

external validity

falsification

generalizability

idiographic

longitudinal case study

member checking

method

methodology

nomothetic

parallel case study

participant checking

praxis

theoretical generalization

theory generation

theory testing

transferability

REVIEW QUESTIONS

1. How can a study based on N=1 produce useful research results? What role does generalizability play in evaluating the quality of such results?
2. What is the historical role of multiple methods in case study research?
3. Outline some of the key advantages and challenges of temporal or spatial comparative case studies.
4. What are the roles of induction and deduction in case study research? How do they relate to nomothetic and idiographic research?

Useful Resources

Colorado State University. 1993–2009. 'Writing guides—Case studies'. http://writing. colostate.edu/guides/research/casestudy. A helpful guide to the history and conduct of case studies as well as to the presentation of their results.

Flyvbjerg, B. 2006. 'Five misunderstandings about case-study research'. *Qualitative Inquiry* 12 (2): 219–45.

George, A., and A. Bennett. 2005. *Case Studies and Theory Development in the Social Sciences*. Cambridge, MA: MIT Press.

Hamel, J., S. Dufour, and D. Fortin. 1993. *Case Study Methods*. Newbury Park: Sage.

Hesse-Biber, S., and P. Leavy. 2004. *Approaches to Qualitative Research: A Reader on Theory and Practice*. New York: Oxford University Press.

Platt, J. 1992. '"Case study" in American methodological thought'. *Current Sociology* 40: 17–48.

Stake, R. 2006. *Multiple Case Study Analysis*. New York: Guilford Press.

Yin, R. 2003. *Case Study Research: Design and Methods*. Los Angeles: Sage.

Notes

1. At the time, structured surveys were by far the most popular means of collecting information in social research.

2. The example is modelled on my own research largely to avoid any misinterpretations about research intentions and interpretations. See Baxter and Lee 2004.

Part II

'Doing' Qualitative Research in Human Geography

Interviewing

Kevin Dunn

6

CHAPTER OVERVIEW

This chapter provides advice on interview design, practice, transcription, data analysis, and presentation. I describe the characteristics of each of the three major forms of interviewing and critically assess what I see as the relative strengths and weaknesses of each. I outline applications of interviewing by referring to examples from economic, social, and environmental geography. Finally, some of the unique issues regarding Internet-based interviews are reviewed.

INTERVIEWING IN GEOGRAPHY

Interviewing in geography is so much more than 'having a chat'. Successful interviewing requires careful planning and detailed preparation. A recorded hour-long interview would have involved days of preliminary background work and question formulation. It would have required diplomacy in contacting **informants** and negotiating 'research deals'. A 60-minute interview will require at least four hours of transcription if you are a fast typist, and verification of the record of interview could stretch out over a couple of weeks. After all that, you have still to analyze the interview material. These are time-consuming activities. Is it all worth it? In this chapter, I outline some of the benefits of interviewing and provide a range of tips for good interviewing practice.

An interview was once defined as 'a face-to-face verbal interchange in which one person, the interviewer, attempts to elicit information or expressions of opinion or belief from another person or persons' (Maccoby and Maccoby 1954, 499). An interview is a data-gathering method in which there is a spoken exchange of information. While this exchange has traditionally been face-to-face, researchers have also used telephone interviews (Groves 1990). However, the new millennium has seen the emergence of **computer-mediated communications (CMC) interviewing**, a mode from which direct access has disappeared.

Types of Interviewing

There are three major forms of interviewing: structured, unstructured, and semi-structured. These three forms can be placed along a continuum, with the **structured interview** at one end and the **unstructured interview** at the other (see also Chapter 1, Box 1.1). Structured interviews follow a predetermined and standardized list of questions. The questions are asked in almost the same way and in the same order in each interview. At the other end of the interviewing continuum are unstructured forms of interviewing such as **oral histories** (discussed fully in Chapter 7). The conversation in these interviews is actually directed by the informant rather than by set questions. In the middle of this continuum are **semi-structured interviews.** This form of interviewing has some degree of predetermined order but maintains flexibility in the way issues are addressed by the informant. Different forms of interview have varying strengths and weaknesses that should be clear to you by the end of this chapter.

Strengths of Interviewing

Research interviews are used for four main reasons (see also Krueger 1994; Minichiello et al. 1995, 70–4; Valentine 1997, 110–12):

1. To fill a gap in knowledge that other methods, such as observation or the use of census data, are unable to bridge efficaciously.
2. To investigate complex behaviours and motivations.
3. To collect a diversity of meaning, opinion, and experiences. Interviews provide insights into the differing opinions or debates within a group, but they can also reveal consensus on some issues.
4. When a method is required that shows respect for and empowers the people who provide the data. In an interview, the informant's view of the world should be valued and treated with respect. The interview may also give the informant cause to reflect on their experiences and the opportunity to find out more about the research project than if they were simply being observed or if they were completing a questionnaire.

Interviews are an excellent method of gaining access to information about events, opinions, and experiences. Opinions and experiences vary enormously between people of different class, ethnicity, age, and sexuality. Interviews have allowed me to understand how meanings differ among people. Geographers who use interviewing should be careful to resist claims that they have discovered the *truth* about a series of events or that they have distilled *the* public opinion (Goss and Leinbach 1996, 116; Kong 1998, 80). Interviews can also be used to counter the claims of those who presume to have discovered *the* public opinion. This can be done by seeking out the opinions of different groups, often marginalized or subaltern groups, whose opinions are rarely heard.

Most of the questions posed in an interview allow for an open response as opposed to a closed set of response options such as 'yes' or 'no'. In this way, each informant can describe events or offer opinions in their own words. One of the major strengths of interviewing is that it allows you to discover what is relevant to the informant.

Because the face-to-face verbal interchange is used in most interviewing, the informant can tell you if a question is misplaced (Box 6.1). Furthermore, your own opinions and tentative conclusions can be checked, verified, and scrutinized. This may disclose significant misunderstandings on your part or issues that you had not previously identified (Schoenberger 1991, 187).

Asking the Wrong Question: A Tale from Cabramatta   Box 6.1

On 23 June 1990, I began my first formal research interview. The informant was a senior office-bearer from one of the Indo-Chinese cultural associations in New South Wales, Australia. My research interest was in the social origins of the concentration of Indo-Chinese Australians around Cabramatta and the experiences of these immigrants. The political context of the time was still heavy with the racialized and anti-Asian overtones of the 1988 'immigration debate' in which mainstream politicians and academics such as John Howard (later prime minister from 1996 to 2008) and Geoffrey Blainey (professor of history) had expressed concern about 'Asian' immigration and settlement patterns in Australian cities. Specifically, Vietnamese immigrants were accused of congregating in places like Cabramatta (Sydney) and Richmond and Springvale (both in Melbourne) and of purposefully doing so in order to avoid participating with the rest of Australia. I had hypothesized that Vietnamese Australians did not congregate voluntarily but that they were forcibly segregated by the economic and social constraints of discrimination in housing and labour markets. Indeed, the geographic literature supported my assertion at the time. I was a somewhat naive and colonialist investigator who saw his role as 'valiant champion' of an ethnic minority.

But back to my first interview. One of my first questions was: 'Please explain the ways in which discrimination has forced you, and members of the community you represent, to reside in this area?' The answer: 'I wouldn't live anywhere else.' This informant, and most subsequent informants, described the great benefits and pleasures of living in Cabramatta. They also explained how residing in Cabramatta had eased their expanding participation in Australian life (Dunn 1993). I had asked the wrong questions and had been told so by my informants. I decided to focus the project on the advantages and pleasures that residence in 'Cab' brought to Indo-Chinese Australians. The face-to-face nature of the exchange, and the informed subject, makes interviews a remarkable method. The participants can tell the researcher, 'You're on the wrong track!'

Interview Design

It is not possible to formulate a strict guide to good practice for every interview context. Every interview and every research issue demands its own preparation and practice. However, researchers should heed certain procedures. Much of the rest of this chapter focuses on strategies for enhancing the credibility of data collected using rigorous interview practice. In the next section, we look at the organization of **interview schedules** and the formulation of questions.

The Interview Schedule or Guide

Even the most competent researcher needs to be reminded during the interview of the issues or events they had intended to discuss. You cannot be expected to recall all of the specific questions or issues you wish to address, and you will benefit from some written reminder of the intended scope of the interview. These reminders can take the form of interview schedules or **interview guides**.

An interview guide or **aide-mémoire** (Burgess 1982c) is a list of general issues you want to cover in an interview. Guides are usually associated with semi-structured forms of interviewing. The guide may be a simple list of key words or concepts intended to remind you of discussion topics. The topics initially listed in a guide are often drawn from existing literature on an issue. The identification of key concepts and the isolation of themes is a preliminary part of any research project (see Babbie 1992, 88–164).

One of the advantages of the interview guide is its flexibility. As the interviewer, you may allow the conversation to follow as 'natural' a direction as possible, but you will have to redirect the discussion to cover issues that may still be outstanding. Questions can be crafted in situ, drawing on themes already broached and from the tone of the discussion. The major disadvantage of using an interview guide is that you must formulate coherent question wordings 'on the spot'. This requires good communication skills and a great deal of confidence. Any loss of confidence or concentration may lead to an inarticulate and ambiguous wording of questions. Accordingly, a guide is inadvisable for first-time interviewers. It is more appropriate for very skilled interviewers and for particular forms of interview, such as oral history.

Interview schedules are used in structured and sometimes semi-structured forms of interviewing. They are also called question schedules or 'question routes' (Krueger 1994). An interview schedule is a list of carefully worded questions (see Box 6.2).

I have found that a half-hour interview will usually cover between six and eight primary questions. Under each of these central questions I nest at least two detailed questions or **prompts**. In some research, it may be necessary to ask each question in the same way and in the same order to each informant. In others, you might ask questions at whatever stage of the interview seems appropriate. The benefits of using interview schedules mirror the disadvantages of interview

Formulating Good Interview Questions   **Box 6.2**

- Use easily understood language that is appropriate to your informant.
- Use non-offensive language.
- Use words with commonly and uniformly accepted meanings.
- Avoid ambiguity.
- Phrase each question carefully.
- Avoid leading questions as much as possible (i.e., questions that encourage a particular response).

guides. They provide greater confidence to the researcher in the enunciation of their questions and allow better comparisons between informant answers. However, questions that are prepared before the interview and then read out formally may sound insincere, stilted, and out of place.

A mix of carefully worded questions and topic areas capitalizes on the strengths of both guides and schedules. Indeed, a fully worded question can be placed in a guide and yet be used as a topic area. The predetermined wording can be kept as a 'fall-back' in case you find yourself unable to articulate a question 'on the spot'. I find it useful to begin an interview with a prepared question. It is damaging to one's confidence if an informant asks, 'What do you mean by that?' or 'I don't know what you mean' in response to your first question.

Interview design should be dynamic throughout the research (Tremblay 1982, 99, 104). As a research project progresses, you can make changes to the order and wording of questions or topics as new information and experiences are fed back into the research design. Some issues may be revealed as unimportant, offensive, or silly after the initial interview. They can be dropped from subsequent interviews. The interview schedule or guide should also seek information in a way that is appropriate to each informant.

While the primary purpose of interview schedules and guides is to jog your memory and to ensure that all issues are covered as appropriately as possible, it is also useful to provide informants with a copy of the questions or issues before the interview to prompt thought on the matters to be discussed. Interview guides and schedules are also useful note-taking sheets (Webb 1982, 195).

TYPES OF QUESTIONS

Interviews utilize **primary** (or original) **questions** and **secondary questions**. Primary questions are opening questions used to initiate discussion on a new theme or topic. Secondary questions are prompts that encourage the informant to follow up or expand on an issue already discussed. An interview schedule, and even an interview guide, can have a mix of types of original questions, including descriptive questions, storytelling prompts, structural questions, contrast and opinion questions, and devil's advocate propositions (see Box 6.3). Since different types of primary questions produce very different sorts of responses, a good interview schedule will generally comprise a mix of question types.

Primary Question Types   Box 6.3

Type of Question	Example	Type of Data and Benefits
Descriptive (knowledge)	What is the full name of your organization? What is your role within the organization? How many brothers or sisters do you have?	Details on events, places, people, and experiences. Easy-to-answer opening questions.
Storytelling	Can you tell me about the formation and history of this organization and your involvement in it?	Identifies a series of players, an ordering of events, or causative links. Encourages sustained input from the informant.
Opinion	Is Canadian society sexist? What do you consider to be the appropriate size for a functional family?	Impressions, feelings, assertions, and guesses.
Structural	How do you think you came to hold that opinion? What do you think the average family size is for people like yourself?	Taps into people's ideology and assumptions. Encourages reflection on how events and experiences may have influenced opinions and perspectives.
Contrast (hypothetical)	Would your career opportunities have been different if you were a man? Or if you grew up in a poorer suburb?	Comparison of experience by place, time, gender, and so forth. Encourages reflection on (dis)advantage.
Devil's advocate	Many practising town planners are voicing concern about the lack of transparency in the new development consent process.	Controversial/sensitive issues broached without associating the researcher with people who are not prepared to make their opinions public.

See also Box 6.4.

| Asking the Tough Questions without Sounding Tough | Box 6.4 |

My work with Indo-Chinese communities in Cabramatta occurred within a political context in which Vietnamese Australians were being publicly harangued by academics and politicians (Dunn 1993). I felt it was important to get the informants to respond to the views of their critics. My interview schedule had the following two devil's advocate propositions. I also used a preamble to dissociate myself from the statements.

> In my research so far, I have come across two general explanations for Vietnamese residential concentration. I would like you to comment on two separate statements that to me represent these two explanations:
>
> First: that Vietnamese people have concentrated here because they don't want to participate in the wider society.
>
> Second: that the Vietnamese are segregated into particular residential areas through social, economic, and political forces imposed upon them by the wider society.

The aim was to gather people's responses to both statements. In most cases, informants were critical of both views. Some informants took my question as a request for them to select the explanation that they thought was the most appropriate. Those people selected the second statement. Others were in no doubt that I disagreed with both views. Either way, devil's advocate propositions are often leading. My political views were noticeable in the question preamble and wording as well as in the preliminary discussions held to arrange the interviews. It is fairer that the researcher's motives and political orientation are obvious to the informant rather than hidden until after the research is published (see Chapters 2, 4, and 17 of this volume).

On an interview guide or schedule, you might have a list of secondary questions or prompts (Box 6.5). There are a number of different types of prompts, ranging from formal secondary questions to nudging-type comments that encourage the informant to continue speaking (Whyte 1982, 112). Sometimes prompts are listed in the interview guide or schedule, but often they are deployed, when appropriate, without prior planning.

Ordering Questions and Topics

It is important to consider carefully the order of questions or topics in an interview guide or schedule. Minichiello et al. (1995, 84) advise that the most important consideration in the ordering of questions is preserving **rapport** between you and

---

Types of Prompt   Box 6.5

Prompt Type	Example	Type of Data and Benefits
Formal secondary question	Primary Q: What social benefits do you derive from residing in an area of ethnic concentration?	Extends the scope or depth of treatment on an issue.
	Secondary Q: What about informal child care?	Can also help to explain or rephrase a misunderstood primary question.
Clarification	What do you mean by that?	Used when an answer is vague or incomplete.
Nudging	And how did that make you feel? Repeat an informant's last statement.	Used to continue a line of conversation.
Summary (categorizing)	So let me get this straight: your view, as just outlined to me, is that people should not watch shows like 'Survivor'?	Outlines in-progress findings for verification. Elicits succinct statements (for example, 'quotable quotes').
Receptive cues	Audible: Yes, I see. Uh-huh. Non-audible: nodding and smiling.	Provides receptive cues, encourages an informant to continue speaking.

---

your informant. An interview might begin with a discussion of the general problem of homophobia or racism: how widespread it is, how it varies from place to place, and how legal and institutional responses to those forms of oppression have emerged. Having broached these general or macro-level aspects of oppression, the interview might then turn to the particular experiences of the informant.

In an interview with a **pyramid structure**, the more abstract and general questions are asked at the end. The interview starts with easy-to-answer questions about an informant's duties or responsibilities or their involvement in an issue. This allows the informant to become accustomed to the interview, interviewer, and topics before they are asked questions that require deeper reflection. For example, to gather views on changes to urban governance, you might find it necessary to first ask an informant from an urban planning agency to outline their roles and duties. Following that, you might ask your informant to outline the actions of their own agency and how those actions may have changed in recent times. Once the 'doings and goings-on' have been outlined, it may then make sense to ask the informant why agency actions and roles have changed, whether that change has been resisted, and how they view the transformation of urban governance.

A final question-ordering option is to use a hybrid of **funnel** and pyramid structures. The interview might start with simple-to-answer, non-threatening questions, then move to more abstract and reflective aspects before gradually progressing towards sensitive issues. This sort of structure may offer the benefits of both funnel and pyramid ordering.

When thinking about question and topic ordering, it can be helpful to have key informants comment on the interview guide or schedule (Kearns 1991a, 2). Key informants are often initial or primary contacts in a project. They are usually the first informants, and they often possess the expertise to liaise between the researcher and the communities being researched. Key informant review can be a useful litmus test of interview design, since these representatives are 'culturally qualified'. They have empathy with the study population and can be comprehensively briefed on the goals and background of the research (Tremblay 1982, 98–100).

STRUCTURED INTERVIEWING

A structured interview uses an interview schedule that typically comprises a list of carefully worded and ordered questions (see Boxes 6.3 and 6.5 and the earlier discussion on ordering questions and topics). Each respondent or informant is asked exactly the same questions in exactly the same order. The interview process is question-focused.

It is a good idea to **pre-test** a structured interview schedule on a subset (say 3 to 10) of the group of people you plan to interview for your study to ensure that your questions are not ambiguous, offensive, or difficult to understand. Though helpful, pre-testing is of less importance in semi-structured and unstructured interviews in which ambiguities (but not offensive questions!) can be clarified by the interviewer.

Structured interviews have been used with great effect throughout sub-disciplines of geography, including economic geography (Box 6.6).

Interviewing—An Economic Geography Application	Box 6.6

In 1991, Schoenberger argued that most industrial geography research had been on the outside looking in, deducing strategic behaviour from its locational effects rather than investigating it directly (1991, 182). One of the assumptions challenged by the use of structured interviews was that the location of firms was strongly associated with the location preferences of the industries in question. Using structured interviews with managers, Schoenberger was able to show that the location of foreign chemical firms in North America was as much, if not more, related to historical and strategic contingencies than to contemporary location preferences.

For example, one of Schoenberger's case studies was a German-owned chemical firm. Her interviews revealed that the firm's board of directors had decided on a major expansion in the US market. But the board had been split between establishing a 'greenfields' site, which would be purpose-built to company needs, and acquiring an already established chemical plant, which would hasten their expanded presence in the market. Plans to establish a greenfields facility were foiled by organized community opposition. The directors who argued for an acquisition then gained the upper hand, and at about the same time a US chemical firm came up for sale. For many decades, German chemical firms had agreed among themselves to specialize in certain parts of the chemical sector. These agreements were about to end, and Schoenberger's case-study firm was keen to expand horizontally into another speciality. The US firm that came up for sale happened to specialize in that area. A host of historical and strategic events had combined to produce a particular location result.

The historical and strategic contingencies that accounted for the location of the German chemical company were revealed through structured interviews. Their location was in fact quite at odds with the apparent preferences of the firm and reveals nothing about the firm's location preferences (Schoenberger 1991, 185). The US chemical sector has a high level of foreign ownership, and most of it was established through acquisition. Replicated interviews with managers and directors across the chemical sector revealed the prevalence of location choices being determined by historical and strategic contingency. Interviewing was therefore an essential method for unravelling the location determinants of chemical plants in the North America.

SEMI-STRUCTURED INTERVIEWING

Semi-structured interviews employ an interview guide. The questions asked in the interview are content-focused and deal with the issues or areas judged by the researcher to be relevant to the research question. Alternatively, an interview schedule might be prepared with fully worded questions for a semi-structured interview, but the interviewer would not be restricted to deploying those questions. The semi-structured interview is organized around ordered but flexible questioning. In semi-structured forms of interview, the role of the researcher (interviewer or facilitator) is recognized as being more interventionist than in unstructured interviews. This requires that the researcher redirect the conversation if it has moved too far from the research topics (see also Chapter 16).

UNSTRUCTURED INTERVIEWING

Various forms of unstructured interviewing exist. They include oral history, **life history**, and some types of group interviewing and in-depth interviewing. Unstructured interviewing focuses on personal perceptions and personal histories. Rather than being question-focused like a structured interview or content-focused as in a semi-structured format, the unstructured interview is informant-focused. Life history and oral history interviews seek personal accounts of significant events and perceptions, as determined by the informants and in their own words (see Chapter 7 and also McKay 2002). Each unstructured interview is unique. The questions you ask are almost entirely determined by the informant's responses. These interviews approximate normal conversational interaction and give the informant some scope to direct the interview. Nonetheless, an unstructured interview requires as much if not more preparation than its structured counterpart. You must spend time sitting in musty archive rooms or perched in front of dimly lit microfiche machines gaining a solid understanding of past events, people, and places related to the interview. But through these interviews we can 'find out about' events and places that had been kept out of the news or that had been deemed of no consequence to the rich and powerful (Box 6.7).

Oral Environmental Histories	Box 6.7

Oral history interviews can collect data about environmental history. This type of interviewing helps produce a more comprehensive picture of the cause and process of environmental change than is available through physical methods of inquiry. Data collected might include people's memories of changes in local land use, biodiversity, hydrology, and climate.

Lane (1997) used oral history interviews to reveal changes in watercourses, weeds, and climate in the Tumut Region high country of the Australian Alps. Interviews were conducted with five main informants, first in their homes and then while driving and walking through the countryside where they had resided. The informants told of the waterholes and deep parts of creeks where they would fish and swim and where they and their children had learned to swim. One informant commented that one of the creeks used to be almost a river—and now you could step over it (Lane 1997, 197). The same informant noted the change in colour and quality of the water. Lane's informants described how the water level and quality had steadily degraded since pine plantations had been planted in the 1960s. This description was consistent with 'scientific' understandings of the impact of pine plantations in which there is an ever-decreasing level of run-off as the pines grow.

> Such specific observations from local residents may often be the only detailed evidence on environmental change that is available. Oral history can fill gaps in the 'scientific record', or it can be used to complement data gathered using physical or quantitative methods. More important, with the use of oral history, environmental change can be set in a human context and related to the history of people who lived in the region (Lane 1997, 204).

INTERVIEWING PRACTICE

Rapport with another person is basically a matter of understanding their model of the world and communicating your understanding symmetrically. This can be done effectively by matching the perceptual language, the images of the world, the speech patterns, pitch, tone, speed, the overall posture, and the breathing patterns of the informant (Minichiello et al. 1995, 80).

Achieving and maintaining rapport, or a productive interpersonal climate, can be critical to the success of an interview. Rapport is particularly important if you need to have repeat sessions with an informant. Even the first steps of arranging an interview are significant, including the initial contact by telephone and other preliminaries that might occur before the first interview. Interviews in which both the interviewer and informant feel at ease usually generate more insightful and more valid data than might otherwise be the case. In the following paragraphs, I outline a set of tips that can help you to enhance rapport before, during, and while closing an interview.

Contact

Informants are usually chosen purposefully on the basis of the issues and themes that have emerged from a review of previous literature or from other background work (see Chapters 5 and 16). This involves choosing people who can communicate aspects of their experiences and ideas relevant to the phenomena under investigation (Minichiello et al. 1995, 168). Once you have identified a potential informant, you must then negotiate permission for the interview. This means getting the consent of the informants themselves, and in some circumstances, it will also involve gaining the sanction of 'gatekeepers' like employers, parents, or teachers. This might occur, for example, if you wanted to interview schoolchildren, prisoners, or employees in some workplaces.

Your first contact with an informant will often be by telephone or by some form of correspondence such as e-mail. In this preliminary phase, you should do at least four things (Robertson 1994, 9):

1. Introduce yourself and establish your bona fides. For example: 'My name is Juan Folger, and I am an honours student from Java State University.'
2. Make it clear how you obtained the informant's name and telephone number or address. If you do not explain this, people may be suspicious and are likely to ask how you got their name or e-mail address. If you are asked this question, rapport between you and your informant has already been compromised.
3. Outline why you would like to conduct the interview with this particular informant. Indicate the significance of the research, and explain why the informant's views and experiences are valued. For instance, you may believe that they have important things to say, that they have been key players in an issue, or that they have experienced something specific that others have not. On the whole, I have found that most people are flattered to be asked for an interview, although they are often nervous or hesitant about the procedure itself.
4. Indicate how long the interview and any follow-up is likely to take.

Making an informant feel relaxed involves dealing with all of the issues mentioned above and in addition spelling out the mechanics of the interview and negotiating elements of the interview process. All of these matters can be outlined in a 'letter of introduction', which may be sent to an informant once they have agreed to an interview or while agreement is still being negotiated. This formal communication should be under the letterhead of your organization (for example, your university) and should spell out your bona fides, the topic of the research, the manner in which the interview will be conducted, and any rules or boundaries regarding confidentiality. You must, of course, seek permission from your supervisor to use the letterhead of an organization such as a university, although ethics procedures within your institution (see Chapter 2) are likely to have made this mandatory. In the absence of a letter of introduction, informants should be made aware of their rights during the interview. This procedure is sometimes referred to as brokering a 'research deal' or a 'research bargain'. The research deal may be agreed to over the telephone or by e-mail or orally just before an interview begins. The deal can be set out in written form. (See Box 6.8 for some of the rights of informants that can be established. Chapter 2 of this volume includes material relevant to the ethics of interviewing. See also Hay [1998] and Israel and Hay [2006].) These preliminary discussions are important to the success of an interview. Indeed, they set the tone of the relationship between interviewer and informant.

The Interview Relation

The relationship established between interviewer and informant is often critical to the collection of opinions and insights. If you and your informant are at

Codifying the Rights of Informants

Box 6.8

In their research on the Carrington community in Newcastle, New South Wales, Winchester, Dunn, and McGuirk (1997) decided to codify informants' rights in the oral histories and semi-structured interviews that were to be conducted. They included the following list of informants' rights on university letterhead, and they gave a copy to each of the informants:

- Permission to record the interview must be given in advance.
- All transcribed material will be anonymous.
- Tapes and **transcripts** will be made available to informants who request them.
- Informants have the right to change an answer.
- Informants can contact us at any time in the future to alter or delete any statements made.
- Informants can discontinue the interview at any stage.
- Informants can request that the audio recorder be paused at any stage during the interview.

To this list one might add further statements (for example, that informants can expect information about the ways in which their contributions to the research would be used). A codification of rights was deemed necessary for two reasons. First, it was done to empower the informants and assure them that they could pause or terminate the interview process whenever they deemed it necessary to do so. Second, the researchers had employed an articulate local resident to conduct the interviews, and so it was important that the interviewer was also constantly reminded of the informants' rights.

ease with each other, then the informant is likely to be communicative. However, there are competing views on the nature of the interviewer–informant relationship. On the one hand is an insistence on **professional interviewing** and on the other 'creative' or empathetic interviewing. Goode and Hatt (in Oakley 1981, 309–10) warn that interviewers should remain detached and aloof from their informants: 'the interviewer cannot merely lose himself [*sic*] in being friendly. He must introduce himself as though beginning a conversation, but from the beginning the additional element of respect, of professional competence, should be maintained. . . . He is a professional in this situation, and he must demand and obtain respect for the task he is trying to perform'.

A very different, indeed opposite, sort of relationship was proposed by Oakley (1981) and Douglas (1985). In their view, a researcher who remains aloof would undermine the development of an intimate and non-threatening relationship

(Oakley 1981, 310). Rather than demanding respect from the informant, Douglas's model of **creative interviewing** insists that each informant must be treated as a 'Goddess' of information and insight. Douglas recommends that researchers humble themselves before the Goddess. The creative or empathetic model of interviewing thus advocates a very different sort of relationship between the informant and interviewer than that recommended for professional interview relations. Overall, there is a range of interview practice that lies between the poles of professional and creative interview relationships.

Decisions about the interview relationship will vary according to the characteristics of both the informant and the interviewer. The cultural nuances of a study group will at times necessitate variations in the intended interaction. However, it is wise to remember that despite any empathy or relationships that are established, the interview is still a formal process of data gathering for research. Furthermore, there is usually a complex and uneven power relationship involved in which information, and the power to deploy that information, flows mostly one way: from the informant to the interviewer (see Chapter 2 and also McKay 2002, Stacey 1988).

Rapport may increase the level of understanding you have about the informant and what they are saying. There are a number of strategies for enhancing rapport. The first is through the use of respectful preliminary work. The second involves the use of a **warm-up** period just before an interview commences. Douglas (1985, 79) advises that rather than getting 'right down to business', it is better to engage in some 'small talk and chit-chat [which] are vital first steps'. This warm-up discussion with an informant could be a chat about the weather, matters of shared personal interest, or 'catching-up' talk. In their surveys and interviews of Vietnamese Australians in Melbourne, Gardner, Neville, and Snell (1983, 131) found that '[t]he success of an interview (when measured by the degree of relaxation of all those present and the ease of conversation) generally depended on the amount of "warm-up" (chit-chat, introductions, etc.).'

My own warm-up techniques for face-to-face interviews have included giving the informant an overview of the questions I plan to ask and presenting relevant diagrams or maps, as well as discussing historical documents (see also Tremblay 1982, 99, 103). Maps, diagrams, tables of statistics, and other documents can also be used as references or stimuli throughout an interview. If an informant offers you food or drink before an interview, it would be courteous to accept. As we shall see in more detail later in this chapter, many of these forms of warm-up are not available in CMC interviews.

You should also have acquainted yourself with the cultural context of the informants before the interview. As Robin Kearns pointed out, 'If we are to engage someone in conversation and sustain the interaction, we need to use the right words. Without the right words our speech is empty. Language matters' (Kearns 1991a, 2). For instance, you must be able to recognize the jargon or slang and frequently used acronyms of institutions or corporations as well as the language of particular professions or cultural groups.

Listening strategies can improve rapport and the productivity of the interview. Your role as interviewer is not passive but requires constant focus on the information being divulged by informants and the use of cues and responses to encourage them. Your role as an active participant in the interview extends well beyond simply asking predetermined questions or broaching predetermined topics. You must maintain an active focus on the conversation. This will help to prevent lapses of concentration. You must also avoid 'mental wandering'—otherwise, you may miss unexpected leads. Moreover, it is irritating to the informant, and a threat to rapport, if you ask a question they have already answered (Robertson 1994, 44).

Adelman (1981) advises researchers to maintain a **critical inner dialogue** during an interview. This requires that you constantly analyze what is being said and simultaneously formulate the next question or prompt. You should be asking yourself whether you understand what the informant is saying. Do not let something pass by that you do not understand with the expectation that you will be able to make sense of it afterwards. Minichiello et al. (1995, 103) provide a demonstration of how critical inner dialogue might occur: 'What is the informant saying that I can use? Have I fully understood what this person is saying? Maybe, maybe not. I had better use a probe. Oh yes, I did understand. Now I can go on with a follow-up question.'

Strategies to enhance rapport continue throughout the interview, through verbal and non-verbal techniques that indicate that the responses are valued. Informants may sometimes recount experiences that upset them or stir other emotions. When an informant is becoming distressed, try pausing the interview or changing the topic and possibly returning to the sensitive issue at a later point. If the informant is clearly becoming very distressed, you should probably terminate the interview.

There may be a stage in an interview when your informant does not answer a question. If there is a silence or if they shake their head, the informant may be indicating that they have not understood your question or simply do not know the answer. They might be confused as to the format of the answer expected: is it a 'yes no' or something else (Minichiello et al. 1995, 93)? In these cases, try restating the question, perhaps using alternative wording or providing an example. You should always be prepared to elaborate on a question. It is important to remember, however, that choosing not to respond is the informant's right. If the informant refuses to answer and says so, you should not usually press them. They may have chosen not to answer because the question was asked clumsily or insensitively (or for some other reason—if the question dealt with sensitive commercial matters, for instance). If you prepare your questions carefully, you should avoid this sort of problem and the consequent loss of empathy and data.

As an interviewer, you should also learn to distinguish between reflective silence and non-answering. Robertson cautions, 'Do not be afraid of silences.

Interviewers who consciously delay interrupting a pause often find that a few seconds of reflection leads interviewees to provide the most rewarding parts of an interview. . . . There is no surer way of inhibiting interviewees than to interrupt, talk too much, argue, or show off your knowledge' (Robertson 1994, 44).

It is important to allow time for the informant to think, meditate, and reflect before they answer a question (see Box 6.9). It is also important to be patient with slow speakers or people who are not entirely host language–fluent. Resist the temptation to finish people's sentences for them. Supplying the word that an informant is struggling to find may seem helpful at the time, but it interrupts them and inserts a term they might not have ordinarily used.

Finishing Sentences, Interrupting, and 'Rushing-on'	Box 6.9

During February 2001, Minelle Mahtani (then of the University of British Columbia) joined me in Sydney to undertake joint, and comparative, research on the media and representations of ethnic minorities in Australia and Canada. This involved interviews with managers and employees within newsrooms. They were powerful and confident informants. Dr Mahtani had a wealth of expertise in such environments, having been a producer with the flagship Canadian Broadcasting Corporation's 'The National', a news television program. Our first field interview was with a network news editor for one of the commercial networks in Australia. Our questions included themes such as media representations of ethnic minorities, attempts by the organization to improve the portrayal of ethnic minorities, the presence of 'minority' journalists, and circumstances in which they or their staff had challenged stereotypical storylines. The questions had been developed and agreed to in advance, but what very different styles we had! The informants would sometimes provide very short and dismissive responses to some questions. When it was clear that they had answered, I would probe or move to another question. Dr Mahtani would wait, however. The silence would hang heavy over the interview. I felt uncomfortable, but these powerful informants got the idea that we wanted a fuller response. They would then justify the view they had briefly given, or they would admit that there were alternative viewpoints to the one they had expressed. My rushing-on was a strategy vastly inferior to the 'sounds of silence' for uncovering richer insight into ethnic minority representations and the dynamics of the newsroom (Dunn and Mahtani 2001).

Closing the Interview

Do not allow rapport to dissipate at the close of an interview. It is critical to maintain rapport—especially if you intend to re-interview the informant. You must prepare for the closure of an interview—otherwise, the ending can be clumsy. Because an interview establishes a relationship within which certain expectations are created, it is better to indicate a sense of continuation and of feedback and clarification than to end the interview with an air of finality.

Try not to rush the end of an interview. At the same time, do not let an interview 'drag on'. There is an array of verbal and non-verbal techniques for closing interviews (Box 6.10). Of course, non-verbal versions should be accompanied by appropriate verbal cues; otherwise, you could appear quite rude. The most critical issue in closing an interview is to express not only thanks but also satisfaction with the material that was collected. For example, you might say, 'Thanks for your time. I've got some really useful, insightful information from this interview.' Not only is gratitude expressed this way, but the informant is made aware that the process has been useful and that their opinions and experiences have been valued.

Techniques for Closing Interviews **Box 6.10**

Four types of verbal cue:

- direct announcement: 'Well, I have no more questions just now.'
- clearinghouse questions: 'Is there anything else you would like to add?'
- summarizing the interview: 'So, would you agree that the main issues according to you are . . . ?'
- making personal inquiries and comments: 'How are the kids?' or 'If you want any advice on how to oppose . . . just ring me.'

Six types of non-verbal cue:

- looking at your watch
- putting the cap on your pen
- stopping or unplugging the audio recorder
- straightening your chair
- closing your notebook
- standing up and offering to shake hands.

Source: Adapted from Minichiello et al. 1995, 94–8.

RECORDING AND TRANSCRIBING INTERVIEWS

Interview recording, transcription, and field note assembly are referred to as the mechanical phases of the interview method. These are the steps through which the data are collected, transformed, and organized for the final stages of analysis.

Recording

Audio recording and note-taking are the two main techniques for recording face-to-face and telephone interviews. Other less commonly used techniques in geography include video recording, compiling records of the interview after the session has ended, and using cognitive maps. Both audio recording and note-taking have associated advantages and disadvantages, as will become clear in the discussion to follow. Therefore, a useful strategy of record-keeping is to combine note-taking and audio recording.

The records of an interview should be as close to complete as possible. An audio recorder will help to compile the fullest recording (Whyte 1982, 117–18). Interviewers who use note-taking need excellent shorthand writing skills to produce verbatim records. However, the primary aim in note-taking is to capture the gist of what was said.

Audio or video recording can allow for a natural conversational interview style because the interviewer is not preoccupied with taking notes. Instead, you can be a more attentive and critical listener. Audio recording is also preferable to note-taking because it allows you more time to organize the next prompt or question and to maintain the conversational nature of the interview. The note-taking researcher can be so engrossed in taking notes that they may find themselves unprepared to ask the next question. Note-takers can also miss important movements, expressions, and gestures of the informant while they are hunched over, scribbling at a furious pace (Whyte 1982, 118). This undermines rapport and detracts from attentive listening.

On the other hand, an audio recorder may sometimes inhibit an informant's responses because the recorder serves as a reminder of the formal situation of the interview (Douglas 1985, 83). Informants may feel particularly vulnerable if they think that someone might recognize their voice if the recording were aired publicly. Opinions given by the informant on the 'spur of the moment' become fixed indelibly on tape (or disc or memory stick) and have the potential to become a permanent public record of the informant's views. This may make the informant less forthcoming than they would have been if note-taking had been used. Some informants become comfortable with an audio recorder as the interview progresses, but others do not. If you find the latter to be the case, consider stopping the recorder and reverting to note-taking.

If you use an audio recorder, place it somewhere that is not too obvious without compromising the recording quality. Modern digital audio recorders have long recording capacities and can be turned on and then left alone. But take care when using an audio recorder not to be lulled into a loss of concentration by the feeling that everything is being recorded safely. There may be a technical failure. You can maintain concentration and avoid the problem associated with recorder failure by taking some written notes. If you are taking notes, there is little likelihood of mental wandering. Everything is being listened to and interpreted, and parts of it are being written down, demanding that you maintain concentration. I find this particularly important when I am conducting the second or third interview in a long day's fieldwork.

Because an audio recorder does not keep a record of non-verbal data, non-audible occurrences such as gestures and body language will be lost unless you are also using a video recorder or taking notes. If an informant points to a wall map and says, 'I used to live there', or if they say, 'The river was the colour of that cushion', then the audio recording will be largely meaningless without some written record. These written notes can be woven into the verbal record during the transcription phase (described later in this chapter). Written notes also serve as a back-up record in case of technical failures. Overall then, a strategic combination of audio recording and note-taking can provide the most complete record of an interview with the least threat to the interview relationship.

Transcribing the Data

The record of an interview is usually written up to facilitate analysis. Interviews produce vast data sets that are next to impossible to analyze if they have not been converted to text. A transcript is a written 'reproduction of the formal interview which took place between researcher and informant' (Minichiello et al. 1994, 220). The transcript should be the best possible record of the interview, including descriptions of gestures and tone as well as the words spoken (although see Box 6.13 below). The name or initials of each speaker should precede all text in order to identify the interviewer(s) and informant(s). Counter numbers at the top and bottom of each page of the transcript enable quick cross-referencing between the transcript file and tapes or digital records of the interview. Converting interview to text is done either through a reconstruction from handwritten notes, a transcription of an audio or video recording, the use of **voice recognition computer software** (See Box 6.11), or editing CMC interview correspondence.

Interview notes should be converted into a typed format preferably on the same day as the interview. If there were two or more interviewers, it is a good idea to compile a combined reconstruction of what was said using each researcher's notebook. This will improve the breadth and depth of coverage. The final typed record will normally comprise some material recalled verbatim as well as summaries or approximations of what was said.

Recorded interviews should also be transcribed as soon as possible after the interview. Transcription is a very time-consuming and therefore resource-intensive task (Whyte 1982, 118). On average, most interviews take four hours of typing per hour of interview. Transcription rates vary according to a host of variables,

Voice Recognition Computer Packages and Interviews   **Box 6.11**

Computing packages have been developed that convert the spoken word into computer text. Packages such as NaturallySpeaking by Dragon Systems can help you convert text at rates above 160 words per minute. However, these systems will only convert the speech of a single speaker. Each system has to be 'trained' to understand a single 'master's voice'. The success of these packages for converting interview data has been mixed. The researcher has to simultaneously listen to a recording and orally repeat the informant's contributions to enable the system to convert the data. Gestures and indications of intonation have to be typed into the **word processing** document manually. Nonetheless, typists can train the software and then re-speak the interview, typing corrections (using the 'correct that' command) and inserting notations as they go along (using hot-keys for such things as speaker initials). Fatigue, illness, and alcohol have all been reported as reducing recognition accuracy.

Listserve reports lodged by researchers have claimed accuracy rates as high as 95 per cent once the program has been 'trained'. I have read a claim that the new age of software requires only 15 minutes of training, although my experience was that even after four hours of coaching, the software was still getting every third word 'wrong' and some of the conversions were hilarious. It is critical that you save your speech files (the training) after each use, and it is advisable that you use a very good quality microphone. It should also be noted that serious investigation of the methodological issues that may surround voice recognition software has barely begun.

such as typist skill, the type of interview, the informant, and the subject matter. You can facilitate transcription by using a purpose-built transcribing recorder. You should transcribe your own interviews for two main reasons. First, since you were present at the interview, you are best placed to reconstruct the interchange. You are aware of non-audible occurrences and therefore know where such events should be inserted into the speech record. You are also better able to understand the meaning of what was said and less likely to misinterpret the spoken words. Second, transcription, although time-consuming, does enable you to engage with the data again. Immersion in the data provides a preliminary form of analysis.

There is no accepted standard for symbols used in transcripts, but some of the symbols commonly used are set out in Box 6.12. CMC interviews do not require transcription, since each informant's answers are already in a text format. The text may well include some **emoticons** that are popular in SMS (short message service) and Internet-based communication.

Once completed, the transcript should be given a title page stating the informant's name (or a code if there are confidentiality concerns), the number of the

Symbols Commonly Used in Interview Transcripts

Box 6.12

Symbol	Meaning
//	Speaker interrupted by another speaker or event: // phone rings//
:	Also used to indicate an interruption
KMD	The initials of the speaker, usually in CAPS and bold
—	When used at the left margin, refers to an unidentified speaker
Ss	Several informants who said the same thing
E	All informants made the same comment simultaneously
. . .	A self-initiated pause by a speaker
. . . . or	Longer self-initiated pauses by a speaker
-	Speech that ended abruptly but without interruption
()	Sections of speech, or a word, that cannot be deciphered
(jaunty)	A best guess at what was said
(jaunty/journey)	Two alternative best guesses at what was said
*	Precedes a reconstruction of speech that was not taped
(. . .)	Material that has been edited out
But I didn't want to	Underlined text indicates stressed discourse
I got nothing	Italicized text indicates louder discourse
[sustained laughter]	Non-verbal actions, gestures, facial expressions
[hesitantly]	Background information on the intonation of discourse
Emoticon	(see http://en.wikipedia.org/wiki/List_of_common_emoticons)
:) or ☺ or :-) or :-D	Smiling, joke marker, happy, laughing hard
:(or :-(	Frowning, sad
;) or *)	Wink
:'(or :: or :,(	Tears, shedding a tear, crying
lol	Laughing out loud

interview (for example, first or third session), the name(s) of the researcher(s) (i.e., who carried out the interview), the date of the session, the location, duration of the interview, and any important background information on the informant or special circumstances of the interview. Quotations that demonstrate a particular point and that could be presented as evidence in a final report on the research might be circled or underlined.

The transcript can be given to the informant for vetting or authorizing. This will normally improve the quality of your record (see Box 6.13). This process of **participant checking** continues the involvement of the informants in the research process and provides them with their own record of the interview.

Debates about Changing the Words: Vetting and Correcting   Box 6.13

In general, it has been thought that a transcription should be a verbatim record of the interview. This would include poor grammar, false starts, 'ers', and 'umms'. There are a number of good reasons advanced for this position. A verbatim record will include the nuances of accent and **vernacular**, it will maintain any sense of hesitancy, and it could demonstrate an embarrassment that was present. For example, Sarah Nelson (2003, 16) reflected on how the 'humming and hawing' of Ulster politicians when asked about sectarian killings was reflective of their hesitancy and hypocritical stances on sectarianism. Transcripts that are not exact textual replications of an interview will lose the ethnographic moment of the interview itself. However, it may be difficult to search for key terms if they are 'misspelled' in a transcript (misspelled as a means of indicating accent or mispronunciation).

A range of researchers working in different disciplines and countries have expressed some concern about the political effects of exact transcription. Many have reflected on the embarrassment that many informants articulate when they receive the transcript of their interview. They express anxiety about the grammar, the false starts to their sentences, repetition, and the 'ers' and 'umms' and 'you knows'. This anxiety is even more strongly felt by informants who live in societies where the dominant language is not their first language. Informants might be so concerned as to withdraw their interview and avoid any future ones (McCoyd and Kerson 2006, 397). Moreover, research reports on the less powerful in society (the poor, single mothers, youth groups) that use the real language of informants and are largely sympathetic to those people can often portray them in a way that reproduces negative images and stereotypes. Nelson (2003) reflected on the way such quoted material reconstructs images of illiteracy, powerlessness, and inferiority. As stated earlier, transcription is a transformation of verbal encounter into text; it is a constructed document that is of the researcher's making (Green, Frauquiz, and Dixon 1997). As bell hooks (1990, 152) famously stated, 'I want to know your story. And then I will tell it back to you in a new way . . . . Rewriting you, I write myself anew. I am still author, authority.' Informants are more interested in the impact of their words than in the nuances of expression. Many researchers recommend sending informants summaries or interpretations of the interview rather than transcripts. It is certainly a good idea to send informants the eventual publications and reports.

Assembling Field Note Files

Assembling interview records marks the beginning of the analysis proper. It begins with a critical assessment of the interview content and practice and is followed by formal preparation of interview logs. To my mind, the best advice on assembling field note files comes from Minichiello et al. (1995, 214–46). In the wide margins of the transcript file, you can make written annotations. Comments that relate to the practice of the interview, such as the wording of questions and missed opportunities to prompt, should be placed in the left margin. These annotations and other issues concerned with contact, access, ethics, and overall method should be elaborated upon in a **personal log** (Box 6.14). The right margin of the transcript file can be used for annotations on the substantive issues of the research project. These comments, which generally use the language and jargon of social science, are then elaborated upon in the **analytical log**. The analytical log is an exploration and speculation about what the interview has found in relation to the research question (Box 6.14). It should refer to links between the data gathered in each interview and the established literature or theory.

Field Note Files   Box 6.14

Transcript File	*Personal Log*	*Analytical Log*
Includes the record of speech and the interviewer's observations of non-audible data and intonation. Also includes written annotations in the margins on the practice and content of the interview.	Reflection on the practice of the interview. Includes comments on the questions asked and their wording, the appropriateness of the informant, recruitment and access, ethical concerns, and the method generally.	Exploration of the content of the interview. A critical outline of the substantive matters that have arisen. Identification of themes. Reference to the literature and theory. In-progress commentary on the research aims and findings.

Source: Adapted from Minichiello et al. 1995, 214–46.

ANALYZING INTERVIEW DATA

Researchers analyze interview data to seek meaning from the data. We construct themes, relations between variables, and patterns in the data through content analysis (see Chapter 14). Content analysis can be based on a search of either manifest or latent content (Babbie 1992, 318–19). **Manifest content analysis** assesses the visible, surface content of documents such as interview transcripts. An example would be a tally of the number of times the words 'cute' and 'cuddly' are used to describe koalas in interviews with members of the public. This might be important to understanding public opinion and the politics of culling in areas of koala overpopulation (for example, Muller 1999). Searching interview data for manifest content often involves tallying the appearance of a word or phrase. Computer programs such as **NVivo** or QSR N6 are particularly effective at undertaking these sorts of manifest searches (see Chapter 15).

Latent content analysis involves searching the document for themes. For example, you might keep a tally of each instance in which a female has been portrayed in a passive or active role. Latent content analysis of interview texts requires a determination of the underlying meanings of what was said. This determination of meanings within the text is a form of coding.

A coding system is used to sort and then retrieve data. For example, the text in transcripts of interviews with urban development authorities could be coded based on the following categories: structures of governance (for example, legislation, party political shifts), cultures of governance (with sub-codes like 'managerialist perspective' and 'entrepreneurial perspective'), coalitions and networks (of various types and agendas), the mechanisms through which coalitions operate, and the various scales at which power and influence emanate and are deployed (see McGuirk 2002). Once the sections of all the interviews have been coded, it is then possible to retrieve all similarly coded sections. These sections of text can be amalgamated and reread as a single file (Box 6.15). This might allow a researcher to grasp the varying opinions on a certain issue and to begin to unravel the general feeling about an issue.

Not every section of text needs to be coded. An interview will include material that is not relevant to the research question, particularly warm-up and closing sections and other speech focused on improving rapport rather than on gathering data. Some sections of text will be multiple-coded. For example, in one sentence an informant may list a number of causes of fish kills, including opencut mine run-off, super phosphates, acid sulphate soils, and town sewage. This may require that the sentence is given four different coding values. Coding is discussed more fully in Chapter 14.

| Coding Interview Data: Five Suggested Steps | Box 6.15 |

Coding Step	Specific Operations: Computing/Manual Versions
Develop preliminary coding system	Prepare a list of emergent themes in the research. Draw on the literature, your past findings, as well as your memos and log comments. Amend throughout.
Prepare the transcript for analysis	Meet the formatting requirements for the computing package being used. / Print out a fresh copy of the transcript for manual coding.
Ascribe codes to text	Allocate coding annotations using the 'code text' function of computing packages. / Place handwritten annotations on transcript.
Retrieve similarly coded text	Use the 'retrieve text' function of computing packages to produce reports on themes. / Extract and amalgamate sections of text that are similarly coded.
Review the data by themes	Assess the diversity of opinion under each theme. Cross-referencing themes allows you to review instances where two themes are discussed together. Begin to speculate on relations between themes.

Presenting Interview Data

Material collected from interviews is rarely presented in its entirety. Most interview data must be edited and (re)presented selectively in research publications. While it is difficult to locate a 'genuinely representative' statement (see Connell 1991, 144–5; Minichiello et al. 1995, 114–15), it is usually possible to indicate the general sense and range of opinion and experience expressed in interviews. One way to indicate this is to present summary statistics of what was said. Computing packages such as NVivo can help you to calculate the frequency with which a particular term or phrase appeared in a document or section of text (see Chapter 15). However, the more common method is through a literal description of the themes that emerged in the interviews (see, for example, Boxes 6.6, 17.1, and 17.2 and the discussion in Chapters 17 and 18 on presenting results).

When describing interview data, you must cite transcript files appropriately. For example, in her interview-based honours research on the changing identity of the Australian industrial city of Wollongong, Pearson (1996, 62) noted that 'Several respondents asserted that elements excluded by the new identity were of little significance to the overall vernacular identity of Wollongong (Int. #1, Int. #6, and Int. #7).' The transcript citations provided here indicate which of the informants expressed a particular type of opinion. In research publications, the transcript citations can indicate the informant's name, number, code, or recorder count. Whenever a direct quotation from an informant is presented, then a transcript page reference or recorder count can be provided.

Transcript material should be treated as data. A quotation, for example, ought to be treated in much the same way as a table of statistics. That is, it should be introduced and then interpreted by the author. The introduction to a quotation should offer, if it has not already been provided, some background on the informant. It is important that readers have some idea of where an informant is 'coming from'; information about their role, occupation, or status is important in this regard. Also important, as Baxter and Eyles (1997, 508) point out, is 'some discussion of why particular voices are heard and others are silenced through the selection of quotes'. Quotations should be discussed in relation to, and contrasted with, the experiences or opinions of other informants. Statements of opinion by an informant should also be assessed for internal contradiction. Finally, a quotation cannot replace a researcher's own words and interpretation. As the author, you must explain clearly what theme or issue a quotation demonstrates.

Knowledge is a form of power. The accumulation and ownership of knowledge is an accumulation of power: power to support arguments and power to effect change. In most interviews, information and knowledge flow from the informant to the researcher. The researcher accumulates this knowledge and ultimately controls it. There is a host of strategies and guidelines to which researchers can adhere to reduce the potential political and ethical inequities of this relationship (see Chapters 2, 16, and 17). In terms of data presentation, it will sometimes be important that an informant's identity be concealed.

Pseudonyms or interviewee numbers have been used by geographers to disguise the identity of their informants when it has been thought that **disclosure** could be harmful. Informants can be given the opportunity to select their own pseudonym. Robina Mohammad (1999, 238–9) used this technique with some success in her interviews with young Pakistani Muslim women in England. The interviews included discussion of patriarchal authority, 'English cultures', and the cultures and dynamics of the Pakistani Muslim community in Britain. Some informants selected Pakistani Muslim pseudonyms for themselves, others chose very English names. These selections were themselves very interesting and provided further insight into the cultural perspectives and resistances of these women. Other researchers allocate pseudonyms that reflect the ethnicity of the informants (McCoyd and Kerson 2006, 396, 404). Gill Valentine (1993) felt it necessary to disguise the name of the town in which her interviews with lesbians had taken place. Similarly, Mariastella Pulvirenti (1997, 37) disguised the street names that were mentioned by female Italian Australians when discussing their housing and settlement experiences in Melbourne.

Naming an informant (or locating them in any detailed way) and directly associating them to a quotation could be personally, professionally, or politically harmful. Researchers must be very careful when they deploy data they have collected. Interviewers are privileged with insights into people's lives. Some researchers recommend instituting an alias or pseudonym for informants very early in the mechanical phase so that no electronic records will bear the informant's real identity. However, it can prove difficult to remember who the real people behind the aliases are, and some researchers only impose the pseudonym in the presentation phase of the research.

The presentation of interview-based research must contain an accessible and transparent account of how the data were collected and analyzed (Baxter and Eyles 1997, 518). This account should outline the subjectivity of the researcher, including their biases or 'positioned subjectivity' (see Chapters 2, 16, and 17). As we have already seen from the discussion in Chapter 3, it is only through transparent accounts of how interview-based research was undertaken that the trustworthiness and wider applicability of the findings can be assessed by other researchers.

INTERVIEWS USING COMPUTER-MEDIATED COMMUNICATION

Computer-mediated communication (CMC) interviewing can include interviews with individuals and groups. They can be either **asynchronous** or **synchronous** (see Meho 2006, 1284; Mann and Stewart 2002, 604). The most common form of CMC interviewing has been asynchronous, using the to-and-fro of e-mail exchanges. Synchronous CMC interviewing has mirrored the environment of on-line chat. Meho's (2006, 1284–5) review of literature on e-mail interviewing concluded that since the early 2000s, this mode had moved from being experimental and is now considered an established format for research-oriented interviewing. Most early uses of CMC for social research involved attaching questionnaires to e-mails (and sometimes interview schedules). However, e-mail interviewing now commonly takes the form of an ongoing set of exchanges between a researcher and an informant. In the next few pages, I review advantages and disadvantages of CMC interviewing and then provide a series of tips for better practices specific to this mode.

Advantages of CMC Interviewing

CMC-delivered interviews offer five general sets of advantages: (1) an expanded sample; (2) reduced **interviewer effects**; (3) enhanced convenience; (4) more reflective informant responses; and (5) cost savings (see Chapter 10 for additional discussion of some of these issues).

First, Internet delivery of questions allows a researcher to overcome spatial, temporal, and social barriers that would restrict access to informants for face-to-face interviews. Interviews can be more easily facilitated with people living overseas or in inaccessible locations (remote places or war zones) or who have mobility limitations (e.g., differently abled), and e-mailed exchanges can be much more convenient for people who are shift workers or those who are based at home with small children (Bampton and Cowton 2002; Mann and Stewart 2002, 604–6; Meho 2006, 1288). E-mail interviews can also allow researchers to transgress social hurdles to gain access to informants: they will suit people who are shy, who are cautious about their identity being revealed, or whose cultural context is too disparate from that of the researcher; they are also appropriate when the topic is very sensitive. Good examples in the literature include interviews with political and religious dissidents, criminals and criminalized people, oppressed minorities, and deviant subcultures. McCoyd and Kerson (2006) found from their research with women in North America who had had medical terminations (after learning of a foetal anomaly) that e-mail interviews were very effective for discussions on a topic that generated strong emotions among the informants and broached issues that are not normally discussed in public. Meho's (2006, 1286–7) review of the uses of e-mail interviewing indicates that response rates among these 'difficult to reach' samples were on average about 40 per cent.

The absence of 'face-to-face' in e-mail interviewing means that researchers can interview people who dress and look very differently from themselves—such as subcultural groups like Goths or bikers (Mann and Stewart 2002, 606). More broadly, a second advantage of e-mail interviews is a reduction in interviewer effects because of the researcher's visual anonymity. Many of the cues we use to make judgments about people are based on visual appearance (e.g., dress, body shape, skin colour, jewellery, hair styles), and these cues are much more limited in CMC interviews in which informants can adopt pseudonyms and even de-gender themselves. This anonymity comes at the cost of the non-audio data that are usually gathered in a face-to-face interview. Contextual effects also dissipate. At the instrumental level, this means that the interview is not interrupted by telephone calls, colleagues, small children, or spouses. A more complicated matter is the absence of the researcher from the informant's own ethnographic setting and the loss of those observations. However, e-mail interviewing offers great convenience to informants, allowing them to choose the time of their responses, to consider their answers at their leisure, and to do so in the comfort of their own home (Bampton and Cowton 2002; Mann and Stewart 2002, 607; Meho 2006, 1290). An informant in McCoyd and Kerson's (2006, 397) research with women who had medical terminations reflected that 'I'm looking forward to doing the interview . . . it is a much more relaxed and productive way to do it [through e-mail]. This way, I can do it when things are quiet and I'm in the right frame of mind.' The informant has much more control over the pace and flow of the interview, more so than if it were a telephone or face-to-face interview during which they might feel rushed to offer an answer.

A fourth advantage of e-mail interviewing is that the answers that informants offer can be more detailed, reflective, and well-considered than those in other formats. James and Busher's (2006) e-mail interviews with people in university settings found that there was a 'richness of reflection among the participants'. Informants who took their time in responding to a question 'tended to generate more thoughtful answers' (2006, 414). James and Busher gave the example of an informant who began one of their answers with 'I didn't email you straight back, because I was thinking about my answer. So my responses were more carefully thought through and probably longer than if I'd tackled the whole thing in a face-to-face interview' (university-based informant, quoted in James and Busher 2006, 415). Informants can rethink, proofread, and re-craft their responses so that they most accurately represent their views and experiences (Bampton and Cowton 2002). Many of the above advantages apply to one-on-one e-mail interviews rather than to group discussions.

A final set of advantages of CMC interviewing is the reduced cost relative to face-to-face interviewing. The obvious savings are associated with travel costs and time—and carbon footprints. Another saving comes in the mechanical phase, since there is no need for a conversion of spoken word to text. The answers are already in text form, and this also removes issues of transcription error and interpretation (Chen and Hinton 1999; Mann and Stewart 2002, 608–9). This obviates the need to engage one of the debates about transcription referred to earlier (Box 6.13). Informants have 'cleaned' their own responses to a level they are satisfied with before they post them (McCoyd and Kerson 2006, 397). Of course, researchers do have to edit out the unnecessary e-mail symbols, signature sections, and line returns.

Challenges of CMC Interviewing

The weaknesses or limitations of CMC interviews, relative to face-to-face interviewing, stem mostly from the spatial and temporal displacements between the informants and the researcher. These issues include concerns about the authenticity of the informant, the loss of visual cues that assist rapport-building, and the 'clunkiness' of the interview relationship. There are also issues of uneven Internet access and comfort with the medium, as well as ethical issues having to do with privacy and anonymity.

The advantages that stem from the visual anonymity and the use of pseudonyms have the negative effect of reducing our ability to know who we are interviewing. The identity dynamism facilitated in Internet correspondence means that people can make misleading claims about who they are (James and Busher 2006). However, Mann and Stewart (2002, 210) do note that maintaining a false identity in a substantive correspondence is not so easy to do, and researchers ought be wary of inconsistencies and contradictions that reveal such false personae. More subtly, informants are more able to embellish or be bombastic in e-mailed answers, because the researcher does not have the visual cues to help detect and address such tendencies (Meho 2006, 1289). Similarly, prompts are delayed until the next e-mail interchange, at which point they may be considered less important than new primary questions, crowding out space for such clarifications (Chen and Hinton 1999; Meho 2006, 1289–90). Researchers using e-mail interviews must instead rely heavily on 'reading between the lines' of answers.

The absence of **paralinguistic clues** entails broader issues. It is much more difficult for researchers to tell whether an informant is becoming distressed or uncomfortable when reflecting on or writing an answer to a question (Bampton and Cowton 2002). Their answers are in a 'narrower bandwidth', and this has dramatic consequences on the potential for rapport. It is difficult to communicate empathy and sympathy (e.g., in regard to grief) to informants via e-mail without it sounding banal or insincere (Mann and Stewart 2002, 617–19). The generally truncated process of e-mail interviewing means that interviewing a single informant can extend over weeks and months. There will be substantial gaps at times between responses. These issues are related to the processes of reflection and response consideration discussed earlier, but they are also linked to the personal availability of the informants and to the ability of the researcher to properly assimilate previous answers and to consider necessary probes and relevant primary questions. These gaps can be interpreted, incorrectly, as disinterest on the part of the informant or as a sign that the researcher is underwhelmed or disgusted by previous responses. It is difficult for researchers to know why an informant might be delaying their response (Bampton and Cowton 2002). But the resulting form of rapport that develops is very different from what has been described for face-to-face interviews.

CMC interviewing is really only appropriate for study groups with widespread Internet access and literacy. This means that it would be inappropriate for homeless people or those who lack basic information technology skills and familiarity (Bampton and Cowton 2002; Mann and Stewart 2002, 605). Participating in an interview through e-mail or on-line in a chat format, is quite physically demanding, requiring a lot of time spent hunched over a computer screen and typing (Chen and Hinton 1999). It is more onerous for the informant than answering a question orally over the telephone or face-to-face. Interviews that require informants to download materials and open attachments shift a good deal of technical work to the informant, raising the concern that it may divert attention from the questions and research topic to the technology. One general lesson is that

CMC interviews are most appropriate for technically savvy study groups, including, for example, 'on-line communities', which are natural groups that already exist within the Internet environment (Mann and Stewart 2002, 615).

Decisions about good ethical practices surrounding CMC interviews are still evolving. One concern surrounds the privacy of informant comments. On the one hand, people often feel an inflated sense of anonymity within the Internet, using pseudonyms to avoid attribution. But e-mailed messages and posted comments can be traced (Cho and LaRose 1999). A bigger issue concerns the usual promises that researchers offer to informants about the confidentiality of their comments. Internet communications are intercepted, and server hosts do retain copies of e-mails, as do institutions like universities. This means that informants' answers are not only kept by or accessible to the researcher. As well, many universities and most research corporations have 'My Documents' directories on all researchers' computers through which files are stored centrally on a common server. E-mails should not be considered ephemera that disappear as soon as you delete them from your own inbox: they have potential longevity and circulation that is almost limitless, and the ethical risks entailed in that require careful consideration (Crystal 2006, 132; Spinks, Wells, and Meche 1999, 148–9). Although many universities currently require informants to sign consent forms before they participate in interviews, the Internet environment has different protocols for signatures and approvals, and some researchers consider that an e-mail from the informant detailing their informed consent should be sufficient. Some argue that such correspondence must occur before and separate from any discussion of substantive matters; others are more sanguine about the separation, suggesting that consent statements can immediately precede the first set of answers (Cho and LaRose 1999, 429; McCoyd and Kerson 2006, 394).

A final set of concerns regarding CMC interviews has to do with the way they add to Internet clutter. Requests that a person be involved in e-mail interviews or post responses in an on-line setting can be confused with corporate and commercial correspondence with which Internet users are increasingly bombarded. Forecasts by the Radicati Group (2005, 14) determined that in 2009, those who use e-mail for work will receive 109 messages per day on average (as opposed to 99 per day in 2005). This Internet clutter could pose substantial problems for engaging CMC interview informants in the future.

Strategies for Good CMC Interviews

As with face to-face interviews, rapport between researchers and informants can be enhanced or jeopardized in various phases of CMC interviews. In the contact phase, you should make quite clear how you obtained the potential informant's e-mail address. Internet users are increasingly suspicious of how people obtained their address and fed-up with the e-mail spam they receive. The sweeping up of e-mail addresses has been referred to as **trolling**, and any research that carries a sense that informant contacts were collected in that way will be treated with suspicion. E-mails also need to be sent in an individualized way to informants, not as a group e-mail (James and Busher 2006, 408). Contacting on-line groups or networks and asking them participate in on-line interviews should be preceded by some interaction with the group and some sanction from gatekeepers. A researcher who launches into a group unannounced will likely receive antipathetic reactions. Similarly, a long period of non-participation in a group followed by a request for interviews that demonstrates a good deal of knowledge about the group will create the impression that the researcher has been **lurking**—another form of Internet anti-social behaviour (Cho and LaRose 1999, 422–4, 430). Further, informants should receive individually tailored responses and follow-ups, not formally worded e-mails that anaemically state, 'Here is the next question'. This recursiveness means that most CMC interviews take a semi-structured or unstructured form.

One of the threats to rapport once an e-mail interview is underway is the silences and gaps in communication. The longer the gaps between exchanges, the higher the rates of drop-out and non-completion (Meho 2006, 1288). Researchers will need to remind informants about questions for which an answer has not yet been forthcoming. Bampton and Cowton (2002) advise that this asynchronous aspect (clunkiness) of interviews (and other differences from face-to-face format) should be made clear to the informants at an early stage. In general, informants need to be given a clear sense of the time frame of the whole process, the expected amount of response, the number of parcels of questions they will receive, and that there will be courteous reminders at times (see also Mann and Stewart 2002, 616). Nonetheless, patience is likely to be a virtue when one is waiting respectfully for answers, and pestering could spoil the field for future researchers (Cho and LaRose 1999, 425). When an answer from an informant is received, it is wise to send a prompt acknowledgment, and considered follow-ups can be delayed until you have been through a process of inner critical dialogue. James and Busher (2006, 412) recommend 'strategies of visibility' to maintain contact while informants are considering their answers: 'Haven't heard from you for while', 'How is it going?' These types of individualized prompts will remind your informants that you are still keen to hear from them, and it can provide an opportunity for informants to raise any issues they have about the question or make clear their desire to skip that question. These are ways to be vigilant to potential drop-outs and distress (Mann and Stewart 2002, 616–19).

Some researchers recommend that initial e-mails from the researcher should contain some element of personal disclosure relevant to the topic of the interview. This has the effect of building up a sense of trust and sharing (Mann and Stewart 2002,

615–17). The Internet is an environment with a spirit of information-sharing, so CMC interviews should commit to releasing the general findings and recommendations to the world wide web (Cho and LaRose 1999, 432).

On-line cultures may include the use of acronyms, abbreviations, and emoticons. Some researchers suggest that these textual cultures should be encouraged (Meho 2006, 1293), but others think that they should be discouraged in formal correspondence (Dumbrava and Koronka 2006, 62), and still others believe that they should be tolerated. Emoticons, for example, could be used by informants to indicate the kind of tone and emotion that would be apparent in face-to-face interviews (Bampton and Cowton 2002). McCoyd and Kerson (2006, 396) reported an informant placing the emoticon for tears '::' at the end of a sentence to indicate her emotional context as she was telling the researchers about her feelings regarding the medical termination of her foetus. Common emoticons could enhance rapport between researchers and informants once an interview is well underway, and the use of extra letters and dots can be a good indication of pauses ('hmmmmmm, I'm not sure about that', 'weeellll maybe') (see Box 6.12). There are debates regarding the extent to which researchers should ensure proper grammar and spelling in CMC interviewing. In general, some latitude is granted for contractions that imitate speech (e.g., I'm, don't) as well as some abbreviations (txt, fwd) (Crystal 2006, 113–17). However, rich and well-expressed formal text provides a clearer and more usable form of data than text that is full of emoticons and SMS-like abbreviations (Mann and Stewart 2002, 15, or see Spinks, Wells, and Meche 1999, 149, on business communications).

Many of the rules and cultures of Internet communication carry over into CMC interviews. Researchers should be cautious about attempting to use humour and sarcasm. E-mails should have appropriate subject lines, and university e-mail addresses and URLs should be used to provide informants with a stronger sense of the project's credibility (Chen and Hinton 1999; Cho and LaRose 1999, 431). Capitalized and bold text should be avoided, since it is considered on-line shouting. Each e-mail should have only enough text to fit into a single screen, or the first screen should have a summary of all that follows (see Bampton and Cowton 2002; Crystal 2006, ch. 4; Hassini 2006, 33; Spinks, Wells, and Meche 1999, 147–9) These rules of 'netiquette' flow from long-standing principles established in the mid-1990s (see Scheuermann and Taylor's [1997, 270] reference to the Ten Commandments of Etiquette or Rinaldi's [1994] often-cited guidelines).

The ethical issues surrounding CMC interviews mean that the threats to privacy should be prominent in consent agreements. Researchers can make the compiled records of interviews anonymous and store all such files off-line on external hard drives, and they can attempt to erase all e-mails and destroy all print-outs (Chen and Hinton 1999; McCoyd and Kerson 2006, 394; Mann and Stewart 2002). These actions can provide the informant with some assurance of confidentiality.

Because of the limitations and challenges of CMC interviewing, some have concluded that it is best seen as a complementary mode and that face-to-face interviewing remains the gold standard method (see Meho 2006). According to this view, CMC interviewing is appropriately used for informants who are difficult to physically reach or for those who already have a presence within some form of on-line community or network. However, the mode has special advantages of its own, such as the opportunity for more reflective responses, and it represents an appropriate method for the Internet age.

CONCLUSION

The rigour of interview-based research is enhanced through adequate preparation, diverse input, and verification of interpretation. Being well-informed and prepared will give you a deeper understanding of the 'culture' and discourse of the group(s) you study. You can then formulate good questions and enhance levels of rapport between you and your informants. You should also purposely seek out diversity of opinion. By interviewing more than one informant from each study group, you can begin to draw out and invite controversy or tensions. An opinion from one informant should never be accepted as demonstrative of group opinion unless it is shown to be so. Finally, some means of verifying your interpretations of interview data are necessary (for example, participant checking, peer checking, and cross-referencing to documentary material).

Interviews bring people 'into' the research process. They provide data on people's behaviour and experiences. They capture informants' views of life. Informants use their own words or vernacular to describe their own experiences and perceptions. Kearns (1991a, 2) made the point that 'there is no better introduction to a population than the people themselves.' This is what I find the most refreshing aspect of interview material. Transcribed interviews are wholly unlike other forms of data. The informant's non-academic text reminds the researcher and the reader of the lived experience that has been divulged. It reminds geographers that there are real people behind the data.

KEY TERMS

aide-mémoire

analytical log

asynchronous interviewing

computer-mediated communications (CMC) interviewing

creative interviewing

critical inner dialogue	pre-testing
disclosure	primary question
emoticon	professional interviewing
funnel structure	prompt
informant	pyramid structure
interview guide	rapport
interview schedule	secondary question
interviewer effects	semi-structured interview
latent content analysis	structured interview
life history	synchronous interviewing
lurking	transcript
manifest content analysis	trolling
NVivo	unstructured interview
oral history	vernacular
paralinguistic clues	voice recognition computer software
participant checking	
personal log	warm-up
	word processing

REVIEW QUESTIONS

1. Select one of the four questions below, and spend about 15 minutes constructing an interview schedule for a hypothetical five-minute interview with one of your colleagues. Use a mix of primary question types and prompts. Think about the overall structure of your schedule, and provide a sense of order in the way the issues are covered. Try to imagine how you will cope if the interviewee is aggressive, very talkative, or non-communicative. Will your schedule still work?

   a. Most of us would agree that a greater use of public transport is an environmentally and economically sound goal. However, most of us would personally prefer to use a private car and only pay lip service to such noble goals. Why?

   b. Beach activity is decidedly spatial. Performances are expressive, and behaviour is at times territorial.

   c. The re-integration of the differently abled into 'normal society' is a noble ideal. However, this integration will always be confounded by the organization of public space and the reactions of the able-bodied when the differently abled are in public space.

d. The local environment plan (LEP) of every local council should allocate a specific area for sex industry uses.

2. Conduct two semi-structured in-depth interviews with someone of an older generation than yours. It could be an older relative (however, do not interview a sibling or parent). Limit both interviews to approximately 30 minutes. Construct an interview guide that operationalizes key concepts in the following research question: 'Ours is a patriarchal society. We are often told, however, that the society of our parents and grandparents was structured by an even more restrictive and oppressive system of sexism and compulsory heterosexuality. Investigate how the opportunities, resources, and experiences differed according to gender for earlier generations. Pay particular attention to gender variations in the use of and access to space.'

3. Devise a list of rapport strategies you could use if you were to conduct a face-to-face interview with an older relative not well known to you. Consider the preliminary, contact, warm-up, and closing phases of the interview.

4. The absence of direct contact in CMC interviewing has implications for rapport. Map out a list of strategies specifically tailored to maintain rapport in an interview comprised of e-mail exchanges. You can imagine that the topic of the interview is one of those outlined in Questions 1 and 2. In making your list, reflect on the means by which you could maintain 'listening strategies' over the web, as well as 'strategies of visibility'.

Useful Resources

Baxter, J., and J. Eyles. 1997. 'Evaluating qualitative research in social geography: Establishing "rigour" in interview analysis'. *Transactions of the Institute of British Geographers* 22 (4): 505–25.

Bennett, K. 2002. 'Interviews and focus groups'. In P. Shurmer-Smith, ed., *Doing Cultural Geography*. London: Sage.

Blunt, A. 2003. 'Home and identity'. In A. Blunt et al., eds, *Cultural Geography in Practice*. Euston: Arnold.

Cloke, P., et al. 2004b. 'Talking to people'. In *Practising Human Geography*. London: Sage.

Douglas, J.D. 1985. *Creative Interviewing*. Beverley Hills, CA: Sage.

Edwards, J.A., and M.D. Lampert, eds. 1993. *Talking Data: Transcription and Coding in Discourse Research*. Hillsdale, NJ: Lawrence Erlbaum Associates.

Findlay, A.M., and F.L.N. Li. 1997. 'An auto-biographical approach to understanding migration: The case of Hong Kong emigrants'. *Area* 29 (1): 34–44.

Kearns, R. 1991. 'Talking and listening: Avenues to geographical understanding'. *New Zealand Journal of Geography* 92: 2–3.

Mann, C. 2000. *Internet Communication and Qualitative Research: A Handbook for Researching Online*. London: Sage.

Minichiello, V., et al. 1995. *In-Depth Interviewing: Principles, Techniques, Analysis*. 2nd edn. Melbourne: Longman Cheshire.

Oakley, A. 1981. *From Here to Maternity: Becoming a Mother*. Harmondsworth: Penguin.

Robertson, B.M. 2000. *Oral History Handbook*. 4th edn. Adelaide: Oral History Association of Australia SA Branch Inc.

Schoenberger, E. 1991. 'The corporate interview as a research method in economic geography'. *Professional Geographer* 43 (2): 180–9.

Tremblay, M.A. 1982. 'The key informant technique: A non-ethnographic application'. In R.G. Burgess, ed., *Field Research: A Sourcebook and Field Manual*. London: Allen and Unwin.

Valentine, G. 1997. 'Tell me about . . . : Using interviews as a research methodology'. In R. Flowerdew and D. Martin, eds, *Methods in Human Geography: A Guide for Students Doing a Research Project*, 110–26. Harlow: Longman, Harlow.

Whyte, W.F. 1982. 'Interviewing in field research'. In R.G. Burgess, ed., *Field Research: A Sourcebook and Field Manual*. London: Allen and Unwin.

Oral History and Human Geography

7

Karen George and Elaine Stratford

CHAPTER OVERVIEW

This chapter describes how oral history can be used in geographical research. After defining oral history, we outline the aspects that make it unique and distinguish it from other forms of interviewing. These aspects include establishing rapport, dealing with sensitive issues, understanding the ethics of interviewing, ways to ask questions, and the importance of sound quality.

INTRODUCTION

> It was a blinking dust storm. Every time you come up to Loxton there was you got off the track oh well that's where you was until you got yourself out again and it was always blowing dust. I thought 'Gawd' I always used to say, 'Fancy living up in this hole.' [Ruth Scadden, in George 1999a, 161]

These are the words of Ruth Scadden, the wife of a soldier settler who was eye-witness to a wave of major environmental change in the horticultural town of Loxton in South Australia's Riverland after World War II. Over the past 60 years, this 'dust bowl' farming area has been gradually transformed into an irrigated settlement producing citrus fruits, grapes, and stone fruits for Australian and overseas markets. When interviewed in 1999, Ruth had begun to witness a further wave of change as long-term irrigation revealed its impact on the River Murray, a situation that has now become critical. Her perceptions, understanding of rural life, and representations of that life to others are threads of everyday and ordinary existence whose cumulative weavings constitute a rich tapestry of local geographical knowledge.

Ultimately what has always struck me as being remarkably interesting about how one is influenced is that there is a local geography involved. For example, when I was working in Bougainville, there was one other academic working on the island, an anthropologist, and simply because we met fairly frequently we managed to produce two or three joint articles together, and that situation seems to me to have always continued. So there are these local factors which no one can actually build into an intellectual trajectory or even practical planning, have been incredibly important at how one actually shapes what it is one does. [John Connell, geographer]

These are the recollections of a scholar in the field, recorded during an interview for the Institute of Australian Geographers' Millennium Project on Australian Geography and Geographers. His words trace just some of the complex lines that ultimately form the web of an individual's life experiences and locales.

This chapter outlines the basic scope of oral history as a technique to gather information, insights, and knowledge from participants in social research—people such as Ruth Scadden and John Connell. It describes how oral history can be a powerful source of **situated learning** and how it can facilitate enhanced understandings of space, place, region, landscape, and environment—the five central filaments of human geography. Importantly, the chapter also summarizes a range of ethical, technical, and communicative guidelines for the effective conduct of oral history.

WHAT IS ORAL HISTORY, AND WHY USE IT IN GEOGRAPHICAL RESEARCH?

The practice of oral history involves a prepared interviewer recording a particular kind of interview. The interview is usually conducted in an informal question-and-answer format with a person who has first-hand knowledge of a subject of interest. Background preparation allows the interviewer to follow up responses and prompt further information. Oral history interviews may concern a very specific subject, or cover an entire lifespan, or trace a complex issue that unfolds over time.

Historian Allan Nevins first used the term 'oral history' in the 1940s to describe a project at Columbia University in which the memories of a group of *significant* Americans were recorded (Robertson 2006, 3). While 'oral *tradition*' as a method of passing stories down through generations has existed for centuries, oral *history* was defined differently because its aim was to record the firsthand knowledge and experience of participants. During the 1960s and 1970s, the value of oral history in discovering and preserving the experiences of *ordinary* people was recognized. Since then, oral history has become an important tool for studying hidden histories and geographies, the place-based lives and memories of disadvantaged people, minority groups, and others whose views have been ignored or whose lives pass quietly, producing few if any written records. In short, there are insights to be gained from oral histories to better understand space, place, landscape, region, and environment in ways that are sensitive to context (see, for example, Andrews et al. 2006; Stratford 1997).

American geographer Isaiah Bowman suggested that 'Geography tells what is where, why and what of it' (in Rivera 1997). Acknowledging that this definition no longer fully represents the complexity of the discipline but borrowing Bowman's phraseology in any case, oral history tells what happened, how, why, and what it was like from a personal perspective. For this reason, it has become a useful tool in human geography, illuminating how recollections and representations are *placed* over extended periods and enabling researchers and participants to track and understand changes across spatial scales as well as temporal ones. In this respect, oral history has been described as the voice of the past (Thompson 2000) and as 'a picture of the past in people's own words' (Robertson 2006, 2). As a research method, it provides a means to step back to the mix of past times and places *as they are mediated* through the words and memories of another person in the present.

Another way of thinking about oral history in relation to human geography is to acknowledge that people witness and engage in all manner of change, including environmental change—and here we mean 'environment' broadly as 'that which surrounds'. While documents and photographs may tell part of the story, eyewitness accounts can deepen the image and provide unique or specific detail from many different perspectives (Sackett 2005). Take, for example, how oral history helps to uncover people's experiences of the built environment and to trace the narratives of their geographical engagements there. As oral historian for the Adelaide city council, Karen George learned from statistical records in annual reports that during the 1950s and 1960s the population of the city declined markedly. Many buildings were declared unfit for human habitation, and residents were forced out of the city as it was being transformed into a business district. Only when she interviewed such former residents did Karen understand the significance of this phenomenon. The city was the home of individual families *and* also a community and a support network; destruction of homes resulted in the breakdown of both. People described meeting places that no longer existed and reminisced about people they used to see every day. From such stories Karen reconstructed an image of the city before the exodus. Interviews with a health inspector who had declared many of the houses unfit allowed her to see the event from another perspective. Ironically, events have now gone full circle, and the council has vigorously encouraged people to move back to the city to live and to build new community networks and sense of place (City of Adelaide Oral History Collection, Adelaide City Archives).

A second example shows that oral history can uncover the way that geographers themselves understand their professional contributions to how space, place, region, landscape, and environment are constituted (Matless, Oldfield, and Swain 2007; Powell 2008). As coordinator of the Institute of Australian Geographers Millennium Project between 1996 and 2001, Elaine Stratford encouraged members of the profession to undertake oral histories with eminent geographers (Stratford 2001; see also Hay 2003b; Rugendyke 2005; Sheridan 2001). Participants were asked to give some thought to the contributions that geography has made or may make to Australian society. Two responses begin to hint at the wealth of disciplinary knowledge that can be gained by the extended and in-depth interview style of oral history.

> I think that geography could continue its contribution to the development of Australian society by expanding the public imagination and . . . values about [the public's] . . . relationship with the environment and so on. Because it is a very special place here, a very special environment with special needs and so on. [Joseph Powell]

> I think geography has made an enormous contribution. It would be difficult to look at that in its totality because different geographers are obviously looking at that component from the perspective of the work that they themselves have done. I think geography has contributed or could contribute again enormously to the understanding of the habitation of this country and what this means for the future. [Elspeth Young]

How Is Oral History Different from Interviewing?

In Chapter 6, Kevin Dunn describes in detail how to conduct research interviews. Most of his guidelines hold true for oral history. However, oral history practice does differ from interview practice in a number of ways, which we set out in the sections below.

Perhaps one of the clearer differences between interviewing for research and conducting oral history interviews is that many elements in the oral history process place particular emphasis on the role of the participant. Through preparations and techniques that make a participant comfortable with recording an interview, oral historians aim to record as natural, rounded, and complete a story as possible.

Starting Ethically

Matt Bradshaw and Elaine Stratford note in Chapter 4 that it is a significant matter to engage with research participants and share, interpret, and represent their experiences. Therefore, these acts require an ethical approach. Oral history work involving researchers from post-secondary institutions must be assessed and approved by those institutions' ethics committees (see Chapter 2). Similarly, private practitioners must be ethical in their approach, and membership in oral history associations demands this. Thus, in describing below the various stages and techniques of oral history, we assume that ethical considerations and/or clearances are in place before research commences—a matter that has parallels with interviewing more generally and on which Kevin Dunn elaborates in Chapter 6 of this volume.

The Oral History Association of Australia (http://www.ohaa.net.au) was established in 1978 to promote the practice and methods of oral history, educate in the ethical use of oral history methods, encourage discussion on all aspects of oral history, and foster the preservation of oral history records. There are branches of the association in each state. As well as providing advice and training in oral history, the association has drawn up 'Guidelines for ethical practice' (http://www. ohaa.net.au/guidelines.htm). It strongly advises that these guidelines, which protect the rights of both participant and interviewer, are followed by anyone involved in oral or life history. A similar organization exists in Canada (http:// www.canoha.ca), and others thrive internationally: in the UK (http://www.ohs. org.uk), the US (http://www.dickinson.edu/oha), and New Zealand (http:// www.oralhistory.org.nz). Many oral historians belong to the International Oral History Association (http://www.ioha.fgv.br).

Getting to Know Your Participant—The Preliminary Meeting

Establishing rapport with a participant is integral to success. Oral historians use a particular approach to help establish rapport, and the **preliminary meeting** is a key part of it. After contact by letter, telephone, Internet, e-mail, or other appropriate means, the interviewer arranges a meeting, usually at the home of the participant or another place of his or her choice. No audio recorder is produced at this orientation session. Rather, the time is used to establish a relationship, gather background information, and 'assess' the participant and interview environment. Some interviewers use an information sheet to record information about their informant, such as where and when they were born, aspects of their school, employment, and/or personal background, and other data that might be pertinent to the interview. Although much of this material might be covered again in the recorded interview, preliminary notes establish context and ensure accuracy—for instance, in spelling and pronunciation.

The preliminary meeting also offers the interviewer opportunities to ask to see materials that may enhance research. News clippings, letters, diaries, or photographs may suggest new questions not previously considered. If a new topic is raised during the interview proper, you may be unprepared to ask questions about it. If it emerges in a preliminary session, you have time to conduct further research about it before the interview.

'Assessing' the participant sounds clinical, but some people remember things and are able to talk about them more readily than others. You may find that someone you thought would provide you with great material remembers very little, is extremely nervous, or is overly wary about their responses. Thus, at a preliminary meeting you have the opportunity to defer or cancel the interview by telling your participant that he or she has already provided what you needed. This strategy can prevent the embarrassment for both participant and interviewer of a stilted recording, filled with clipped responses and phrases like 'I don't remember', 'I don't recall'.

Sensitive Issues

If a project deals with sensitive issues (including personal, political, or professional details that may require the researcher to guarantee the confidentiality of parts of the oral history transcript), a preliminary meeting will allow you to discover how your participant feels about answering particular questions or exploring aspects or phases of their life or the subject under investigation. Even when interviewing in a subject area that seems straightforward and non-contentious, it is good practice to assure participants of their right to refuse to answer a question or to withdraw from the research altogether without prejudice. Discussing these matters in advance saves embarrassment during an interview and allows participants to think in advance about what they might wish to say about difficult subjects.

Sensitive subjects can arise unexpectedly during an interview. If this happens, exercise care and consideration. Ask your participant if she or he wishes to continue or would rather stop. If they wish to continue, let them speak. If the subject material becomes very personal and you think that it should be excluded, let the participant know this. University ethics committees (e.g. research ethics boards, institutional review boards) may make the useful suggestion that if the participant is distressed, you might ask whether they would like you to call a friend or family member or refer them to a helpful counsellor.

After the interview is over, allow time for winding down. Winding down can be as simple as accepting another cup of tea or listening to other stories not related to the research subject.

Multiple Interviews

Unless there are unavoidable constraints, oral history recordings with a participant may be completed over several sessions. When longer recordings are feasible, individual sessions may be confined to an hour or so and second and subsequent occasions used to complete the history.

Multiple interviews can be very valuable. There is time between appointments for you to listen to the initial recording and note responses to enlarge upon later. Participants may also reflect on their answers: remembering triggers further memories to be shared at ensuing meetings. By a third or even fourth meeting, a bond between interviewer and participant has usually developed, which can result in an even better interview.

The Question of Questions

Open questions are integral to effective oral history. These questions begin with words such as who, what, where, when, why, and how. They reveal who was involved in an event, what happened, where and when it happened, how it felt, and why that was so. They yield the details that make oral history such an effective source of nuanced (if always partial) recollections.

In Chapter 6, Kevin Dunn refers to secondary questions or 'prompts', which oral historians often label **follow-up questions**. They often comprise the body of an interview. Most are prompted by a participant's response to an initial question. If a participant says that his or her first day on the job was 'frightening', the logical follow-up question is 'why?' or 'in what way?' If he or she responds by saying that 'fellow-workers were aggressive', a logical follow-up question is 'can you give me an example?' Through follow-up questions, great depth may be added to the detail of information being sought.

Interview Structure

Oral histories can appear to be 'unstructured', but such is not really the case. Interview guides or aides-mémoire are often used. Interviews in the tradition of oral histories can often be divided into a three-part format, comprising orientation, common, and specific questions. As Robertson (2006, 22) points out, this

> three part structure provides an excellent framework for interviews. It helps you to avoid aimless or superficial interviews and it can lead to recordings that are easier to use for research, publication or broadcast because of their well-defined structure and focus.

Orientation questions establish the participant's background. **Common questions** are those asked of each participant in a project. They build up varying views and information about certain themes. **Specific questions** relate to individual experiences and are developed through follow-up work. The flow of an interview is determined by participant responses, so follow-up questions are always different across interviews.

Questioning the Source

As the sections above suggest, oral history is active and shared; you can question the source. For example, as part of Karen George's South Australian research on War Service Land Settlement after 1945, she consulted written records of applicant interviews with the Land Board. Although these documents included board members' notes about applicant responses, she could not ask them, 'What do you mean by that?' When conducting oral histories with people who had appeared before the board, she could. Karen then created as complete a picture of those board interviews as memory allowed. She was able to ask what happened that particular day, what the interviewers were like, how respondents felt about the questions when they were asked and after the experience (George 1999a).

Sound Quality, Interview Sites, and Other Technical Issues

Oral history is *always* recorded via audio, and sometimes audio-visual, technologies and predominantly on digital tape recorders; 'solid state' recorders that use compact flash memory cards, or that record onto a hard disc, are highly recommended (Robertson 2006, 43–44), because a significant aim of recording interviews is to create enduring **sound documents**. Oral history preserves the participants' voices and content of their interviews. High-quality sound allows the emotion, inflections, and tone of each voice to be heard. It is also important because material may be used for broadcast. Background noises, interruptions, substandard recording equipment, and a too-talkative interviewer diminish sound quality.

Because you are recording a unique sound document, it is important to record interviews in a location that is as quiet as possible and to use the best equipment available. Oral history associations in all countries are the best place to approach with inquiries about renting equipment. High-quality recorders are often available on loan from libraries and, in Australia, from branches of the Oral History Association of Australia (OHAA). Some of these organizations run workshops on conducting oral history and the correct use of recording equipment.

Robertson's *Oral History Handbook* is an excellent source of information and advice on all recording equipment, from budget-priced to super high-quality (Robertson 2006, ch. 4). Recording technology is constantly undergoing change, and new digital recording devices are now becoming available. It may be prudent to seek professional advice from organizations such as your national oral history association before investing in new equipment. If you use digital recording technology, you must remember to download sound files after each interview and ensure that back-up copies of work are saved in secure settings, especially if you have promised to do so as part of the ethical commitments of the project. It is often very valuable to work in conjunction with a major university library or state/provincial library. For instance, the University of California at Berkeley has a regional oral history office. In South Australia, if you agree to deposit recordings (with signed conditions of use agreements) into the oral history collection, you are given free access to a top-quality recorder and your recordings will be digitally preserved and copies made on CD for you and your participant. Check relevant libraries to see whether they offer similar arrangements.

No matter what type of recorder you are using, always try to find a quiet place free from interruptions to conduct an interview. Recording high-quality sound means that extracts from your recordings can easily be broadcast or used in museum interpretations, on websites or CDs, or for audio walking tours. This quality also makes any transcription and analysis much easier. Note that workplaces are among the worst locations, whereas private homes usually offer a dining or living room, both of which can be quieter. Avoid kitchens whenever possible. Refrigerators are renowned for droning away in the background or cutting in and out with a thump. Ticking or chiming clocks should be stopped or removed because their regular pulse in the background is distracting. If you cannot avoid background noises completely, such as traffic sounds on a busy road, direct the interviewee's microphone away from the noise. Close doors and windows to minimize background noises. If you talk about these things at your preliminary meeting, participants will usually help out and not think you are rude when you ask them to stop great-grandmother's cuckoo clock or turn off the fridge!

Interpreting non-verbal responses and gestures is common to oral histories and other interviews. The comment 'It was about this big' needs to be translated by the interviewer into 'about a metre high'. It is best to convey these details on tape, since you might forget them later. Recall, too, that an interviewer's verbal responses can be detrimental to a sound recording, particularly one for broadcast purposes. A litany of 'yes, yes', 'mmm', 'oh really', or 'wow' remarks that commonly occur in a conversation interrupts the recorded flow of a story. Respond with a nod or a smile instead. Let participants know you will remain quiet *and* involved. Listen to your own recordings to gauge how silent you actually are. A pause, a moment of quiet may be the instant before the best story (see Chapter 6 for additional discussion). Oral history also can be a demanding process for participants, and they may need time to stop and reflect. Be sure to allow them this time. Silence on tape can also be very emotive. Long seconds of silence recorded in the midst of a painful story reveal a struggle with strong emotions better than words ever could.

Why and How to Make Oral History Accessible

Interviews conducted for research often have very limited circulation. However, in oral history interviewers are encouraged to deposit their recordings in libraries or archives. This step ensures the preservation of master or original recordings, and if the participant has agreed, allows recordings to be made available to other researchers. It is always worthwhile to search oral history collections *before* you begin an interview to make sure that your informant has not already been interviewed and to check whether there are other recordings that might provide data for your project. Even interviews concerning completely different topics may contain useful information. For example, since the majority of the men and women Karen George (1999a) interviewed about soldier settlement in South Australia grew up during the Depression of the 1930s, a researcher interested in that period of history could glean a lot of information from their answers to questions about their backgrounds and childhoods.

Another advantage of depositing recordings in a library is that some larger repositories offer limited assistance with the transcription of interviews. Whether you produce full transcripts, **timed tape logs** (which note subjects discussed at different time points in the recording), or broad interview summaries is normally dependent on the project's aim and on funding. Professional transcription is expensive but worthwhile if material from interviews is to be reproduced in a publication. It is worth noting here that professional transcribers are trained to reflect pauses and the unique cadences of the spoken word through punctuation and layout. Transcripts are rarely completely verbatim because it is common for participants to feel concerned about their poor grammar, repetitions, and crutch phrases (such as 'you know') when they see them in print (see Chapter 6 for additional discussion on this matter).

Making interviews, both recordings and transcripts, accessible to others should only be done with the signed agreement of your participants. It is essential that you draft a **conditions of use form** outlining what will happen to the material they share with you—what their rights are, who will own copyright, where the recorded interview will be stored and for how long, and what it will be used for. Although participants should be encouraged to share their stories with a wide audience, they should also be allowed to add conditions to this agreement, to restrict portions of the recording—or indeed the entire interview—until after their death if they wish.

USES OF ORAL HISTORY—SPREADING THE WORD

As well as depositing oral history recordings into libraries or archives and using information from interviews in published and unpublished writing, there are other ways to share the results of your work. If you have made high-quality recordings, the possibilities are extensive. While quotations from interviews can be presented in displays and on the Internet in written form, it can be even more effective to use sound excerpts. Sound bytes used on a webpage allow users to hear as well as read about your interviews. **Listening posts**—that is, posts with speakers in them—may be used as part of exhibitions, or portions of interviews can be used in audio commentary.

Well-recorded interviews offer much scope for presentation to groups, in radio, film, and video. For example, the voices of long-term employees of Balfours, one of the last city-based factories in Adelaide, were used effectively in conjunction with video recording of the working of the bakery (Starkey and George 2003). Using images and oral history excerpts, the project recorded and preserved images and descriptions of the original factory and of the processes that have not been used since the factory relocated to modern suburban premises. Certainly, narration may be made much more engaging when excerpts from oral histories are played in conjunction with visual materials. Similarly, when Karen George conducted an interview with a long-time printer in 2008, she also recorded the sounds of early linotype and printing machines, some of which may never be operated again after the printer's death. The sounds of the machines have the potential of being used in conjunction with oral history, photographs, and video footage in an interactive museum display—or they could be used as part of an audio walk (see, for example, Butler 2007).

Last Words

Geography's central concern is to understand people in place, spatial relations, landscapes, regions, and environments. It also aims to contribute to research-based outcomes that advance well-being. For many geographers, these composite tasks involve philosophical and political investments in learning about—rather than appropriating—marginal, informal, and otherwise undocumented perspectives (see Chapters 3 and 13) as well as in comprehending those that are central, formal, and documented. Like those of the interview or focus group, oral history techniques allow both researchers and participants to explore the nuances of social and spatial interactions, events, and processes in ways that can make these goals possible. However, in the pursuit of these goals, never forget that your 'source' is another human being, a person sharing with you a distinctive and valuable gift—their memory.

Key Terms

common questions	listening posts
conditions of use form	multiple interviews
follow-up question	open questions

orientation questions   sound document

preliminary meeting   specific questions

situated learning   timed tape logs

REVIEW QUESTIONS

1. What are some of the ethical issues associated with the use of oral history?
2. What are some of the relationships between oral history and human geography?
3. What are some of the ways in which oral history might help you explore an area of human geography in which you are interested?
4. Develop an idea for an oral history project, prepare a list of potential participants, and search existing oral history collections for previously recorded interviews on related topics.
5. Record an interview, paying particular attention to sound quality. Develop a multi-media spoken or web-based presentation using excerpts from the interview combined with other media—such as photographs and documents. You could present this as a talk or on the web.
6. Develop a conditions of use form to be used in conjunction with one of the two projects above.

USEFUL RESOURCES

Baylor University, Institute for Oral History. 2008. 'Introduction to oral history'. http://www.baylor.edu/oral%5fhistory/index.php?id=23566.

Butler, T. 2007. 'Memoryscape: How audio walks can deepen our sense of place by integrating art, oral history and cultural geography'. *Geography Compass* 1: 350–72.

George, K. 1999. *A Place of Their Own: The Men and Women of War Service Land Settlement at Loxton after the Second World War*. Adelaide: Wakefield Press.

Jenkins, A., and A. Ward. 2001, 'Moving with the times: An oral history of a geography department'. *Journal of Geography in Higher Education* 25 (2): 191–208.

Perks, R., and A. Thomson, eds. 2006. *The Oral History Reader*. 2nd edn. London: Routledge.

Powell, R.C. 2008. 'Becoming a geographical scientist: Oral histories of Arctic fieldwork'. *Transactions of the Institute of British Geographers* 33 (4): 548–65.

Robertson, B.M. 2006. *Oral History Handbook*. 5th edn. Adelaide: Oral History Association of Australia SA Branch Inc.

Sheridan, G. 2001. 'Dennis Norman Jeans: Historical geographer and landscape interpreter extraordinaire'. *Australian Geographical Studies* 39 (1): 96–106.

Sommer, B.W., and M.K. Quinlan. 2009. 'Capturing the living past: An oral history primer'. Nebraska State Historical Society. http://www.nebraskahistory.org/lib-arch/research/audiovis/oral_history/index.htm.

Stratford, E. 2001. 'The Millennium Project on Australian Geography and Geographers: An introduction'. *Australian Geographical Studies* 39 (1): 91–5.

Thompson, P. 2000. *The Voice of the Past: Oral History*. 3rd edn. New York: Oxford University Press.

University of California, Berkeley. 2009. 'Regional Oral History Office'. http://bancroft.berkeley.edu/ROHO/index.html.

Focusing on the Focus Group

8

Jenny Cameron

CHAPTER OVERVIEW

An investigation of how consumers in England decide what to purchase in the context of debates about sustainable and ethical food production (Eden, Bear, and Walker 2008), a study of the environmental problems poor communities face and the interventions they develop in a low-income city in Ghana (Osumanu 2007), and an exploration of the experiences of Filipina domestic workers in Canada (Pratt 2002)—all are examples of research projects that employ focus groups to disentangle the complex web of relations and processes, meaning and representation, that comprise the social world. With the shift to more nuanced explorations of people–place relationships in geography, the focus group method has been increasingly recognized as a valuable research tool.

Focus groups can be exhilarating and exciting, with people responding to the ideas and viewpoints expressed by others and introducing you, the researcher, and other group members to new ways of thinking about an issue or topic. This chapter discusses the diverse research potential of focus groups in geography, outlines key issues to consider when planning and conducting successful focus groups, and offers strategies for analyzing and presenting results.

WHAT ARE FOCUS GROUPS?

The focus group method involves a small group of people discussing a topic or issue defined by a researcher. Briefly, a group of between 6 and 10 people sit facing each other around a table; the researcher introduces the topic for discussion and then invites and moderates discussion from group members. A session usually lasts between one and two hours (you might see parallels here with university tutorial group meetings).

The focus group is one of the group techniques used in research. As shown in Figure 8.1, these techniques range from group interviews in which each

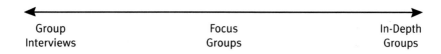

Group   Focus   In-Depth
Interviews   Groups   Groups

Figure 8.1: Relationship between Focus Groups, Group Interviews, and In-depth Groups

participant is asked the same question in turn and there is little or no interaction between participants (Barbour 2007) through to in-depth groups in which the emphasis is on the interaction between participants, with participants sometimes even deciding on discussion topics (Kneale 2001). In-depth groups also meet regularly for extended periods of time (sometimes months).

As with in-depth groups, interaction between participants is a key characteristic of focus groups. The group setting is generally characterized by dynamism and energy as people respond to the contributions of others (see Box 8.1). One comment can trigger a chain of responses. This type of interaction has been described as the **synergistic effect** of focus groups, and some propose that it results in far more information being generated than in other research methods (Berg 1989; Stewart, Shamdasani, and Rook 2007). In the focus group excerpt in Box 8.1, for example, the discussion shifts from family farming practices, to people's commitment to an area, to ways of working with government, to projects that address environmental degradation. Yet as the farmer points out at the conclusion of this excerpt, the speakers all highlight the effect that taking a long-term approach has on economic, environmental, and community practices.

The Synergistic Effect of Focus Groups	Box 8.1

Farmer A: Where we make a mistake in business is in thinking of tomorrow. The family approach is what's happening to the next generation. It's a much longer-term approach. I'm more interested in investing my resources for the next generation and therefore you build a solid business.

Farmer B: My attitude is that I'm the tenant in time.

Farmer A: The custodian.

Farmer B: Yeah, the custodian. My father gave it to me, and I'll hand it on to the next generation. And people say you could sell it and make lots of money, but that just doesn't come into the equation. The thought of selling it and leaving the good life—the kids probably will. And I think there are an awful lot of farmers with that attitude. And I think it has probably in lots of ways been to our detriment. We could use that asset and make more money—as if money is the most important thing.

Consultant: I think that's right. I think one of the important reasons there have been successes and perhaps less problems here is that even though we have all identified lots of problems, we are really committed to this community and making it better. And I think there are an amazing range of people that do choose to live here—they don't have to—but choose to live here and [have] invested huge amounts of time and energy. And I also think this community, just thinking back to my experience, that it's really open to working with whatever government is in at the time and turning the rules or the policies or the dollars that are around for the best here. Like local government saying we don't want yet another regional development board but we will have the money and this is our structure and this is what we'll do. I think there has been some creative use of government money and good partnerships and also just that huge commitment, that energy to make it work.

Manager: You mention our successes and I think one of the unheralded things we've done really well is look towards the sustainability of the whole area from land management which underpins our whole economy. Because we've poured irrigation water onto this country for years and years and we've never really looked at the repercussions: the drainage problems, the salinity problems. And I think in recent years, in the last 15, 20 years, that's really been addressed— the work that's gone into it by some very dedicated people and I think that message has gone across to virtually all land holders in the area. With the advent of some major arterial drains, community drains, the cooperation—the cooperatives virtually that have been formed to bring this into being, really will underpin the future of our economy and the management of our natural resource, which is absolutely vital to the future of our farmers and businesses, etc.

Consultant: And a lot of that work's been voluntary.

Farmer A: It all comes back to the notion that it's the next generation. It's a different approach.

Source: Videotape excerpt from focus group conducted by Katherine Gibson, Jenny Cameron, and Arthur Veno, Shepparton, Victoria, 5 June 1997 (see Gibson, Cameron, and Veno 1999).

The interactive aspect of focus groups also provides an opportunity for people to explore different points of view and to formulate and reconsider their own ideas and understandings. Kitzinger (1994, 113) describes this form of interaction in the following terms: '[p]articipants do not just agree with each other. They also misunderstand one another, question one another, try to persuade each other of the justice of their own point of view and sometimes they vehemently disagree.' For researchers who are interested in the socially constructed nature of knowledge, this aspect of focus groups makes them an ideal research method; the multiple meanings that people attribute to places, relationships, processes, and events are expressed and negotiated, thereby providing important insights into the practice of knowledge production.

As in group interviews, the researcher plays a pivotal role. He or she promotes group interaction and focuses the discussion on the topic or issue. The researcher draws out the range of views and understandings held within the group and manages—sometimes even encourages—disagreement among participants (Myers 1998).

Initially, focus groups can be extremely challenging for researchers who are new to the process. They are, however, well worth it. In focus groups, the diversity of processes and practices that make up the social world and the richness of the relationships between people and places can be addressed and explored explicitly. Furthermore, group members almost invariably enjoy interacting with each other, offering their points of view, and learning from each other. Researchers also find the process refreshing (for example, see the discussion by two sceptical anthropologists in Agar and MacDonald 1995).

USING FOCUS GROUPS IN GEOGRAPHY

Focus groups were originally used by sociologists in the US during World War II to examine the impact of wartime propaganda and the effectiveness of military training materials (Merton 1987; Morgan 1997). Although this work resulted in several sociological publications on the technique, focus groups were neglected by social scientists in the post–World War II period in favour of one-to-one interviews and participant observation (Johnson 1996). It was in the field of market research that the focus group method found a home. Since the 1980s, there has been renewed interest in the technique among social scientists, and this has led to considerable diversity in focus group research (Lunt and Livingstone 1996; Morgan 1997). Focus groups can be a highly efficient data-gathering tool, but they are also appropriate in 'more critical, politicized, and more theoretically driven research contexts' (Lunt and Livingstone 1996, 80), exploring, for instance, the discourses that shape practices of everyday life, the ways in which meanings are reworked and subverted, and the creation of new knowledges out of seemingly familiar understandings. The range of uses and purposes of focus groups is evident in geographic research employing the technique.

Geographers have used focus groups to collect information. Zeigler, Brunn, and Johnson (1996) used them to find out about people's responses to emergency procedures during a major hurricane. They claim that the focus group technique provided insights that might not have been revealed through methods like questionnaires or individual interviews. As a consequence, they were able to recommend important refinements to disaster plans. J. Burgess (1996) has also used focus groups, in combination with participant observation, to obtain information about factors that inhibit visits to and use of woodlands. Her findings have contributed to the development of landscape design and management strategies to enhance the use of woodlands. As a data-gathering technique, focus groups are ideal for investigating not just *what* people think and do but *why* people think and behave as they do. Barbour (2007, 24) notes that focus groups are particularly useful for understanding people's beliefs and practices. In focus groups, 'participants are given scope to justify and expand on their views in a non-judgemental environment', giving researchers a chance to learn how people rationalize what might seem illogical to researchers.

One concern of some researchers involved in data gathering is that because of the relatively small number of participants, findings are not applicable to a wider population (for a discussion of this issue, see Chapter 4). Combining focus groups with quantitative techniques is an extremely useful way of dealing with this issue. A survey questionnaire, for instance, might be administered to a random sample of the population from which the focus group was drawn to test the generalizability of the insights gained from the group discussions. Quantitative methods can supplement focus groups in other ways.

Preliminary surveys are sometimes helpful in identifying focus group members or the topics for detailed focus group discussion.

Conversely, focus groups can supplement quantitative research. They have been used to generate questions and theories to be tested in surveys (Goss and Leinbach 1996; Pratt 2002), to refine the design of survey questionnaires (Jackson and Holbrook 1995), and to follow up the interpretation of survey findings (Goss and Leinbach 1996; Wrigley et al. 2004), particularly when there seem to be contradictory results (Morgan 1996). It is, however, entirely appropriate to use focus groups as the sole research method rather than in combination with other research techniques.

For geographers interested in the process of knowledge production, the focus group is an excellent research tool. Robyn Longhurst (1996) employed focus groups as a forum in which pregnant women could converse and interact. The narratives, accounts, anecdotes, and explanations offered by these women provided Longhurst (1996) with insights into a new discursive landscape of pregnancy. Similarly, Gibson, Cameron, and Veno (1999) have been concerned not just with reproducing knowledge of the problems and difficulties confronting rural and non-metropolitan communities in Australia but with reshaping understandings so that new responses might be engendered. The seemingly isolated instances of innovation that several focus group members could readily recall provoked other participants to think of additional examples. The beginnings of a body of knowledge on regional initiative began to emerge through these discussions.

In a report of their Indonesian research on individual and household strategies related to the allocation of land, labour, and capital, Goss and Leinbach (1996) also highlight the collective rather than individual nature of knowledge production. By interacting with other focus group members, Javanese villagers developed new understandings of their social conditions. Indeed, Goss and

Leinbach argue that 'the main advantage of focus group discussions is that both the researcher and the research subjects may simultaneously obtain insights and understanding of particular social situations *during* the process of research' (Goss and Leinbach 1996, 116–17, emphasis in original). For geographers who are committed to the idea that research can be used to effect social change and empower 'the researched', the potential for focus groups to create and transform knowledges and understandings of researchers and participants is compelling (see also Johnson 1996; Swenson, Griswold, and Kleiber 1992).

The focus group method has an important contribution to make to geographic research. It is a highly effective vehicle for exploring the nuances and complexities associated with people–place relationships. The material generated in focus groups can add important insights to work that seeks to describe, document, and explain the social world. But focus groups serve not just to 'mine', 'uncover', and 'extract' existing knowledges (Gibson-Graham 1994); they can also provide insights into how people construct their world views through interactions with others (Pratt 2002), and they can provide opportunities for researchers and participants to jointly develop new knowledges and understandings.

Planning and Conducting Focus Groups

Given that the focus group method can be used for a range of research purposes in geography, there will be some variation in how groups are organized and conducted. There are, however, basic principles and methodological issues that need to be considered. To be sure, the success of a focus group depends largely on the care taken in the initial planning stage.

Selecting Participants

Selecting participants is critically important. Generally, participants are chosen on the basis of their experience related to the research topic. J. Burgess's (1996) study is a good example of this **purposive sampling** technique (see Chapter 4 for a discussion of participant selection). In work intended to ascertain the perceptions of crime and risk in woodlands among different social and cultural groups, she selected women and men of varying age, stage in the lifecycle, and ethnicity to participate in focus groups. In another study (Casey et al. 1996) of local perspectives on potential strategies to address agricultural pollution in the Minnesota River Basin, people involved in different aspects of farming were invited to participate. Groups were made up of farmers—who varied in age and gender and in size and type of farm—and local staff from agriculturally based government agencies and non-profit groups.

Composition of Focus Groups

Should people with similar characteristics participate in the same group, or should groups comprise members with different characteristics? This decision will be largely determined by the purpose of your research project.

Holbrook and Jackson (1996), for example, sought to address issues of identity, community, and locality by grouping together people with characteristics like age and ethnicity in common. In their research on environmental responsibility, Bedford and Burgess (2002) had people with similar experiences in each focus group but a range of different focus groups—suppliers, retailers, regulators, consumers, and advocates. They describe this as 'ensur[ing] homogeneity within the group and heterogeneity between them' (2002, 124). Other researchers have noted that discussion of sensitive or controversial topics can be enhanced when groups comprise participants who share key characteristics (Hoppe et al. 1995; O'Brien 1993). In some projects, it may be more appropriate to have groups made up of different types of people. Goss and Leinbach (1996) were interested in the social relations involved in family decision-making and deliberately chose to conduct mixed-gender groups. The different knowledges, experiences, and perspectives expressed by women and men became an important point of discussion.

Another consideration is whether people already known to each other should participate in the same group. Generally, it is best not to have people who are acquainted in the same group, but in some research, particularly place-based research, it may be unavoidable. Researchers need to be aware of the limitations that this can produce. One is peer pressure, with participants not wanting to appear 'out of step' with their acquaintances. Similarly, some participants may **under-disclose** or selectively disclose details of their lives, as Pratt (2002) found in her research.

A different problem arises when participants **over-disclose** information about themselves. One strategy for dealing with this is to outline fictional examples and ask group members to speculate on them. In groups they ran in Indonesia, Goss and Leinbach (1996) provided details of three fictional families and asked group members to discuss which of the families would be most likely to accumulate capital. Participants did not have to disclose information about their own situations but could still discuss family strategies. Participants can also be asked to treat discussions as confidential. Since this confidentiality cannot be guaranteed, it is appropriate to remind people to disclose only the things they would feel comfortable about being repeated outside the group.

Of course, you should always weigh whether a topic is too controversial or sensitive for discussion in a focus group and would be better handled through another technique such as individual in-depth interviews. (Most universities now have ethics committees, institutional review boards, or research ethics boards to ensure that researchers carefully manage material from focus groups and other qualitative research methods. For more on this, see Chapter 2.)

Size and Number of Groups

The size of each group and the number of groups are other factors to be considered. Too few participants per group—fewer than four—limits the discussion, while too many—more than 10—restricts the time available for individual participants to contribute.

In terms of the number of groups, one rule of thumb is to hold three to five groups, but this guideline will be mediated by factors such as the purpose and scale of the research and the heterogeneity of the participants. A diverse range of participants is likely to require a larger number of groups. J. Burgess (1996), for instance, conducted 13 focus groups with people of varying age, stage in the life-cycle, and ethnicity, while Secor (2003) held four groups with women migrants to Istanbul. Likewise, Le Heron et al. (2001) held four focus groups with dairy and sheep meat farmers in New Zealand.

The structure of the focus group is also a factor to consider. When less **standardized questions** are used and when there is a relatively low level of researcher intervention and moderation, more groups are needed, since both these factors tend to produce greater variability among groups (Morgan 1997). Time, cost, and availability of participants may also limit the number of groups that can be held. The overall research plan—especially whether focus groups are the sole research tool or one of a number of tools—will also affect decisions about the number of groups convened. Finally, another guide to the number of focus groups is to use the concept of **saturation** (Krueger 1998, 72). This means that you continue to conduct focus groups until you can gather no new information or insights.

Recruiting Participants

The strategy used to recruit participants will depend on the type of participants you require for your study. Gibson, Cameron, and Veno (1999) recruited business and community leaders in two regions by initially contacting local people who featured in local newspapers and targeting managers of key government and non-government agencies. These initial contacts were asked to suggest other people who would make interesting contributions to the study, including those who would have a different point of view from them (this snowball **recruitment** technique is also discussed in Chapter 4). A preliminary phone conversation quickly established whether nominees were interested and able to attend. This was followed by a letter with more information about the project. A few days before the focus groups were held, participants were telephoned again to reconfirm their participation. Twelve people were invited to attend each group to allow for cancellations due to illness, last-minute change of plans, and so on (several people from each group did drop out).

After an unsuccessful attempt to recruit participants by advertising in local newspapers and writing letters to local organizations, Holbrook and Jackson

(1996) went directly to places where potential participants were likely to meet and socialize, such as community centres, homes for the elderly, play groups, and clubs. Managers or convenors of the centres helped to set up the groups, or the researchers visited venues and invited people to participate. Once people had been involved in a focus group, news of the project spread by word of mouth, and other people were recruited easily. When Eden, Bear, and Walker (2008) wanted a more diverse range of food consumers, they made contact with a whole-food retailer and a vegan group, among others. Like Eden, Bear, and Walker and Holbrook and Jackson, researchers need to think strategically about how best to locate potential participants (see also Burgess, Limb, and Harrison 1988).

Questions and Topics

Before conducting focus groups, give thought to the questions or topics for discussion. This involves not only the general content of questions or selection of issues for discussion but also the wording of questions and issues, identification of key phrases that might be useful, the sequencing and grouping of questions (see Chapter 6 for additional material on question order), strategies for introducing issues, and the links that it might be important to make between different questions or issues. Also give thought to using others sorts of stimuli to generate discussion. Barbour (2007, 84–8), for example, discusses the use of cartoons, snippets from television shows, photographs, newspaper clippings, advertising materials, and vignettes. However, it is important to select such material carefully and, where possible, to check that it will not offend participants.

One way to proceed is to devise a list of questions. Swenson, Griswold, and Kleiber (1992) developed a list of 20 questions to act as **probes** for focus groups comprising rural journalists. Another list of 20 questions was used in separate focus groups with community development leaders. Holbrook and Jackson (1996) identified six themes related to the experience of shopping and then used these themes to develop questions that were raised spontaneously and that fitted with the flow of the discussion. J. Burgess (1996) first took the participants in each focus group on a walk through a woodland and then introduced for discussion five primary themes related to elements of the walk. As part of the recruitment process, Gibson, Cameron, and Veno (1999) asked each participant to prepare a brief two-minute summary of their perception of social and economic changes in the region over the previous 20 years. The similarities and differences among these statements provided the basis for discussion.

Take care when letting people know in advance what the questions or topics will be. If attendance or discussion is likely to be enhanced by providing this information, then it may be appropriate. Sometimes, however, it might be necessary for you to paint a very broad picture. For example, it might be more judicious to let a group of men know that you are interested in how they manage the interrelation between work, recreation, and home than to tell them that you are interested in contemporary negotiations of masculinity (provided, of course, that you do want to know about masculinity in work, recreation, and home environments). (See Chapter 2 for a consideration of the ethical dimensions of this sort of approach.) This example is also a reminder that you must be sure to use language that participants will understand when you are providing them with information on the themes of your research.

Generally, questions or topics should be suited for a discussion of between one and two hours. With very talkative groups, it may be necessary to intervene and move the discussion on to new topics. Alternatively, if you have planned a hierarchy of questions or themes, then it may be appropriate to allow the group to focus on the more important areas of discussion. With less talkative groups, you may need to introduce additional or rephrased questions and prompts to help draw information out and open up the discussion. You should think about such questions and prompts during the preparation stage.

Another issue to consider is whether questions and topics should be standardized across all focus groups involved in your study or whether new insights from one group should be introduced into the discussions of the next. In many qualitative research situations, it may be appropriate to incorporate material from earlier groups, but this issue should be determined with reference to the project aim. Information that might identify people who attended earlier groups should not be revealed to subsequent groups.

As well as conducting meetings with several groups, you may find it useful and appropriate to have each group meet more than once. Multiple focus groups may be a particularly useful strategy when participants are being asked to explore new and unfamiliar topics or to think about an apparently familiar topic in a new way. Multiple groups may also be appropriate as a way of developing trust between the researcher and research participants. For instance, when researching the experiences of single mothers, I met several times with one group of teenagers who were very wary of talking with people associated with educational, medical, and media institutions (Cameron 1992).

Conducting Focus Groups

Generally, focus groups are best held in an informal setting that is easily accessible to all participants. The rooms of local community centres, libraries, churches, schools, and so on are usually ideal. The setting should also be relatively neutral: for example, it would not be advisable to convene a focus group about the quality of service provided by an agency in that agency's offices. Food and drink can be offered to participants when they arrive to help them relax, but alcohol should never be provided. It is also helpful to give out nametags as participants arrive.

Much has been written about the ideal focus group **facilitator** or **moderator** (for example, Morgan 1997; Stewart, Shamdasani, and Rook 2007). In academic research, the researcher, who is familiar with the research, is often best positioned to fill this role. To gain some confidence and familiarity with the process, a less experienced researcher might initially take the role of note-taker while a more experienced researcher facilitates the first groups. Focus groups can also be run with more than one facilitator, and a less experienced researcher might invite a more experienced researcher to take the lead. But beginning researchers should not be afraid to try facilitation; as Bedford and Burgess (2002, 129) note, 'the desired qualities are those possessed by the average undergraduate—the ability to listen, the ability to think on your feet, and a knowledge of and interest in the subject the group is discussing.'

When a note-taker is present, that person should sit discreetly to one side. The notes, particularly a list of who speaks in what order and a brief description of what each talks about, can be helpful when transcribing audio recordings of the discussion. A seating plan is also essential. Because the facilitator has to attend to what participants are saying and monitor the mood of the group, he or she should not take extensive notes, although the facilitator may want to jot down a point or two to return to in discussion.

It is highly advisable to audio-record focus groups. The group will usually cover so much material that it will be impossible to recall everything that was discussed. In addition, because presentation of focus group results generally includes direct quotes to illustrate key points, a transcribable audio recording can be very helpful. The quality of the recorder and microphone is crucial. (See Chapters 6 and 7 for a fuller discussion.) Most recorders come with a built-in microphone, but several flat 'desk' microphones placed around the table will ensure much better sound quality and that quieter voices are recorded. The audio-visual departments of universities can be excellent places for you to get advice. Ensure that you test the equipment long before the focus group as well as later in the room before group members arrive. Spare batteries (and tapes, if you are using them) are essential pieces of equipment for focus group researchers. Take care that the setting for the group meeting is quiet enough for discussion to be recorded clearly.

The facilitator usually initiates discussion by giving an overview of the research and the role of the focus group in the project. The themes or questions for discussion can then be introduced. Since group members may be unfamiliar with the focus group technique, a brief summary of how focus groups operate should also be given. Box 8.2 provides an example of a focus group introduction.

The facilitator moderates discussion by encouraging exploration of a topic, introducing new topics, keeping discussion on track, encouraging agreement and

A Sample Introduction to a Focus Group Session   Box 8.2

In this focus group, three researchers acted as facilitators. As people arrived, they were greeted by one of the research team, introduced to the other researchers and group members, and offered tea or coffee. When all participants had arrived, the group was invited to sit around a table. The primary facilitator for this focus group session explained the consent form that had already been placed in front of each person. Participants were asked to read and sign it. The consent forms were passed to one of the researchers, and the session was ready to begin. The researcher acting as the primary facilitator introduced the project:

> *Well, I'd like to thank you all for making the time and coming along today and contributing and sharing your knowledge. In this particular project that we're working on, we're looking at how communities negotiate change, and the reason that we particularly wanted to look at this community was that it seemed there was a lot of change going on and the community, ummm, seemed to be, ummm—we were interested in how you saw your community handling that change and specifically what we're trying to get is to—in the long term is to generate a set of suggestions for other communities on how to manage and negotiate change. So we're hoping to learn from both the mistakes and the right things you've done. So what we're looking for today is a frank and open discussion about how you see change occurring in your community over the last 20 years and how that has been handled. And a little later on in the session we'll get you to—as we go on through the session we'll ask you specific things to help flesh out answers and issues that might be raised. And anything you feel like contributing or adding to just jump in and have that because these focus groups are to get at what the ideas and issues are as you see them. So I think we mentioned in the initial contact with you that we'd like to start out with a two-minute presentation from each of you as to how you see the critical features of change in your community. So we might start around this way. And if you would introduce yourself and your affiliation as you start.*

Once all the group members had made their presentations, the primary facilitator opened the discussion up:

> *Great, thanks very much. Well, that's been really informative to get all those different perspectives. What we'd like to do now is to explore some of these issues. But from here on in the process should change and you should feel free enough to ask, agree, disagree, jump in, put your opinion forward, and so on. And if things get a bit noisy, then we'll just jump in and try and get some semblance of order. One of the common themes that runs through what you've all said is that the community fabric has been affected in a really negative way by all the changes that*

have occurred. And what I'm trying to get at is what could have been done to improve that. So what do you think?

From this point on, different group members responded to questions from the researchers, asked each other questions, agreed, and disagreed with each other. The topics for discussion flowed as people each contributed, adding a slightly different perspective and introducing new ideas. The researchers also asked questions, made points of clarification, and introduced new areas of discussion.

Source: Videotape excerpt from focus group conducted by Katherine Gibson, Jenny Cameron, and Arthur Veno, La Trobe Valley, Victoria, 19 June 1997 (see Gibson et al. 1999). An example of an introduction is also provided by Myers (1998, 90).

disagreement, curbing talkative group members, and encouraging quiet participants. Examples of the sorts of phrases used by facilitators are outlined in Box 8.3.

Examples of Phrases Used in Focus Group Facilitation   Box 8.3

- Encouraging exploration of an idea:
  'Does anyone have anything they'd like to add to that?'
  'How do you think that relates to what was said earlier about . . . ?'
  'Can we talk about this idea a bit further?'
- Moving onto a different topic:
  'This is probably a good point to move on to talk about . . .'
  'Just following on from that, I'd like to bring up something we've not talked about yet.'
  'This is an important point because it really picks up on another issue.'
- Keeping on track:
  'There was an important point made over here a moment ago, can we just come back to that?'
- Inviting agreement:
  'Has anyone else had a similar experience?'
  'Does anyone else share that view?'
- Inviting disagreement:
  'Does anyone have a different reaction?'
  'We've been hearing about one point of view, but I think there might be other ways of looking at this. Would anyone like to comment on other sorts of views that they think other people might have?'
  'There seem to be some differences in what's been said, and I think it's really important to get a sense of why we have such different views.'

- Clarifying:
  'Can you give me an example of what you mean?'
  'Can you say this again, but use different words?'
  'Earlier you said that you thought . . . now you're saying . . . can you tell us more about what you think/feel about this topic/issue?'
- Curbing a talkative person:
  'There's a few people who've got something to add at this point. We'll just move onto them.'
  'We need to move onto the next topic. We'll come back to that idea if we have time.'
- Encouraging a very quiet person:
  'Do you have anything you'd like to add at this point?'

Source: Drawn from discussions in Carey (1994), Krueger (1998), and Myers (1998) and from personal experience.

Some aspects of facilitation require special comment. Expressing and exploring different points of view is important in focus groups, yet research shows that groups have a preference for agreement (Myers 1998). The facilitator plays a central role in creating the context for disagreement. This can be done by stating in the introduction that there is no correct answer and that disagreement is normal and expected, by asking directly for different points of view, and by making explicit implied disagreement and introducing it as a topic for discussion (Myers 1998, 97). Watch for non-verbal signs of disagreement such as folded arms, movement away from the table, and a shaking or downcast head. You might ask the whole group or target the disagreeing member to give a different point of view. Of course, as facilitator, never state that someone is wrong or display a preference for one position. In the unlikely event that the discussion becomes heated, then intervene immediately, suggest that there is no right answer, and move the group on to the next question.

Very talkative or very quiet participants can be a challenge. Talkative people need to be gently curbed, while quiet ones need to be encouraged to participate. Along with the sorts of phrases listed in Box 8.3, your non-verbal signals can be useful. Pointing to someone who is waiting to speak indicates to the talkative person that there are others who need to have a turn. Making frequent eye contact with the quieter person and offering encouraging signs, like nodding and smiling when they do speak, is important. Remember, though, that silence gives people time to reflect and gather their thoughts. Do not feel that you have to fill silences; give people time to respond.

At the conclusion of a focus group, you might review key points of the discussion, providing a sense of completion and allowing participants to clarify and correct your summary. Group members should always be thanked for taking the time to attend and for their contributions. You can do this again with a personal letter to each participant.

On-line Focus Groups

With developments in on-line technology, some researchers find **on-line focus groups** useful, particularly for bringing together people who are extremely busy or geographically dispersed (even in different parts of the world). They also save time, since 'saved' discussions eliminate the need for transcribing. Some researchers have used on-line focus groups to broach sensitive topics because anonymity is possible—either participants devise their own pseudonyms or in some computer programs, participants have the option of submitting both anonymous and identified postings (Kenny 2005; O'Connor and Madge 2003). However, participants need on-line access and some familiarity with computers.

There are two forms of on-line focus groups—**synchronous** (or real-time) **groups** such as chat rooms and **asynchronous groups** such as **bulletin boards** or blogs that may run for days, weeks, or even months. Like face-to-face focus groups, synchronous groups tend to have 6 to 10 participants, whereas asynchronous groups can be much larger. Sweet (2001) recommends 12 to 20 participants, but Kenny (2005) reports successfully running one with 40. In both settings, the role of the facilitator is critical. They post questions for participants to respond to, they seek clarification, they probe for more information, they bring the discussion back 'on track', and they can even use computer logs to monitor when someone is not contributing or is dominating. Facilitating can be demanding, particularly in synchronous settings. The discussion tends to be as lively as it is in face-to-face groups; however, because participants cannot be seen (although the use of web-cam technology could change this), the facilitator cannot 'read' the visual cues: is a participant disagreeing, bored, asleep, out of the room making a cup of tea, distracted by someone else in the room? Nevertheless, those who have used on-line focus groups find them a useful addition to the qualitative researcher's toolbox.

ANALYZING AND PRESENTING RESULTS

Krueger (1998, 46) importantly reminds us that 'analysis begins during the first focus group.' Listen carefully to responses, and clarify any unclear or contradictory contributions, since this information may be critical later when you present the results. For example, if young people say that they would watch television news and current affairs if the coverage were more relevant to them, it is probably important to get them to explain or give examples of how news items could be made more relevant (see also Box 8.4).

Since there is always a richness of material, analyzing focus group discussions can be as time-consuming as it is interesting. The first step involves transcribing the audio recording. A complete transcript of the entire discussion takes a great deal of time, since one hour of recording usually requires more than four hours to transcribe. When a detailed comparison of groups is to be undertaken, full transcripts may be necessary. Generally, a partial or abridged **transcription** (which involves transcribing only key sections of the discussion) will suffice. This is best done as soon as possible after the focus group, with the facilitator(s) and note-taker working in collaboration to decide which sections should be transcribed. A record of the seating plan and running order of speakers and a brief description of what was said are extremely helpful at this point. If you as the researcher have the time, it is also advantageous for you to transcribe the audio recording, because it will give you an opportunity to become more fully immersed in the content (and if you are new to focus group research, you will be able to reflect on your facilitation style and identify your strengths and weaknesses). (For a full discussion on transcribing interviews, see Chapter 6.)

It is advisable to transcribe and undertake a preliminary analysis of the first focus group before conducting any others. This is a way of checking that your questions are understood by participants and are eliciting the type of information you need for your research. It is also a way of checking that you understand and can interpret the responses of participants. For example, in an initial focus group, you might not think to ask young people to clarify what they mean by relevant news coverage, but by carefully reading the transcript, you are likely to pick up this omission.

Once you have the complete set of focus group transcripts available, read the material over several times to help make yourself very familiar with the discussion. One relatively straightforward strategy for proceeding draws from the questions or themes on which the discussion focused. Write each question or theme on the top of a separate sheet of paper, and then list on each sheet the relevant points made. Finally, make a note of key quotes that might be used in written material (Bertrand, Brown, and Ward 1992). This is an approach that works well when the discussion did not deviate widely from the questions or themes set by the researcher or when comparisons are to be made between focus groups (Bertrand, Brown, and Ward 1992). For example, in a research project comparing the land management strategies for dealing with salinity preferred by farmers, policy-makers, and researchers, the sheets with the responses of the different groups to each question or theme can be easily compared.

When the purpose of the research project is to identify key themes or processes associated with a particular issue or topic, it may be more appropriate to use

margin coding (Bertrand, Brown, and Ward 1992). To do this, read through the transcripts, identify key themes or categories, and devise a simple colour-, number-, letter-, or symbol-based coding system to represent the themes or categories. You should then reread the transcripts and highlight the words, sentences, and paragraphs related to each category or theme by writing the appropriate code in the margin. Once transcripts have been coded, a cut-and-paste technique—completed either on a computer or manually—can be used to group the discussion related to each theme or process (see Chapters 6, 14, and 15 for more information on this). Always keep an original of the transcripts for future reference. A variation of this thematic analysis is to develop a list of key words and, in a word-processing package, type two or more key words beside each comment. Using the search function, you can then locate related points of discussion. Computer programs specifically designed for qualitative analysis, like NVivo, can also be used and are particularly helpful when you have a large amount of transcribed material to analyze (see Chapter 15 for a discussion of this).

The ability to find material quickly is an important consideration, since analysis and writing rarely proceed in a linear fashion. During the writing process, new insights unfold (see Chapter 17), and frequently you may find it necessary to return to the original transcripts to refine and reformulate ideas. Sometimes it will be necessary to listen to and make additional transcriptions of sections of the recordings.

When reporting on focus group research, present your results only in terms of the discussion within the groups. As noted earlier, focus groups do not produce findings that can be generalized to a wider population. Focus group results are also expressed in impressionistic rather than numerical terms. Instead of precise numbers or percentages, the general trends or strength of feeling about an issue are typically given. As Ward, Bertrand, and Brown (1991, 271) have noted, focus group reports are 'replete with statements such as "many participants mentioned . . .", "two distinct positions were observed among the participants", and "almost no one had ever . . .". Reporting on their study into people's responses to emergency procedures, Zeigler, Brunn, and Johnson (1996), for example, noted that the people in their focus groups generally responded with either compliant behaviour or under-reaction. Zeigler, Brunn, and Johnson then used direct quotes to illustrate the different ways that the responses were expressed (see Box 8.4 for an example of a focus group analysis, and see also the ways that Jackson and Holbrook (1995), Jarrett (1994), and Myers and Macnaghten (1998) discuss focus group findings).

In some projects, it might not be the general trends but the ambiguous or contradictory remarks that the researcher particularly wants to explore. The development and presentation of an argument may refer not only to what was talked about but the way it was talked about in the group setting. This became

Example of How Focus Group Results Can Be Written Up   Box 8.4

The following is an extract from a journal article reporting on findings from Australian focus group research on the impact of television news and current affairs on young people's political participation and active citizenship. The analysis of the focus group discussion starts by highlighting the way that young people find the reporting of news and current affairs too complex, particularly because of the sophisticated language and the absence of background information. The analysis continues:

> Political current affairs television was viewed by respondents as too complex to incorporate into their everyday viewing habits, but young people also feel it is not worth investing time in television current affairs because any political information received from the programs is usually trivialized and played for entertainment value. For example, 'A Current Affair' [a news magazine show] was described by Debra, a nineteen-year-old university student, as 'Hey Hey It's Ray' after the celebrity of its host Ray Martin, and was seen by her focus group as a form of populist emotional exploitation. As the following responses suggest, there was a strong feeling amongst the groups that television current affairs portrays politicians as being 'full of it'.
>
> > Bianka: The whole politics thing. They're all liars; they're all full of it.
> > Craig: All the media carry on with is stuff like when they asked Hewson [a former politician] if [the] GST [Goods and Services Tax] would be applied to a birthday cake and they just blew that up. Who gave a shit?
>
> The respondents felt news and current affairs did not help them develop a political identity. They also expressed distrust in politicians who attempt to 'persuade' them to choose a lesser evil. As Bianka points out: 'They all change their minds when they get what they want. I mean what's the point?' What eventuates is a distrust of not only politicians but also the media that is supposed to decipher the positive and negative elements of each candidate's actions.

Note how the findings are reported only in terms of the focus groups and in tentative terms with the use of phrases such as 'responses suggest', 'respondents felt', and '[w]hat eventuates'. Main themes that emerged from the focus groups are summarized by the authors, and quotes from participants are used to illustrate and elaborate these themes.

Source: Evans and Sternberg 1999, 105.

a significant aspect of the focus group research conducted by Gibson, Cameron, and Veno (1999, 29):

> The stories of success and hope that emerged when the discussion
> was shifted onto the terrain of community strengths and innovations
> were numerous. They came stumbling out in a disorganised manner
> suggesting that these stories were not readily nor often told. In the face
> of dominant narratives of economic change perhaps such stories are
> positioned as less important or effective. It is clear that there is a lack
> of a language to talk about this understanding of community capacity;
> yet, as we will argue, this understanding has the potential to contribute
> to the ability of a region to deal effectively and innovatively with the
> consequences of social and economic change.

One important element in the process of writing up (or 'writing-in', as Mansvelt and Berg call it in Chapter 17) is to find a balance between direct quotes and your summary and interpretation of the discussion. When too many quotes are included, the material can seem repetitive or chaotic. Too few quotes, on the other hand, can mean that the vitality of the interaction between participants is lost to the reader. Morgan (1997, 64) recommends that the researcher should aim to connect the reader and the original participants through 'well-chosen' quotations (see Chapter 18).

CONCLUSION

Focus groups demand careful preparation on the part of the researcher. The selection and recruitment of participants, the composition, size, and number of groups, and the questions and topics to be explored are all key points to consider during the planning stage. Even the apparently mundane details of appropriateness of venue, provision of refreshments, and quality of audio equipment are critical to the success of focus groups. A well-prepared researcher also gives thought beforehand to the process of facilitation, including the points to cover in the introduction, the wording of key questions, topics, and phrases, the probes and prompts that might be useful in further exploring a theme or topic, and strategies for drawing out different points of view, keeping the discussion on track, and dealing with both the more talkative and quieter members of the group. As soon as possible after the focus group, start the process of analysis, beginning with transcribing the audio recordings, followed by reading and rereading the transcripts, summarizing main points, and identifying central themes.

Although they require careful planning beforehand and a great deal of reflection afterwards, focus groups are an exciting and invaluable research tool for geographers to use. Participants almost invariably enjoy interacting with each other, and the discussion can generate insights and understandings that are new to both participants and researchers. The interactive element makes focus groups ideally suited to exploring the nuances and complexities of people–place relationships, whether the research has a primarily data-gathering function or is more concerned with the collective practice of knowledge production.

KEY TERMS

asynchronous groups

bulletin board

facilitator

margin coding

moderator

on-line focus groups

over-disclosure

probe

purposive sampling

recruitment

saturation

standardized questions

synchronous groups

synergistic effect

transcription

under-disclosure

REVIEW QUESTIONS

1. Find a research project from a recent issue of a geographical journal that you think could have been conducted using focus groups. Why do you think focus groups would be appropriate? Discuss the participants you would select, the composition of the focus groups, the size and number of groups you would use, the questions you would ask or themes you would use, and strategies for recruiting participants.

2. In a few days time, you will be facilitating a series of focus groups. Describe the final steps you would take to prepare for the groups. What are some of the issues that you anticipate might arise when conducting the groups? Discuss the strategies you would use to manage these issues. You might want to consider the following issues: too much agreement among participants; over-disclosure of personal information; groups that are overly talkative or overly silent.

3. The University of Pacifica is upgrading its on-campus library facilities. Your research company has been commissioned by the university to conduct a focus group study on the sorts of changes that students think are most important. Devise a series of questions that you would use to canvass students' views. Select two students to co-facilitate the focus group, and select seven to eight students to participate in the focus group. Other students observe, noting how the content of the discussion flows, and the verbal and non-verbal interactions between participants and the facilitators.

Useful Resources

Barbour, R. 2007. *Doing Focus Groups*. Los Angeles: Sage.

Barnett, J. 2009. 'Focus groups tips for beginners'. TCALL Occasional Research Paper no. 1. Texas Center for Adult Literacy and Learning. http://www-tcall.tamu.edu/orp/orp1.htm#1.

Dick, B. 2000. 'Structured focus groups'. http://www.scu.edu.au/schools/gcm/ar/arp/focus.html.

Morgan, D.L. 1997. *Focus Groups as Qualitative Research*. 2nd edn. Thousand Oaks, CA: Sage.

Rezabek, R. 2000. 'Online focus groups: Electronic discussions for research'. *Forum Qualitative Social Research* 1 (1). http://www.qualitative-research.org/fqs-texte/1-00/1-00rezabek-e.htm.

Stewart, D., P. Shamdasani, and D. Rook. 2007. *Focus Groups: Theory and Practice*. 2nd edn. Thousands Oaks, CA: Sage.

Historical Research and Archival Sources

Michael Roche

9

CHAPTER OVERVIEW

Despite heightened levels of interest in qualitative research methods, human geographers have tended to overlook one of the oldest of qualitative techniques: historical approaches based on the study of primary documents. The interrogation of archival sources is an essential technique for most historical geographers, and while other human geographers have been willing to incorporate a historical dimension into their work (Driver 1988), they do not necessarily have any archival research skills. Human geographers more generally can benefit from having some appreciation of these skills, for they can be deployed for contemporary as well as historical research.

INTRODUCTION

After 30 years of working as a historical geographer, I still relish the opportunity to undertake archival research. There is a continuing sense of delving into the unknown, of engaging in academic detective work trying to understand inevitably fragmentary and partial surviving **records**, of striving to make sense of the evidence you are scrutinizing. My own introduction to **archival research** was orchestrated only at the graduate level. To a large degree, I was able to learn by trial and error, following the tendency of historical geographers at the time to regard archival research as part of their craft, something to be acquired on the job. Good archival scholarship was to be inferred from reading journal articles or books by leading historical geographers and from discussions with supervisors. In many ways, this was a laudable model, one that allowed me to develop my skills and understanding at my own pace, but geography students of today wishing to use archival sources can benefit from a more overt discussion of the fundamentals of archival research. This is also the case because archival research needs to engage with larger disciplinary theory and the research ethics that are also a part of historical inquiry. Even so, like other methods, archival skills can to some extent only be learned by doing archival research. Expertise improves with experience, and this is not easily reduced to a checklist of best practice.

WHAT IS ARCHIVAL RESEARCH?

> Archival scholarship at its best, it seems to me, is an ongoing, evolving interaction between the scholar and the voices of the past embedded in the documents. [Harris 2001, 332]

Archival sources are a subset of what historical geographers and historians refer to as **primary sources**. They include non-current records of government departments held in public **archives** but can be extended to include company records and private papers. As well as **documents**, handwritten and typed, these sources can embrace personal letters, diaries, logbooks, and minutes of meetings, as well as reports, plans, maps, and photographs. More recently, they have included records created in electronic format, which brings with it new challenges (Davison 2003). This chapter concentrates on official papers, including manuscript and typescript files.[1] Most of the comments are also applicable to company archives and private papers. With the target readership of this book in mind, the chapter concentrates largely on government archives, on the past century or so, and on the 'New World'. This focus simplifies the discussion, since much of this documentation is type-written and the language of the more recent past is relatively easy to comprehend today. As a collection of unique, single documents, created contemporaneously with the events they discuss, the materials lodged in archival repositories provide a particular window on the geography of earlier times. As such, they are a major source of valuable information for geographers. Historical approaches applied to archival sources will not allow all of the research questions of human geography to be addressed; however, they do provide a means of answering questions about the recent as well as the more distant past that are not recoverable by the other techniques or from other sources available to human geographers.

More often than not, the researcher will make use of public archives housed in a government agency charged with the preservation of non-current records. On other occasions, small regional collections, such as those associated with some museums, may be targeted. Sometimes, access to the records of private organizations may be sought. Michael Williams (1992) offers a concise summary of a range of archives from the national to the local. Increasingly, the web provides the initial contact point with research and archival collections. Box 9.1 lists some major repositories and their 'www' addresses. In addition, various specialized

Identifying the Archives	**Box 9.1**

Archives New Zealand:
  http://www.archives.govt.nz/index.html
Library and Archives Canada:
  http://www.collectionscanada.ca/index-e.html
National Archives of Australia:
  http://www.naa.gov.au
National Archives of Ireland:
  http://www.nationalarchives.ie
National Archives of Scotland:
  http://www.nas.gov.uk
National Archives of South Africa:
  http://www.national.archives.gov.za
Public Record Office (England and Wales):
  http://www.nationalarchives.gov.uk
US National Archives and Records Administration:
  http://www.archives.gov

archives, such as those dealing with women (Mason and Zanish-Belcher 2007) and others, capitalize on new technologies, taking the form of digital archives such as the Aluka Project, which brings together material on liberation struggles in southern Africa (see http://www.aluka.org) (Isaacman, Lalu, and Nygren 2005). The latter contains published and oral material and in that sense reflects a broader definition of 'archive' than I have adopted in this chapter.

For Mayhew (2003), historical geography has a two-fold significance for the discipline as a whole that lies beyond increasing the understanding of the geography of the past: to re-evaluate taken-for-granted concepts and to develop a comparative perspective so that as geographers we might more fully appreciate what is distinctive about today's world and how we understand it in disciplinary terms. Historico-geographical research based on archival research underpins both of these objectives.

Advice on Conducting Good Historical and Archival Research

The first step in reconstruction of past stages of a cultural area is mastery of its written documents. [Sauer 1941, 13]

It is an oversimplification to believe that the study of change through time by means of either historical documents or field evidence does not require special training and skills. [Perry 1969, 96]

Approaching Archival Research　　　　　　　　　**Box 9.2**

The first point to make about archival research is that it cannot be contained within a single methodology. Any sizable archive holds a vast array of material, and even if one's research questions are fairly specific, the chances are good that there will be far more potentially relevant documents than there will be time to examine them . . .

. . . [A]rchival research tends to gravitate towards one of two polar reactions — neither, I think particularly helpful. It is easy enough to be taken over by the archives, to attempt to read and record all their relevant information. In this way months and perhaps years go by, and eventually the investigator has a vast store of notes and usually, rather weak ideas about what to do with them. A fraction of the archives have been transferred from one location to another, while the challenges of interpretation have been postponed . . .

In effect the archives have swallowed the researcher. At the other pole are those who come to the archives with the confidence that they know precisely what they want. They have conceptualized their research thoroughly in advance. They pretty much know how they will argue their case and what their theoretical position is. But they do need a few more data, which is why they return to the archives. As long as they cleave to their initial position, either they will find that data they need and leave fairly quickly or they will not find them and also leave. Fair enough for certain purposes. But they are imposing their preconceptions on the archives. They have solved the problem of archival research by, in effect, denying the complexity of the archives and the myriad voices from the past contained in their amorphous record (Harris 2001, 330–1).

So where to begin? Good archival research is difficult to reduce to a checklist of points. However, it is useful to explore sequentially the sorts of things you might need to do and the obstacles that you might encounter in undertaking archival research. At the start, like any other research project, your work ought to be informed by prior in-depth reading on current scholarship around the topic but accompanied by an openness about the ultimate direction of the research. This point is well made by the highly accomplished Canadian historical geographer Cole Harris (Box 9.2).

Archival research begins before you arrive at the archive. You ought to be familiar with the existing **secondary literature** before beginning any search for archival material. For instance, for a project involving state agencies, it is important to read any previously published institutional histories as well as the annual reports of that agency and to look at parliamentary debates (or their equivalent). You should always write to any archives you intend to visit well beforehand and, if possible, obtain a letter of support from an academic supervisor. Outline your research topic in reasonable detail, indicate whether it is for a small project or for a thesis, and signal what you are seeking to do and how much time you have at your disposal. With larger archives, you may be able to do this on-line. This preparatory work will help you to make best use of what is inevitably too little time. The **archivist**, with an intimate knowledge of both the ways the materials are organized and their contents, will be able to help you identify relevant files.

Archives are not like libraries, although there may be some similarities. They may operate on quite restricted hours. Some will require that you sign up for a reader's ticket (you may be able to do this electronically). You may also need to have someone vouch for you and to show some form of identification. If you can, you should check out the specific characteristics and requirements of the archives before you visit.

What should you take to the archives? Take related research notes, pencils, and paper. Most archives operate on a **'pencil only' rule** to minimize damage to the original documents should any be accidentally marked, although many researchers now bring their own laptop computers. Although most public archives will have supplies of scrap paper, I would also advise you not to depend on the archivists to supply stationery. Do not expect a small regional archive to supply pencils, to have a pencil sharpener, or to provide a convenient power outlet for your computer.

Sometimes ingenuity is called for in that you may only be able to address research questions obliquely. The archivists can sometimes provide helpful, expert advice about record sets that you may not have considered useful. Typically, as a new user you will be given the opportunity to explain what you are researching and why. The archivist will tell you how the **finding aids** work and can offer suggestions about where to start looking. Their experience and expertise can often prove invaluable, but it is important to remember that they may have limited time available to offer help to individual researchers.

A crucial difference between a library and an archive lies in the way that each stores material and in the nature of the finding aids. Libraries typically catalogue books and journals by either the 'Dewey Decimal' or the 'Library of Congress' classification systems that group together all books on similar subjects. In contrast, public archivists seek to maintain the integrity of the record sets they obtain from government departments or other agencies in terms of preserving specific files' place in the broader record set, maintaining the original ordering of documents in the file, providing storage conditions that will ensure the long-term survival of the records, and making them available to the public. Archivists place great emphasis on the **provenance** of the **files**; the actual order of the material within the files in itself tells the researcher something about the situation that prevailed when the file was being created.[2] Thus, whereas in a library you can refer to a catalogue to find a book on a particular subject on an open shelf, in an archive basic paper-based finding aids take the form of sequential **series lists** of all the files held by particular agencies. These lists itemize all the files created by an organization using the original description system (usually numerical but some time alpha-numerical).

In many major archives, electronic searching of the collections is now possible. This means that you can search for specific items in the same way, superficially at least, that you would use a library catalogue. In large archives, this procedure can reduce the amount of time you spend hunting for files and release extra time for actually reading the retrieved materials. However, this time-saving may come at the expense of your having any sense of the overall structure of the records. Do not forget about provenance. What survives in the file is likely to be only a fragment, and it may be quite partial in terms of providing any insights about the past. In many cases, however, particularly with small archives, you will have only the series list of files to guide you. The files remain organized according to the system that the original creating agency devised. Inevitably, you will find that some of the originating records staff have been more thorough and less idiosyncratic than others. The name of a file may not always be a clear guide to its contents, material may have been misfiled, and some files may have been lost or destroyed. For instance, I recently found that files marked 'railways accommodation' had nothing to do with housing, the topic I was working on, but actually referred to the number of passenger carriages and freight wagons that could be 'accommodated' at the railway station yards. Similarly, a file (mis)labelled as 'hoses' was actually about houses (here the file number provided the clue).

In some national collections, precious and fragile originals may have been electronically scanned or photographed and made available on-line or as microfilm copies (e.g., Hackel and Reid 2007). Again, it is important to check in advance as to whether you will be reading originals or copies of the documents that you plan to consult. Some sources have been much studied and are available in published form (e.g., Powell 1973). However, remember to use these sources critically, because they may have been edited to reflect the conventions and morals of a later age. For example, well-known New Zealand historian Bill Oliver had suspicions that the published correspondence of two prominent nineteenth-century New Zealand political families had been inadequately edited: 'upon inspecting the originals I found that the editor had not only made mistakes in the transcription (a venal sin) but has defaced the manuscript with overwriting and instructions to his typist a (a mortal sin or if it isn't it should be)' (Oliver 2002, 107). Just because some primary source material has been published in printed form does not mean that you can relax your critical judgment.

You may find that there is restricted access to some files. Personnel files typically fall within this category. The period during which restricted access applies varies from one country to another, but 30 years after the closing of the file is typical. In some instances, a lesser degree of restriction applies, and permission to look at files may be granted by a senior archivist, government official, or someone associated with the organization that created them. A formal written request outlining your research project may result in the granting of access; however, some conditions may be attached—for example, you may be permitted to read only a particular portion of the material while the remainder of the file remains physically sealed. In other cases, the researcher has no choice but to wait patiently until the material is released.

Unlike material in a library, archived files are not kept on open shelves. They are not necessarily even kept at the same site as the reading room and may only be delivered from storage on request on an hourly basis or less frequently, so be prepared to have other tasks to occupy your time while you wait (for example, searching the finding aids for other files to call up). Photocopying material is usually possible, but it can be comparatively costly, and you may have to pay in advance. Some material may be deemed too fragile or, if bound, too difficult to photocopy. It is therefore advisable for you to find out what the policy is beforehand. Plans, maps, and charts larger than A3 size can be copied by means other than photocopying, but this is sometimes quite expensive. However, it may be the only means of obtaining a copy of an essential document. Some archives now permit researchers to make their own digital copies of documents. Various conditions apply, including registering your camera and completing associated documentation and agreeing not to use flashes or tripods or to fold documents. The use of digital images also raises new issues regarding labelling and storage if you are to make effective use of such materials (Box 9.3).

Digital Images	**Box 9.3**

It is now becoming more common for archivists to allow researchers to make their own digital images of material. This has obvious advantages in terms of the ability to make images of, for instance, bound volumes that would not have been photocopied because of concerns over preserving their binding. Digital images are also less expensive than photocopying, and they do not involve a lengthy delay in obtaining them, an important point for time-pressed students who cannot wait for an archival turnaround time of several weeks. However, I would be mindful of Harris's (2001) comment about transferring the archive from one location to another. Ease of copying in itself can create other difficulties if reference details are not kept meticulously. As a checklist, I would suggest the following points (nearly all of which I have fallen foul of over the past 12 months . . .):

1. Ensure that the memory card in the camera is clear and the battery charged.
2. When copying lengthy documents, be wary of making blurred images and of missing pages.

3. Recognize that you may require some maps and images reproduced with greater clarity than you can obtain with your hand-held camera, and be prepared to pay for photocopier or high-quality camera images.

4. Ensure that you have a reliable system for linking the digital image to the source file (my low-tech approach to this has been to include a slip of paper with the file details on it alongside the photographed page so that I have a visual reference on each digital image).

5. Store the digital images so that they can be located and retrieved easily.

When the archivist gives you the file to work on, you will find in most cases that new items are on top of the older material, particularly if the material is secured by paperclip. You will probably need to work from back to front. Will the material answer any of your research questions? It may be immediately obvious that the material is relevant to your inquiry, or it may appear only tangentially relevant or even irrelevant. Sometimes it is difficult to make a judgment at first glance, and you may have to recall material you have examined previously but whose significance you did not appreciate at the time. Alan Baker, a British historical geographer, has offered some guiding thoughts on evaluating primary sources, including archival materials (Box 9.4).

Assessing Evidence in Historical Geography   Box 9.4

No source should be taken at face value: all sources must be evaluated critically and contextually. The history and geography of a source needs to be established before it can legitimately be utilized and incorporated into a study of historical geography. The historical sources we use were not compiled and constructed for our explicitly geographical purposes; they were more likely to have been prepared, for example, for the purposes of taxation and valuation, administration and control. We also have to understand not only the superficial characteristics of a specific source but also its underlying motivation, background and ideology of the person(s) who constructed it. In order to make the most effective and convincing use of a source we must be aware of its original purpose and context and thus its limitations and potential for our own project.

Source: Baker 1997, 235.

Baker's words seem to me to be crucial for those using archives as qualitative sources in human geography. It is essential to understand as fully as possible the original purpose of the document, who created it, what position they held, and how and when it was made. Some generic questions to pose when assessing documentary sources are laid out in Box 9.5.

Questions to Ask of Documentary Sources

Box 9.5

1. Can you establish the authenticity of the source—is it genuine? Are you looking at the original?
2. Can you establish the accuracy of the document—how close is it to the source of events or phenomena? How accurately was the information recorded? (Cross-check with other sources.)
3. What was the original purpose for collecting the information? How might it have influenced what information was collected?
4. How has the process of archiving the information imposed a classification and order upon historical events?

Source: After Black 2006.

The questions raised in Box 9.5 provide a useful start, although I would make three qualifying points. First, it is possible to extend 'document' to include maps and plans (see Harley 1992). Second, this approach tends to privilege the ideas behind actions. That is to say, the past is being understood in idealist terms whereby the thought behind the action is regarded as providing the understanding necessary to interpret these events. Historical geography can be written legitimately from a viewpoint other than that of contemporary observers (Baker 1997). Third, the documents themselves cannot be read in isolation but must be understood in their wider context, and even then any conclusions will be provisional rather than absolute.

To some extent, all archival researchers develop individualized approaches to note-taking from archival materials. However, there are two basic strategies. The first involves collecting material by topic, noting specific details and suitably referenced quotations. Classically, historical researchers have made use of large index cards for this purpose, although many now use laptop computers to organize their notes. New topics can be noted on new cards as more files are read and new research questions formulated. The alternative approach is to record chronologically any pertinent information from each file and then subsequently identify themes that emerge across the files. Both strategies have advantages and disadvantages. The former depends on identifying key topics at the beginning of the project within which to collect information. Such an approach offers the ability to add new topics or identify dead ends and see how themes merge or diverge. My personal view is that while this approach means that many diverse sources are brought together, it can blur a researcher's capacity to make good inductive judgments. The latter method is more sensitive to the provenance of files and can give a clearer sense of the role of particular officials or departments. It does, however, involve a degree of double-handling in that evidence that has already been collected by the researcher needs to be reorganized after each visit to the archives and perhaps annotated further. It is important to follow up other research questions that may emerge from this re-sorting process. The latter approach is one that I have used over many years. It suits me, and as a full-time academic, I can incorporate it into my way of working. I would acknowledge that it probably works best when one is working in an area where both the secondary literature and the archival sources are familiar to the researcher, even if the specific contents of the files are unknown. Students with a relatively short period of time available for archival research may prefer to adopt the first strategy and will probably be using laptop computers, particularly when they have the keyboard skills (speed and accuracy) that I lack!

After you have located and extracted archival evidence, it must be adequately cited in the written products of the research. The first step is to carefully record the specific document description and file reference. For example, the personnel file for Edward Phillips Turner, director of forests in New Zealand from 1928 to 1931, is located in the New Zealand Forest Service files at Archives New Zealand in Wellington. The specific reference is 'F Acc W2338 82/113 E.P. Turner'. The 'F' refers to the Forestry files, 'W2338' is the accession number for that collection of documents, 82 is the series, and 113 is the specific file number, while E.P. Turner is the descriptive label (a little confusing in that the individual in question always used the double-barrelled surname). There is considerable variation among filing systems. Many are much simpler than the one used in the example above. You need to quickly become familiar with the system used by the agency whose records you are working on, and here the expert assistance of the archivist can be invaluable. The crucial thing is to record the details carefully. This is important for two reasons. First, it enables you as a researcher to keep track of where you found specific information. Second, it enables a subsequent researcher to relocate the material. The idea is simple enough, but given the nature of archival material, it is somewhat more exacting than, for example, the standard bibliographic requirements of author, date, title, and publisher/place for a book in the reference list of a thesis. Citing archival materials correctly can also pose problems in that human geography has tended to adopt versions of in-text citation systems, such as Harvard (see Hay 2006). Most archival sources sit uncomfortably within this framework and are generally better referenced in footnotes or endnotes, typically used by historians. Students undertaking archival research may need to negotiate a variation from their university's social science–oriented formats for referencing.

The archive does not constitute the only source for historical research. For instance, newspapers, private papers, and unpublished memoirs may provide valuable material for cross-referencing with the archival record. As well, once archival work is completed, the researcher may need to follow up on unfamiliar key actors by checking old editions of *Who's Who* or newspaper obituaries, as well as on unfamiliar organizations or period issues; here contemporary newspaper accounts can be invaluable. This 'post-archive' work can of course help to shape and inform the purpose of subsequent trips to the archive.

Moreover, files are not the end point of research, as US cultural geographer Carl Sauer reminded historical geographers nearly 70 years ago:

> Let no one consider that the historical geographer can be content with what is found in archive and library. It calls, in addition, for exacting fieldwork. One of the first steps is the ability to read the documents in the field for instance of an account of an area written long ago and compare the places and their activities with the present, seeing where the habitations were and the lines of communication ran, where the forests and the field stood, gradually getting a picture of the former cultural landscape behind the present one. [Sauer 1941, 13]

Although Sauer's words may indicate nostalgia for a pioneering rural past while your focus could just as easily be urban and social, his challenge remains pertinent and has only partially been taken up by more recent generations of geographers (for example, Raitz 2001).

CHALLENGES OF ARCHIVAL RESEARCH

There are two types of challenges facing researchers working with archives. The first is intellectual and the second technical. When dealing, for instance, with the file materials contained in an official government archive, it is important to bear in mind the sorts of power relations inherent in the surviving materials. This is rather more than just acknowledging that the surviving files are fragmented and partial. The records are those created by politicians and officials. They reflect the outlooks and understandings of the dominant groups in the national context at the time they were created. Duncan (1999) writes of these concerns in terms of complicity stemming from use of the 'colonial archive'. For much of the nineteenth century and well into the twentieth century, these records were created largely by men in the upper echelons of society, and in colonial situations such as those in Australia, Canada, and New Zealand, they are predominantly the records of colonizing British settlers. Summarizing the contents of files from the archives merely reproduces these uneven power relations rather than interpreting them. The records of nonofficial, community, or sporting groups may provide a way into understanding the concerns and aspirations of those who had no position in the public political sphere. In the same way, oral histories from the recent past may provide insights into gendered and minority concerns. Furthermore, an awareness of the power relations within the archival material may allow the researcher to reinterpret surviving materials. For instance, what I once mapped as examples of illegal felling of forest in New Zealand in the 1870s I would now be inclined to understand as resistance on the part of Maori forest owners to the imposition of authority by the Crown and as the flouting of government regulations by timber-cutters who had limited alternative means of supporting themselves (see Chapter 2 for further discussion of power relations in qualitative research).

The most fundamental technical difficulty relates to the ability to actually read the documents retrieved in the archive. During the first half of the nineteenth century, many official documents were handwritten in **copperplate script**. This script looks elegant, but it can take some time for novices to learn to read it proficiently, a situation that may be exacerbated when officials wrote both across and along a page in order to save paper. Perseverance will pay off. Archives from the latter part of the nineteenth century are generally written in a **modern hand**. They are generally readable with a bit of effort, the main problems occurring with faint letterbook copies. However, original manuscripts concerned, for example, with the early settlement of North America in the period before 1700 may be written in **secretary hand**. This was the script of professional scribes of the time, and it is difficult to read without specialized instruction. Unless the material has been transcribed and printed, its translation requires additional palaeographical skills. In any case, all kinds of handwritten documents made in the past tend to be difficult to decipher, especially when the investigator is trying to read a faint letterbook copy of the original. For instance, in the mid-nineteenth century, the 'long s' written much like an 'f' was frequently used in official correspondence. Words with a double 's'—'lesson', for instance—are rendered as what looks to us like 'lesfon'. Not only do spellings change as you move back in time, but so does the very construction of the English language.[3] This makes it more of a challenge to understand the world view of these earlier times. From around the 1880s, typewritten material becomes more common in government files, but important marginal annotations will be handwritten and often cryptic in meaning. These annotated comments are particularly important for the insight they can give into discussion within an organization about the issue to which the larger document relates.

A good example of some of the challenges posed by handwritten documents is the particularly hard-to-read correspondence of Captain Campbell Walker, the first conservator of forests in New Zealand in 1876–77. Because of the importance of these letters to the topic I was researching, I decided to photocopy and later transcribe them but only managed to do so after painstakingly working out, on a letter-by-letter basis, how he wrote the alphabet. Anyone contemplating the use of archive material as a source for qualitative research in human geography must be prepared to be patient and resourceful; using documentary evidence is rarely easy and generally requires a great deal of time.

The units of land area and currency may also be different from those in use to-day (for example, acres rather than hectares). This raises the issue of whether to convert every measurement to the current system or to give a general conversion factor and use the units of the period (generally, I prefer the latter). Some facility with the original units is useful. Appreciating that there are 640 acres to a square mile makes it possible to recognize, for example, that the apparently precise data on the forest areas in Otago Province in New Zealand in 1867 are actually only estimates to the nearest quarter square mile, or 160 acres. You may also need to understand more specialized measures, depending on your field of research. For example, throughout much of the British Empire in the nineteenth century, quantities of sawn timber are often given in superficial feet (colloquially referred to as 'superfoot')—that is, 12 inches by 12 inches by 12 inches (30 cm by 30 cm by 30 cm), but in North America the equivalent term 'board foot' was used.

If you are working through a file and time runs short, you may find yourself copying whole documents that you think could be of use because you do not have the time to read them carefully and decide whether they are. In the end, you find yourself with page after page of material that may subsequently prove to be marginal to the research. This is particularly important in that when you return to your research material, it is too easy for the photocopied document to overshadow your handwritten notes so that you again end up with the situation Harris refers to as the archive 'swallowing the researcher' (see Box 9.2).

Mistakes in interpretation can and do occur. I once mistook the numbered applications for the position of director of forests in New Zealand for the rank-ings of candidates. The result was an apparently nonsensical list of candidates. On closer subsequent investigation of the date stamps showing the receipt of the applications, it became clear that the numbers related only to the order in which they had been received. Retrospectively, I can draw three points from this epi-sode. First, scrutinize documents carefully. In this case, the answer was there in the documents, but I did not see it the first time around. Second, if you are un-certain about what the documents indicate, acknowledge this, and do not make too definite a claim regarding the surviving evidence. Third, by 'learning the ropes' as an undergraduate or graduate student, you can avoid some of the more obvious pitfalls of interpretation before you have anything published.

Ethics and Archives

It is all too easy for archival researchers to dismiss ethical issues as something relevant only to geographers working on present-day topics using other qualita-tive or quantitative methodologies. Actually, archival researchers also have ethical obligations, accentuated by the fact that the individuals who created—or are the subjects of—the records in question are in all likelihood now deceased and unable to represent themselves. Other ethical dimensions of archival research have less to do with safety, harm, and risk to the researcher or research participants and more to do with retaining the integrity of the material contained in the files and the preservation of the archives themselves. After all, those using archives should regard access to the material as a privilege. Historical records are precious and often irreplaceable. All researchers are under an obligation to look after archival material and to ensure that it is preserved in good order for any subsequent scholars.

The US National Council of Public History identifies three guidelines for using archive material.[4] We can reasonably substitute 'geographer' for 'historian' in each of these guidelines.

1. Historians work for the preservation, care and accessibility of the historical record. The unity and integrity of historical record collections are the basis of interpreting the past.
2. Historians owe to their sources accurate reportage of all information relevant to the subject at hand.
3. Historians favour free and open access to all archival collections (National Council on Public History 2003).

Another situation in which ethical issues may arise is when files that contain classified or otherwise restricted material are issued to you by mistake. While it may be tempting to capitalize on an archivist's error in issuing a file before a time embargo or other restriction has elapsed, in the longer term this is counter-productive; it is equivalent to an unsanctioned questionnaire or an interview in which the participant does not know the true purpose of the research. Such behaviour can result in tighter lending conditions being imposed on all subsequent archival users.

Public archives typically specify conditions to which users must agree to adhere when they sign in or request a reader's card. Not all primary documents are in public archives, however, and having access to such documents can present practical and ethical issues. Finer (2000) recounts an episode in which after she initially received unlimited access to the records of a prominent Italian social reformer, she learned partway through the project that new conditions governing access, the scope of the research, and its objectives were being imposed (Box 9.6).

Researching in a Private Archive   **Box 9.6**

For researchers using public archives, the protocols are fairly well-established and reinforced in documentation that is part of the user registration process.

On occasions, researchers will have access to private papers or records of small organizations. On the basis of a particular research project, Finer (2000) puts forward four 'negotiations' that ought to be undertaken on those occasions to ensure the smooth running of the project. They are:

1. To insist on and ideally participate in the drafting of a detailed written agreement regarding precisely what is to be attempted in the research and to what end.

2. To draft a timetable agreed to in advance by staff members who are in a position to affect access to records or to other facilities such as photocopying.

3. To reach agreement in advance on the handling of sensitive material and the extent to which and on what terms it is to be cited.

4. To reach agreement on matters of faith/ideology—that is, the extent to which it is or is not considered necessary for the researcher to be of the same persuasion as the person(s) being researched.

While Finer's project had a biographical dimension to it rather than much in the way of human geography research, her points have a general utility.

PRESENTING THE RESULTS OF ARCHIVAL RESEARCH

There is no single correct way of presenting the results of archival research. The theoretical foundation of the research project, the sorts of empirical information retrieved, and the writing style of the researcher all shape how the research project or thesis is expressed.[5] Typically, however, archival researchers will make use of direct quotations from key documents to demonstrate their case (see also Chapter 18 for a discussion of this). They will also be mindful of the actions of key actors within organizations (and sometimes the importance of the role of obscure bureaucrats as well) in shaping decisions and policy that may have had far reaching geographical significance. They also make use of case study material to illustrate points. On occasions, good use may be made of cartographic or pictorial material.

Adept researchers are often able to move easily from specific points of detail to sketch a much larger picture and to relate it to what is known about related topics. I would recommend critical perusal of recent issues of the *Journal of Historical Geography* and *Historical Geography* and recent books by recognized figures in the field (e.g., Harris 2003; Colten 2005). However, it is not just a matter of identifying key quotations but rather of building an argument. This obliges you to select ideas in a logical way from the pre-existing literature and then to use them to provide an informed discussion based on what you have found in the archives. The desirable end point, however, is to be in command of the source material. Rather than merely reproducing a chronicle of part of what is contained in the archive, strive to make your writing a synthesis of specific detail and informed interpretation.

CONCLUSION

Although it has much to offer human geography in general, archival research has tended to be neglected by other than historical geographers. As a research method, historical research using archival sources:

- calls for creative thinking in identifying source materials relevant to your research problem;
- needs patience, precision, and critical reflection in collecting and evaluating material;
- requires a sense of historico-geographical imagination in interpreting source material whereby theorization does not outstrip the evidence;
- is partial and requires that you relate archival material to other contemporary sources of a textual and pictorial sort that may be held in other collections;
- asks researchers to continually negotiate between the theoretical and the empirical.

Archival work can be extremely time-consuming and, superficially at least, frustrating in that the information retrieved may offer only partial answers, particularly when you find yourself under time pressure to complete a research project. Archival work done properly takes time and patience. Rarely will the surviving archival material provide 'full' answers to the questions you pose. In the case of public archives, the surviving material typically says more about politics, economics, the concerns of elites, and men than it does about social and private spaces, women, and minority groups. It is, however, still possible to use these records to recreate something about the lives of ordinary people. But surrendering to the temptation to merely summarize the content of files, a trap into which inexperienced archival researchers can fall, is another way of being—as Harris terms it—'swallowed by the archive'.

It is all too easy in discussing archival research to create the impression that there is no room for novices when in fact more human geographers need to be encouraged to incorporate archival work into their research programs. I would simply describe archival research as somewhat akin to confidently accepting the challenge of working on a jigsaw puzzle even though you can be reasonably certain that pieces are missing and that the box cover with the picture of the completed puzzle will never be found. Good archival research can be extremely satisfying, both in learning the skills to conduct it and in the presentation of results.

KEY TERMS

archival research

archives

archivist

copperplate script

document

files

finding aids

modern hand

'pencil only' rule

primary sources

provenance

records

secondary literature

secretary hand

series lists

REVIEW QUESTIONS

1. What sorts of research questions can be addressed using archival sources?
2. What are the problems of a researcher being, as Harris terms it, 'swallowed by the archive'?
3. What is meant by provenance, and why it is important to archival researchers?
4. What steps are required in assessing historical evidence?
5. What approach would you adopt to organize and store digital images from archival collections?

USEFUL RESOURCES

Baker, A.R.H. 1997. 'The dead don't answer questionnaires: Researching and writing historical geography'. *Journal of Geography in Higher Education* 21 (2): 231–43.

Black, I. 2006. 'Analysing historical and archival sources'. In N. Clifford and G. Valentine, eds, *Key Methods in Geography*, 477–500. London: Sage.

Cameron, L. 2001. 'Oral history and the Freud archives: Incidents, ethics and relations'. *Historical Geography* 29: 38–44.

Hackel, S., and A. Reid. 2007. 'Transforming an eighteenth-century archive into a twenty-first century database: The Early Californian Population Project'. *History Compass* 5 (3): 1013–25.

Hall, C. 1982. 'Private archives as sources for historical geography'. In A.R.H. Baker and M. Billinge, eds, *Period and Place: Research Methods in Historical Geography*, 274–81. Cambridge: Cambridge University Press.

Hanlon, J. 2001. 'Spaces of interpretation: Archival research and the cultural landscape'. *Historical Geography* 29: 14–25.

Harris, C. 2001. 'Archival fieldwork'. *Geographical Review* 91 (1 and 2): 328–34.

Harvey, K. 2006. 'From bags and boxes to searchable digital collections at the Dalhousie University Archives'. *Journal of Canadian Studies* 40 (2): 120–38.

Kurtz, M. 2001. 'Situating practices: The archives and the file cabinet'. *Historical Geography* 29: 26–37.

Mayhew, R. 2003. 'Researching historical geography'. In A. Rogers and H. Viles, eds, *The Student's Companion to Geography*, 2nd edn, 260–5. Oxford: Blackwell.

Ogborn, M. 2003. 'Knowledge is power: Using archival research to interpret state formation'. In A. Blunt et al., eds, *Cultural Geography in Practice*, 9–20. London: Arnold.

———. 2006. 'Finding historical data'. In N. Clifford and G. Valentine, eds, *Key Methods in Geography*, 101–115. London: Sage.

Ventresca, M., and J. Mohr. 2005. 'Archival research methods'. In J. Baum, ed., 'The Blackwells Companion to Organizations'. Blackwell Reference Online. http://www.blackwellreference.com/subscriber/tocnode?id=g9780631216940_chunk_g978063121694040.

Notes

1. Much of this chapter also relates to photographs, but this sub-field has a literature of its own—for example, Schwartz and Ryan (2003) and Quanchi (2006).

2. A discussion of the behind-the-scenes work of the archivist in appraising, arranging, and describing records is provided by Harvey (2006).

3. Indeed, you may be dealing with records in another language. For instance, there is good deal of correspondence in Te Reo Maori in Archives New Zealand, much but not all of which is accompanied by an English version prepared by official translators.

4. These guidelines are expanded on in the American Historical Association's 'Statement of standard of professional conduct', which is available on-line at http://www.historians.org/PUBS/Free/ProfessionalStandards.cfm.

5. Given that this volume focuses on qualitative methods, I have omitted discussion of how a researcher might extract and present in tabulated form quantitative information derived from archival sources.

Using Questionnaires in Qualitative Human Geography

10

Pauline M. McGuirk and Phillip O' Neill

CHAPTER OVERVIEW

This chapter deals with questionnaires, an information-gathering technique used frequently in mixed-method research that draws on quantitative and qualitative data sources and analysis. We begin with a discussion of key issues in the design and conduct of questionnaires. We then explore the strengths and weaknesses for qualitative research of various question formats and questionnaire distribution and collection techniques, including on-line techniques. Finally, we consider some of the challenges of analyzing qualitative responses in questionnaires, and we close with a discussion of the limitations of using questionnaires in qualitative research.

INTRODUCTION

Qualitative research seeks to understand the ways people experience events, places, and processes differently as part of a fluid reality, a reality constructed through multiple interpretations and filtered through multiple frames of reference and systems of meaning-making. Rather than trying to measure and quantify aspects of a singular social reality, qualitative research draws on methods aimed at drawing out and interpreting the complexities, context, and significance of people's understanding of their lives (Eyles and Smith 1988). Within this epistemological framework, how can questionnaires contribute to the methodological repertoire of qualitative human geography? This chapter explores the possibilities.

Commonly in human geography, questionnaires pose standardized, formally structured questions to a group of individuals, often presumed to be a **sample** of a broader **population** (see Chapter 4). Questionnaires are useful for gathering original data about people, their behaviour, experiences and social interactions, attitudes and opinions, and awareness of events (McLafferty 2003; Parfitt 2005). They usually involve the collection of quantitative *and* qualitative data. Since such **mixed-method** questionnaires first appeared with the explosion of behavioural geography in the 1970s (Gold 1980), they have been used increasingly to gather more complex data in relation to matters as varied as the environment, social identity, transport and travel, quality of life and community, work, and social networks.

While there are some limitations to the depth and extent of qualitative data that questionnaires are generally capable of gathering, they have numerous strengths. First, they can provide insights into relevant social trends, processes, values, attitudes, and interpretations. Second, they are one of the more practical research tools in that they can be cost-effective, enabling extensive research over a large or geographically dispersed population. This is particularly the case for questionnaire surveys conducted on-line for which printing and distribution costs are minimized (Sue and Ritter 2007). Third, they are extremely flexible. They can be combined effectively with complementary, more intensive forms of qualitative research, such as interviews and focus groups, to provide more in-depth perspectives on social process and context. For instance, Ruming, Mee, and McGuirk's (2004) investigation of the impacts of the social mix policies of the New South Wales Department of Housing on understandings of community combined key informant interviews with housing officials, questionnaires with local residents, and follow-up in-depth interviews with volunteers who had participated in the questionnaire. Data from the questionnaire provided a framework for the in-depth interviews, allowing key themes, concepts, and meanings to be teased out and developed (see Askew and McGuirk 2004, England 1993, and Winchester 1999 for similar examples). In this mixed-method format particularly, questionnaires can be both a powerful and a practical research method. Comparatively, Beckett and Clegg (2007) report on the success of qualitative research into women's experiences of lesbian identity using only carefully designed postal questionnaires to gather rich accounts from respondents. This process was seemingly nurtured rather than constrained by using the questionnaire as a research instrument. It allowed respondents the privacy and time to consider and develop their responses to sensitive questions. This example contradicts the presumption that questionnaires cannot be a powerful means of collecting detailed qualitative data in many research contexts.

QUESTIONNAIRE DESIGN AND FORMAT

While each questionnaire is unique, there are common principles of good design and implementation. Producing a well-designed questionnaire for qualitative research involves a great deal of thought and preparation, effective organization-al strategies, and critical review and reflection, as an array of literature suggests (for example, de Vaus 2002; Dillman 2007; Fink and Kosecoff 1998; Foddy 1993; Fowler 2002; Gillham 2000; Lumsden 2005; see also the relevant chapters in Babbie 2001; Clifford and Valentine 2003; Flowerdew and Martin 2005; Hoggart, Lees, and Davies 2002; and Sarankatos 2005). The design stage is where a great deal of researcher skill is vested, and it is a critical stage in ensuring the worth of the resulting data. Notwithstanding the quality of the questionnaire devised, we are beholden as researchers to ensure that we have sufficient reason to call on the time and energy of the research participants. The desire to generate our 'own' data on our research topic is insufficient justification (Hoggart, Lees, and Davies 2002). As with any study, the decision to go ahead with a questionnaire needs to be based on careful reflection on detailed research objectives, consid-eration of existing and alternative information sources, and appropriate ethical contemplation that is attuned to the particular cultural context of the research (see Chapters 2, 4, and 16).

The content of a questionnaire must relate to the broader research question as well as to your critical examination and understanding of relevant processes, concepts, and relationships. As a researcher, you need to familiarize yourself with relevant local and international work on your research topic. This ensures clarity of research objectives and will help you to identify an appropriate participant group and relevant key questions. You need to be clear on the intended purpose of each question, who will answer it, and how you intend to analyze responses. You also need to be mindful of the limits to what people are willing to disclose, being aware that these limits will vary across different social and cultural groups in different contexts. Public housing tenants, for instance, might be wary about offering candid opinions about their housing authority. Respondents might be cautious about what they are willing to disclose in questionnaires administered via e-mail because of the loss of anonymity that occurs when e-mail addresses can be matched with responses (Van Selm and Jankowski 2006). Every question, then, needs to be carefully considered with regard to context and have a clear role and purpose appropriate to the social and cultural norms and expectations of the participant group (Madge 2007).

Begin by drawing up a list of topics that you seek to investigate. Sarantakos (2005) describes the process of developing questions for a questionnaire as a pro-cess of translating these research topics into variables, variables into indicators, and indicators into questions. Identify the key concepts being investigated, and work out the various dimensions of these concepts that should be addressed. Then identify indicators of the dimensions, and use them to help you formulate specific questions. Doing this will ensure that each question relates to one or more aspects of the research and that every question has a purpose. De Vaus (2002) suggests that it is helpful to think about four distinct types of question content:

1. *attributes:* Attribute questions aim at establishing respondents' characteristics (for example, age or income bracket, dwelling occupancy status, citizenship status).
2. *behaviour:* Behaviour questions aim at discovering what people do (for example, recreation habits, extent of public transport use, food consumption habits).
3. *attitudes:* Questions about attitudes are designed to discover what people think is desirable or undesirable (for example, judgment on integrating social housing with owner-occupied housing, willingness to pay higher taxes to fund enhanced social welfare services).
4. *beliefs:* Questions about beliefs aim at establishing what people believe to be true or false or preferred (for example, beliefs on the importance of environmental protection, beliefs on the desirability of social equity).

A guiding principle for question types, however, is that you need to be sure that your target participant group will understand the questions and have the knowledge to answer them. As is the case in newsprint journalism, it is recommended that unless you are targeting a specialized and homogenized group, you phrase questions to accommodate a reading age of approximately 11 years (Lumsden 2005). Respondents also need to be capable of answering the questions: for instance, it might be beyond their ability to comment with any certainty on whether government planning policies have contributed to local coastal degradation, leading them to abandon the questionnaire.

Apart from the typology of question content, there is a range of question types from which to draw. We commonly make a distinction between closed and open questions, each of which offers its own strengths and weaknesses and poses different challenges depending on the mode through which the questionnaire is being administered (e.g., mail, face-to-face, e-mail). **Closed questions** may seek quantitative information about respondent attributes (for example, level of educational attainment) or behaviour (for example, how often and where respondents buy groceries). You should provide simple instructions on how to answer closed questions (e.g., how many responses the respondent can tick). Some examples are set out in Box 10.1. Closed questions can ask respondents to select categories, rank items as an indicative measure of attitudes or opinions, or select a point on a scale as indicative of the intensity with which an attitude or opinion is held (see Sarankatos 2005, ch. 11). A major benefit of closed questions is that the responses are easily coded and analyzed, a bonus when interpreting a large number of questionnaires. Indeed, for web-based questionnaires, a data file can be assembled automatically as respondents type in their answers. Closed questions are demanding to design, however, since they require researchers to have a clear and full understanding of what the range of answers to a question will be. Respondents' answers are confined to the range of categories designed by the researcher as an exhaustive and exclusive list of possible answers, and this can be a limitation. It has also been found that when respondents are asked to 'tick all appropriate categories' on a list (see the category list question in Box 10.1), they can turn to **satisficing behaviour.** That is, they keep reading (and ticking) until they feel they have provided a satisfactory answer and then stop. This is a particular issue for web-based questionnaires in which limited screen space means that all categories cannot be viewed on one screen (Dillman, Tortora, and Bowker 1998).

A greater potential limitation of closed questions is that they rest on the assumption that words, categories, and concepts carry the same meaning for all respondents and this may not always be the case. For example, how a respondent answers the question 'How often have you been a victim of crime in the past two years?' will depend on what the respondent includes in their definition of a crime (de Vaus 2002). It is worthwhile to be aware, too, that the ways particular questions are posed or how they relate to preceding questions can influence respondents' answers. For instance, Babbie (2001) demonstrates how greater support in questionnaire surveys is indicated habitually for the option worded as 'assistance to the poor' rather than as 'welfare' and for 'halting rising crime rate' rather than 'law enforcement'. A criticism of closed questions, then, is that 'one may learn more about the behaviour of the sample in responding to a set a categories . . . than about the behaviour under investigation' (Cox 1981, 264). This limitation can be lessened by offering an answer option such as 'other (please specify)' or by using **combination questions** that request some elaboration on or explanation of the selection made in a closed question (see Box 10.1).

In general, **open questions** have greater potential to yield in-depth responses in keeping with the thrust of qualitative research: to understand how meaning is attached to process and practice. Open questions offer less structured response options than closed questions, inviting respondents to recount understandings, experiences, or opinions in their own terms. Rather than offering alternative answers, which restrict responses, they provide space (and time) for free-form responses. Open questions also 'give voice' to respondents and allow them to question the terms and structure of the questionnaire itself, demonstrate an alternative interpretation, and add qualifications and justifications. This capacity acknowledges the co-**constitution of knowledge** by researcher and research participant (Beckett and Clegg 2007). For instance, in a questionnaire used by Winchester, McGuirk, and Dunn (1997) in Carrington, New South Wales, concerning urban redevelopment and its impact on the close-knit nature of the stable community being researched, a respondent pointed out that it could not be assumed that a stable community implied a close-knit community, as the questionnaire

Types of Questionnaire Questions

Box 10.1

Closed questions

Attribute information
How often do you shop at this shopping mall? (please tick the appropriate box)
- Less than once a week ☐
- Once a week ☐
- Twice a week ☐
- More than twice a week ☐

Category list
What was the *main* reason you chose to live in this neighbourhood? (please tick the appropriate box)
- Proximity to work ☐
- Proximity to family and friends ☐
- Proximity to schools or educational facilities ☐
- Proximity to shopping centre ☐
- Proximity to recreational opportunities ☐
- Environment ☐
- Housing costs ☐
- Good place to raise children ☐
- Pleasant atmosphere of neighbourhood ☐
- Other (please specify)

Rating
Please rank the reasons for buying your current house (please rank all relevant categories from 1 [most important] to 6 [least important]).
- Price ☐
- Location ☐
- Size ☐
- Proximity to job/family ☐
- Investment ☐
- Children's education ☐

Scaling
Please indicate how strongly you agree/disagree with the following statement (please tick the appropriate box):
Having a mix of social groups in a neighbourhood is a positive feature.
- Strongly disagree ☐
- Disagree ☐
- Neutral ☐
- Agree ☐
- Strongly agree ☐

Grid/matrix question

Think back to when you first got involved in environmental activism. What initially inspired you to get involved? (please tick the appropriate box for *each* reason)

	Very influential	*Fairly influential*	*Not very influential*	*Not influential*
Spirituality/ religious beliefs				
Fear/anxiety about ecological crisis				
Desire to change the world				
Nature/ecology experiences and care for the environment				
Political analysis				
Commitment to justice				
Felt like you could make a difference				
Influential person (please specify)				
Influential book/ film (please specify)				
Key event (please specify)				
Contact with an organization, campaign, or issue (please specify)				
Outreach activities by an organization (please specify)				
Wanted to meet new people				
Want to learn new skills				
Sense of personal responsibility				
Other (please specify)				

Combination question

Have changes in the neighbourhood made this a better or worse place for you to live? (please tick the appropriate box)

Changes have made the neighbourhood better ❑

Changes have not made the neighbourhood better or worse ❑

Changes have made the neighbourhood worse ❑

Please explain your answer below.

_____

_____

_____

Open questions

What have been the biggest changes to the neighbourhood since you moved in?

_____

_____

_____

_____

_____

_____

What, if any, are the advantages for civic action groups of using the Internet, e-mail, and cellular phones?

_____

_____

_____

_____

_____

Please describe any problem(s) you encounter using public transport.

_____

_____

_____

_____

_____

_____

seemed to suggest. She recounted examples of her own sense of detachment from that community. Open questions, then, are capable of yielding valuable insights, many of them unanticipated, and they can open intriguing lines of intensive inquiry in scenarios where extensive research is the main focus or where a more intensive approach is not possible (Cloke et al. 2004b). Such scope, however, has resulted in open questions being characterized as 'easy to ask, difficult to answer, and still more difficult to analyse' (Oppenheim 1992, 113). An open format means that responses are likely to lack consistency and comparability. Certainly, respondents answer them in terms that suit their own interpretations. So open questions and the responses they yield are certainly more challenging to analyze than are their more easily coded closed counterparts (see Chapter 14). They point to 'the rich yet ambiguous and messy world of doing qualitative research' (Crang 2005b, 231).

In summary, using open questions makes it possible to pose complex questions that can reveal, to a greater depth than closed questions, people's experiences, understandings, and interpretations of, as well as their reactions to, social processes and circumstances. Beyond capturing these accounts, answers to open questions can tell us a good deal about how wider processes operate in particular settings. Thus, they enable research that addresses the two fundamental questions that Sayer (1992) poses for qualitative research: what are individuals' particular experiences of places and events? And how are social structures constructed, maintained, or resisted? (see Chapter 1).

Beyond choice of question content and type, general principles of questionnaire wording, sequence, and format are fundamental to a questionnaire's success. These principles are outlined in Box 10.2. Many of them revolve around clarity, simplicity, and logic. In question wording, you need to be sure that questions are sufficiently precise and unambiguous to ensure that the intent of your question is clear and well communicated. It is advisable to be familiar with the vernacular of the participant group. In on-line contexts, this may include becoming familiar with the jargon, abbreviations, and grammatical rules commonly used within the on-line community being approached (for instance, the language styles of specific newsgroups) (Madge 2007). Remember that the language of a questionnaire is textual but often also graphical and numerical. These languages work together to affect respondents' perception of the survey and are perceived in ways that are influenced by cultural context (Lumsden 2005). The web's capacity for global reach also means that on-line questionnaires may target international participants, not all of whom, of course, can communicate in English. There are software programs that will allow the researcher to convert a questionnaire written in English into other languages (e.g., the Inquisite survey software system; see www.inquisite.com). In addition, there are commercial services capable of translating both the survey and the responses on a commercial basis (e.g., Zoomerang; see www.zoomerang.com) (Sue and Ritter 2007, 84).

Beyond issues of logic, clarity, and comprehension, attention should also be given to ensuring that questions do not threaten or challenge respondents by assuming implicit cultural, ethnic, or religious beliefs, which may arise from researchers' insensitivity, ignorance, or lack of preparation. The need for concern about respondents' **'cultural safety'** (Matthews et al. 1998, 316) is part of the researcher's

Guidelines for Designing Questionnaires   Box 10.2

- Ensure questions are relevant, querying the issues, practices, and understandings you are investigating.
- Keep the wording concise (about 20 words maximum), simple, and appropriate to the targeted participant group's vernacular.
- Ensure that questions and instruction text are easily distinguishable in format and font.
- Avoid double-barrelled questions (for example, 'Do you agree that the Department of Housing should cease building public housing estates and pursue a social mix policy?').
- Avoid confusing wording (for example, 'Why would you rather not use public transport?'), and be alert to alternative uses of words (for example, for some people 'dinner' implies an evening meal while for others it implies a cooked meal, even if eaten at midday).
- Avoid leading questions (for example, 'Why do you think recycling is crucial to the health of future generations?'), and avoid loaded words (for example, 'democratic', 'free', 'natural', 'modern').
- Avoid questions that are likely to raise as many questions as they answer (for example, 'Are you in favour of regional sustainability?' raises questions of what sustainability means, how a region is defined, and how different dimensions of sustainability might be prioritized).
- Order questions in a coherent and logical sequence.
- Ensure the questionnaire takes no more time to complete than participants are willing to spend. This will depend on the questionnaire context (for example, whether it is conducted by telephone, face-to-face, or by post). Generally, 20 to 30 minutes will be the maximum, although longer times (45 minutes) can be sustained if the combination of context and research topic is appropriate.
- Ensure a spacious and uncluttered layout with plentiful space for written responses to open questions.
- Use continuity statements to link questionnaire sections (for example, 'The next section deals with community members' responses to perceived threats to their neighbourhood.').
- Begin with simple questions, and place complex, reflexive questions or those dealing with personal information or sensitive or threatening topics later in the questionnaire.

broader ethical obligations.

The flow and sequence of the questionnaire will be fundamental to respondents' understanding of the research purpose and to sustaining their willingness to offer careful responses and, indeed, to complete the questionnaire to its conclusion. Grouping questions into sections of related questions connected by introductory statements will help here. In general, open-ended questions are better placed towards the end of a questionnaire, by which time respondents are aware of the questionnaire's thrust and may be more inclined to offer fluid and considered responses. In terms of layout, aim for an uncluttered and spacious design that is easy and clear to follow. Where you use closed questions, aligning or justifying the space in which the answer should be provided will contribute to clarity and simplify coding responses in the analysis stage. With open-ended questions, you need to be conscious of the need to leave enough space for respondents to answer without leaving so much as to discourage them from offering a response altogether.

All of these basic questionnaire design principles need to be observed regardless of how the questionnaire is being distributed: whether by mail, face-to-face, by telephone, by e-mail, or on-line. However, the on-line environment presents some additional design factors that are important to consider (Dillman 2007). Web-based completion of a questionnaire makes it possible to incorporate a wide range of design features such as split screens, drop-down boxes, images, and sound tracks, but most of these features require powerful computers, particular software, and ample download time. You need to consider whether the participant group has the ability and the capacity to receive and respond to the questionnaire and its mode of questions. A total website content of less than 60 kilobytes of text and graphics is recommended (Lumsden 2005). You also need to remember that on-line questionnaires require respondents to think about how to respond to the questionnaire while simultaneously thinking about how to operate their computer, a matter that is particularly important if your target participant group is less computer-literate. Keeping things simple and limiting the number of actions a respondent has to undertake to complete a question is sensible. Finally, you need to take account of whether you will administer your questionnaire solely on-line or through other modes as well (e.g., by mail), in which case you need to be mindful of how questions will be posed in those other modes. Box 10.3 outlines additional key principles for the design of on-line questionnaires (adapted from Dillman and Bowker 2001).

Finally, whether developing a conventional or on-line questionnaire, you should write a cover letter or e-mail to be included with posted questionnaires or sent as pre-notification for telephone, face-to-face, or on-line questionnaires. Box 10.4 offers examples. The letter or e-mail needs to provide general information about the purpose of the questionnaire as well as information about

Guidelines for Designing On-line Questionnaires   Box 10.3

- Introduce web questionnaires with a welcome screen providing basic instructions and information and encouraging completion.
- Ensure the first question is interesting to most respondents, is easily answered, and is fully visible on the first screen.
- Use conventional formats for questions, similar to those normally used on self-administered paper questionnaires.
- Provide clear instructions on the computer action needed to respond to each question, and position them at the point where they are needed.
- Limit the length of the questionnaire. The typical length of a paper questionnaire may seem excessively long when completed on a website where a typical print page can take up several computer screens.
- Keep the layout, colour, and graphics simple to ensure navigational flow and readability are maintained across different browsers and screen resolutions.
- Allow respondents to move on to the next question without first having to answer a prior question.
- Allow respondents to scroll from question to question without having to change screens.
- Consider displaying answer categories as a double-bank if the number of answer choices exceeds what can be viewed in a single column on one screen.
- Include graphics or words that indicate how much of the questionnaire the respondent has completed.
- Close with a thank-you screen.

confidentiality, how the respondent was selected for inclusion in the research, how long the questionnaire will take to complete, and when relevant, instructions on how and when to return the questionnaire.

SAMPLING

Before administering a questionnaire, you will need to make a decision about the target audience, or sample. In quantitative research, questionnaires are used commonly to generate claims about the characteristics, behaviour, or opinions of a group of people ('the population') based on data collected from a smaller sample of that population. The population might be, for example, tenants in public housing, the residents of a given local government area, or people living with HIV/AIDS. The sample—a subset of the population—is selected carefully to be

Examples of Invitations to Participate in Questionnaire Studies	Box 10.4

Sample cover letter

School of Geography
Geography Building
East Valley University
Kingsland 9222
Telephone: (04) 89889778
Facsimile: (04) 89889779
E-mail: E.saunders@evu.edu.ca

High-density residential living in Port Andrew, East Valley

I am Edith Saunders, a research student with the School of Geography at the East Valley University. As part of my research on high-density residential environments in East Valley, I am investigating how people understand and create feelings of home in high-density neighbourhoods. The research is being conducted in collaboration with East Valley Council and is aimed at informing its policy and planning decision-making. The work is focused on the Port Andrew area, and you have been selected to receive this questionnaire as a local resident.

The questionnaire asks about the ways you understand and use your home and the ways you interact with your local neighbourhood spaces and services. The questionnaire will take approximately 30 minutes to complete, and completion is voluntary. The questions are asking primarily about your experiences and opinions—there are no right or wrong answers. All answers will be treated confidentially and anonymously—individuals will not be identifiable in the reporting of the research.

It would be appreciated if you could complete the questionnaire at your earliest convenience and no later than July 30. Once you have completed it, please return the questionnaire in the reply-paid envelope provided. Return of the questionnaire will be considered as your consent to participate in the survey.

Your participation is greatly appreciated. Your opinions are important in helping to build understanding of high-density residential living and how it can be supported through local government planning and provision of neighbourhood spaces and services.

Questions about this research can be directed to me at the address provided. Thank you in advance for your participation.

Yours faithfully,
Edith Saunders

The university requires that all participants be informed that if they have any complaints concerning the manner in which a research project is conducted, it may be given to the researcher or, if an independent person is preferred, to the university's Human Research Ethics Officer, Research Unit, East Valley University, 9222, telephone (04) 8988 1234.

Sample e-mail invitation to participate in an on-line questionnaire

From: kanchana.phonsavat@EVU.edu.ca
To: [email address]

Subject: Survey on high-density residential college living

Dear Student,

I am a research student with the School of Geography at East Valley University (EVU). As part of my research, I am investigating how students understand and create feelings of home in high-density residential college environments. The research is being conducted in collaboration with EVU and East Valley Council. You have been selected to receive this invitation to participate as a student resident of one of EVU's residential colleges.

We are interested in the ways you understand and use your college accommodation and the ways you interact with your local neighbourhood spaces and services. The questionnaire will take approximately 30 minutes to complete and is completely voluntary and confidential. The data will be used to evaluate university and council policies and their support of high-density residential environments.

To complete the questionnaire, please click on the following link:

http://www.newurbanliving.evu.org.ca/surveys.html

It would be great if you could complete the questionnaire in the next two weeks. If you have any questions or need help, please e-mail me at kanchana.phonsavat@EVU.edu.ca.

Thank you in advance for your participation.
Kanchana Phonsavat

The university requires that all participants be informed that if they have any complaints concerning the manner in which a research project is conducted, it may be given to the researcher or, if an independent person is preferred, to the university's Human Research Ethics Officer, Research Unit, East Valley University, 9222, telephone (04) 8988 1234.

representative of the population such that the mathematical probability of the sample characteristics being reproduced in the broader population can be calculated (May 2001). In such cases, a list of the population in question, the **sampling frame**, is required so that a sample can be constructed (for example, the tenant list of a given public housing authority, a local electoral register, a health register of all people in a given geographical area receiving treatment for HIV/AIDS). The rules surrounding sampling are drawn from the central limit theorem used to sustain statistical claims to representativeness, generalizability, and replicability (see McLafferty 2003; Parfitt 2005; Robinson 1998).

On the other hand, questionnaires used in qualitative research are likely to be used as a part of mixed-method research aimed at establishing trends, patterns, or themes in experiences, behaviours, and understandings as part of analysis of a *specific context*, without seeking to make generalizable claims about whole populations (Robinson 1998, 409). Hence, a more appropriate sampling technique for qualitative research is non-**probability sampling** where generalization to a broader population is neither possible nor desirable and sampling frames may not, in any case, be available. Some web surveys, for instance, involve the self-selection of respondents where anyone who agrees to complete the questionnaire can be included in the sample. For example, Tomsen and Markwell's (2007) research in Australia into the perception and experience of safety at gay and lesbian events included an on-line questionnaire. Respondents were invited to complete the questionnaire through targeted advertising in the gay and lesbian press, a media release, radio interviews, and contacting and giving information about the questionnaire to 25 on-line chat groups and e-mail lists to pass on to their members. A total of 332 people from across the country participated in the questionnaire. Specifically, **purposive sampling** (see Chapter 4) is commonly used where sample selection for questionnaire respondents is made according to some known common characteristic, be it a social category (for example, male single parents), a particular behaviour (for example, women who use public transport), or an experience (for example, people who have been victims of crime). There are no specific rules for this type of sampling. Rather, the determinants of the appropriate sample and sample size are related to the scope, nature, and intent of the research and to the expectations of your research communities.

As in all research, these considerations are overlain by the limitations of resource constraints (time and money). Nonetheless, the lack of hard-and-fast rules and a need for pragmatism do not imply the absence of a systematic approach— quite the opposite. Complex and reflexive decisions need to be made about how to approach sampling. For instance, in research on what motivates 'sea-changers' to abandon city life and relocate to regional, coastal areas, researchers would need to take into account whether they should seek respondents in all age groups, all household types, and all income categories. Research on people living with

HIV/AIDS would need to take into account whether the researchers should target early-stage individuals only, both biological sexes or only one, people of any sexual orientation or a specific sexual orientation, only individuals infected from a particular source, and so on. Each decision is liable to have ramifications for how sample recruitment proceeds and what mode of questionnaire distribution is suitable. Questionnaires administered on-line, for example, may be extremely well suited to research on factors shaping environmental advocacy where the target respondents are highly likely to have web skills and to have access to computers as part of their work. By comparison, this mode of distribution may be poorly suited to research on perceptions of cultural displacement among low-income populations in gentrifying areas. These cases illustrate the fundamental importance of research scope, purpose, and intent in shaping the sampling approach and in determining appropriate sample size. Patton (1990) provides details of various types of purposive sampling, along with a discussion of sample size, and Chapters 4 and 5 in this book provide an extended treatment of further pertinent questions regarding selecting cases and participants in qualitative research. In the end, decisions about samples will be shaped by the compromise between cost, desire for accuracy, the nature of the research, and the limits of possibility.

PRE-TESTING

It is vital to try out a questionnaire before it is distributed. **Pre-testing** is when a questionnaire is piloted or 'road-tested' with a sub-sample of your target population to assess the merits of its design, its appropriateness to the audience, and whether it does in fact achieve your aims. For web-based questionnaires, rigorous testing of the questionnaire on a range of platforms and browsers should be undertaken to identify and weed out potential technical problems. In web-based contexts, technical bugs are very likely to result in the respondent abandoning the questionnaire entirely. Getting feedback through pre-testing from those with extensive questionnaire-design experience and from those who might use the data generated (for instance, in the example in Box 10.4, a local authority and a university) will allow possible problems to be identified or improvements made. Scheduling a pre-testing stage provides the opportunity for post-test revisions that might dramatically increase the questionnaire's effectiveness.

Both individual question items and the overall performance of the questionnaire need attention at this stage. Are individual questions and question instructions easily understood? Would any of them benefit from the addition of written prompts? Do respondents interpret questions as intended? Do any questions seem to make respondents uncomfortable? Discomfort and sensitivity (perhaps the question is considered too intrusive) might be indicated by respondents tending to skip or refuse to answer a specific question or section. Alternatively, such outcomes could mean that respondents do not understand the question or do not have the knowledge or experience to answer it. On the questionnaire overall, consider how respondents react to the order of the questions. Does it seem to them that the questions flow logically and intuitively? Are there parts where the questionnaire seems to drag or become repetitive? Technical aspects can also be tested: Is there enough space for respondents to answer open questions? How long will the questionnaire take to complete? Do the data being generated present particular problems for analysis? If you plan to conduct the questionnaire face-to-face with respondents, the pre-test stage can also be a useful exercise in training and confidence-building.

MODES OF QUESTIONNAIRE DISTRIBUTION

Consideration of the mode of questionnaire distribution should be one of the earliest stages of your questionnaire design. It has significant implications for design, layout, question type, and sample selection. The main distribution modes are mail, face-to-face, telephone, and the Internet-mediated modes of e-mail and the world wide web. Each mode has distinctive strengths and weaknesses, and the choice among them depends on the research topic, type of questions, and resource constraints. The best choice is the one most appropriate to the research context and target participant group, while the success of any particular mode is dependent on a design appropriate to context and participant group. So the question is: what should researchers interested in qualitative research be aware of to guide them in the choice of mode?

Mailed questionnaires have the clear advantages of cost and coverage. They can be distributed to large samples over large areas (for example, an entire country or province) at a relatively low cost. The anonymity they provide may be a significant advantage when sensitive topics are being researched—for example, those dealing with socially disapproved attitudes or behaviours, such as racism or transgressive sexual behaviour, or topics involving personal harm, such as experience of unemployment or experience of crime. Respondents may also feel more able to take time to consider their responses if unimpeded by the presence of an interviewer. Clearly, too, the absence of an interviewer means responses cannot be shaped by how an interviewer poses a question, how they interact with the respondent, or how they interpret cues in the conversation in culturally specific ways.

Nonetheless, mailed questionnaires are generally the most limited of the three modes in terms of questionnaire length and complexity. The scope for complex open questions is particularly limited by the need for questions to be self-explanatory and brief, and this may be a significant consideration for qualitatively oriented research. Once the questionnaire is sent out, there is little control over who completes it or, indeed, over how it is completed; respondents may choose to restrict themselves to brief, unreflective, or patterned responses. A response to the question 'what do you value about living in this community?' might yield a response of several paragraphs from one respondent and the comment 'friends and neighbours' from another. There is no opportunity to clarify questions or probe answers. Nor is there control over the pattern and rate of response. Some parts of the target participant group may respond at a higher rate than others. It is common, for instance, for mailed questionnaires to achieve significantly higher response rates in wealthy neighbourhoods than in less socially advantaged neighbourhoods. Finally, mailed questionnaires can be subject to low response rates unless respondents are highly motivated to participate. Response rates of 30 to 40 per cent are considered good (Cloke et al. 2004b), although effective follow-up steps can increase a rate somewhat (May 2001; Robinson 1998).

Distributing questionnaires electronically is a recent variation on mail distribution. There are three main means of electronic distribution: (1) sending the entire questionnaire to respondents as an e-mail attachment, (2) e-mailing respondents an introductory letter with a hyperlink to a web-based questionnaire, and (3) distributing a general request for respondents (for example, via an on-line newsgroup) to complete a web-based questionnaire. A major benefit of electronic distribution is that it 'compresses' physical distance and enormously expands the reach of the questionnaire. Participant groups that are otherwise difficult to contact with paper questionnaires can be reached. This could include, for example, people with restricted mobility who might find it easier to respond on-line than to mail a completed questionnaire. Furthermore, people practising covert or illegal behaviours—for example, graffitists or drug users—may be more easily recruited through the Internet. The Internet is also a powerful way of gaining access to self-organized groups—for example, those with common interests, lifestyles, or experiences organized into chat-rooms, newsgroups, and on-line forums. Mailing lists or on-line newsgroups can be used for circulating the questionnaire or inviting participants to complete an on-line questionnaire. However, some groups are sensitive to the intrusion of researchers via mailing lists and newsgroups to request participation (Chen, Hall, and Johns 2004). Many discussion groups state their privacy policy when you join, so researchers should check the welcome message of public discussion lists for guidelines before using them to recruit potential participants (Madge 2007).

Regardless of the specific means of electronic distribution used, the recruitment of participants will be affected by the age, class, and gender biases that shape computer use, e-mail and on-line patronage (see Gibson 2003). Careful consideration needs to be given to whether the coverage of who can be reached is problematic for the research in hand.

Other benefits of electronic distribution include the cost-savings and efficiencies of e-mail and on-line questionnaires. One distinct advantage is the ability to incorporate colour images and graphics without associated printing costs, although you should avoid overloading on-line questionnaires with cluttered design features or complex graphics that require excessive download time. Electronic distribution opens up new opportunities for flexibility in question design, for more complex questions to be posed, for incorporating adaptive questions with encoded skip patterns, thus removing the need for complex instructions and filter questions, and for increasing the potential to generate rich qualitative data. Researchers who have taken advantage of electronic distribution report response rates comparable with those of conventionally distributed questionnaires, especially if pre-notification e-mails are used, with respondents characteristically submitting lengthy commentaries on open questions (Hoggart, Lees, and Davies 2002; Sheehan and McMillan 1999; Van Selm and Jankowski 2006), a plus for qualitative research. Apart from saving on print and postage costs, the electronic collection of data offers the major advantage over paper questionnaires of eliminating the need for a separate labour-intensive phase for data entry and coding of closed questions (Van Selm and Jankowski 2006).

Mailed and on-line questionnaires do, however, present a particular set of challenges surrounding their hidden costs, privacy issues, and technical capacities and failures. The cost and labour savings of avoiding coding and data entry through electronic data capture can be offset by the costs of design and programming (Hewson et al. 2003). To run a web-based questionnaire, you need to be proficient in producing HTML documents, to use survey construction software packages, which can be costly, or to use the commercial services of a web survey host (see Sue and Ritter 2007, Appendix A). Costs can vary significantly. When it comes to privacy and confidentiality, the identity of web-based questionnaire respondents can be protected if they withhold their names, although technically adept researchers can collect data about web-based participants using, for example, user log files or Java Applets (Lumsden 2005). Anonymity cannot be provided to e-mail questionnaire respondents when the returned questionnaire attaches the e-mail address. Responses stored on computer files, and on-line, can be accessible to hackers, and this may be a particularly important concern if the study being conducted involves sensitive and personal subjects. Using encryption to increase the security of data on a server and backing up and storing data in a secure off-line location are advisable.

Qualitative research is often very effective if an interviewer administers face-to-face questionnaires, although this is a costly option. The major benefits of this mode flow from the fact that an interviewer's presence allows complex questions to be asked (see Chapter 6). As well, an interviewer can take note of the context of the interview and of respondents' non-verbal gestures, all of which add depth to the qualitative data collected (Cloke et al. 2004b; May 2001). As an interviewer, you can motivate respondents to participate and to provide considered, qualitatively informative responses. Moreover, people are generally more likely to offer long responses orally than in writing. However, as Beckett and Clegg's (2007) work on lesbian identity suggests, this outcome is context-dependent. Perhaps more crucially, face-to-face questionnaires give an interviewer the opportunity to clarify questions and probe vague responses (see Chapters 6 and 7 for related discussions). For example, adding probes like 'why is that exactly?', 'in what ways?', or 'anything else?' can elicit reflection on an opinion or attitude. Long questionnaires can also be sustained because direct contact with an interviewer can enhance respondent engagement. The ability to pose complex questions and elicit more in-depth and engaged responses is a major benefit for qualitative research. Moreover, this high level of engagement can also secure high response rates—Babbie (2001) estimates 80 to 85 per cent—with a minimal number of nil responses and 'don't know' answers. However, the level of interviewer skill and reflexivity required to secure optimal outcomes should not be underestimated.

As Kevin Dunn discusses more fully in Chapter 6, the presence of an interviewer can be a powerful means of collecting high-quality data, but it introduces limitations as well. Interviewer/respondent interaction can produce 'interviewer effects' that shape the responses offered. People filter their answers through a sense of social expectation, especially when interviewed face-to-face (Lee 2000). They may censor their answers according to perceived social desirability. That is, they may avoid revealing socially disapproved behaviours or beliefs (such as racism) or revealing negative experiences (for example, unemployment). Beckett and Clegg (2007) chose postal questionnaires specifically to ensure the *absence* of an interviewer. Their argument was that participants should be allowed to recount their stories in their own terms, without any identification with the researchers' associations with particular geographical spaces or social and cultural attributes and without fear of judgment by the researcher. When interviewers are used, one means of dealing with respondents' self-censoring is to incorporate a self-administered section in the questionnaire or to reassure respondents of a guarantee of anonymity. Moreover, the interviewer's presence (as an embodied subject with class, gender, and ethnic characteristics) can also affect the nature of responses given. For instance, Padfield and Procter (1996) suggest that the gender of the interviewer introduces significant variations. So while distinct benefits arise from using face-to-face distribution, there are drawbacks. Perhaps the most limiting is the practical consideration of cost. Interviewer-administered questionnaires are expensive and time-consuming and tend to be restrictive both spatially and with respect to population coverage. However, as we suggested before, this factor may not be a significant drawback if a particular, localized participant group is targeted.

While the opportunities for personal interchange are more restricted in telephone than in face-to-face questionnaires, the telephone mode still offers the possibility of dialogue between researcher and respondent and can provide some of the associated benefits along with a certain anonymity that may limit problematic interviewer effects. Conducting questionnaires over the phone may encourage respondent participation because it may be seen as less threatening than opening the door to a stranger wanting to administer a questionnaire. However, telephone delivery constrains the scope for lengthy questionnaires, with about 30 minutes being the maximum time respondents are willing to participate (de Vaus 2002). Furthermore, because the mode relies on a respondent's memory, the question format must be kept simple and the number of response categories in closed questions needs to be limited. However, the advent of **computer-assisted telephone interviewing (CATI)** and **voice capture** technology is significantly enhancing telephone questionnaires (see Babbie 2001, 265) and extending their potential in this regard. Moreover, they can be administered with great convenience and at relatively low cost.

Telephone questionnaires may rely on a telephone directory as a sampling frame, and this can introduce class and gender biases among respondents as well as ruling out people whose numbers are not listed. Moreover, as cellular phone use increases, landline directories are becoming less useful as a sampling frame. If telephone numbers are available for a purposefully selected group of people, this may not pose a problem. Historically, telephone surveys have had good response rates (Feitelson 1991), and follow-ups can be conducted much more conveniently than for face-to-face or mailed questionnaires by means of a simple call-back. However, growing public annoyance with unsolicited marketing calls means approaches by telephone face rejection or screening out by answering machines.

MAXIMIZING QUESTIONNAIRE RESPONSE RATES

Questionnaire response rates are shaped by the research topic, the nature of the sample, and the quality and appropriateness of questionnaire design as much as by the mode of distribution. In any case, questionnaire response rates tend to be higher when using a purposive sample—as is common in qualitative research—wherein interest in the research topic may be strong. There is strong evidence that response rates for on-line questionnaires are stronger if the questionnaire is relatively brief, taking no longer than 20 minutes to complete, is not overly complex to complete, is relatively simple in design, and does not require participants to identify themselves (Lumsden 2005). Regardless of the mode of distribution, response rates can be improved by undertaking a series of strategies before questionnaire distribution and as follow-up (Dillman 2007) in order to both maximize participation and minimize non-responses. Box 10.5 summarizes the key strategies and indicates for which modes of distribution they are appropriate.

Strategies for Maximizing Response Rates   Box 10.5

Strategy	Face-to-Face	Telephone	Mail	On-line
Ensure mode of distribution is appropriate to the targeted population and research topic.	√	√	√	√
Send notification letter (or e-mail pre-notification) introducing the research and alerting to the questionnaire's arrival (or posting on-line).		√		√
Place newspaper or on-line advertisement in local community newspaper/ magazines or on-line chat-rooms/newsgroups introducing the research and alerting to the conduct of the questionnaire.	√	√	√	√
Ensure questionnaire is concise.	√	√	√	√
Ensure appropriate location of approach.	√			
Ensure appropriate time of approach.	√	√		
Vary time if no contact is made initially.	√	√		
Pre-arrange time/location for conduct of questionnaire, if appropriate.	√	√		
Ensure reply-paid envelope is included in mail-out.			√	
Print questionnaire on coloured paper to distinguish it from introductory material or other mail.			√	
Send follow-up postcard/e-mail thanking early respondents and reminding others (about one week after initial receipt).			√	√

Send follow-up letter/e-mail and additional copy of questionnaire (two to three weeks after initial receipt).			√	√
Avoid abrasive manner.	√	√		
Dress appropriately to the target population.	√			

ANALYZING QUESTIONNAIRE DATA

Analyzing questionnaires used in mixed-method research that blends qualitative and quantitative data requires an approach that distinguishes between closed questions in which responses are provided in an easily quantified format and open questions that seek qualitative responses. Quantitative data arises primarily from closed questions that provide counts of categorical data (for example, age and income bands, frequency of behaviour) or measures of attitudinal or opinion data (see Box 10.1 for examples). Questions such as these are relatively easy to code numerically and analyze for patterns of response and relationships between the variables that the questions have interrogated (May 2001). Indeed, as noted above, response categories can be pre-coded on the questionnaire, simplifying matters even further (see de Vaus 2002 or Robinson 1998 for more detail), while data can be collected readily and easily collated within the electronic environment. The analysis of qualitative responses is more complex. The power of qualitative data lies in its revelation of a respondent's understandings and interpretations of the social world, and these data, in turn, are interpreted by the researcher to reveal the understandings of structures and processes that shape respondents' thought and action (for elaboration, see Crang 2005b; Robinson 1998, 426–7). Chapters 11, 14, and 17 discuss the techniques and challenges of coding and analyzing qualitative data in detail. Nonetheless, it is worth raising some important points specific to analyzing qualitative data arising from questionnaires.

In qualitative responses, the important data often lie in the detailed explanations and precise wording of respondents' answers. For qualitative research, then, it is best to avoid classifying qualitative responses into simple descriptive categories so as to report on them quantitatively, stating, for example, that '49 per cent of respondents had positive opinions about their neighbourhood'. There are two problems here. First, such reporting gives the misleading impression that findings are quantitative and could be used to make generalizations. It may well be statistically misleading, too, to report in this form the results of what might be a relatively small purposive sample. Second, this approach involves 'closing' open questions so that much of the richness of how respondents constructed, in this example, their positive understandings and experiences of their locality, is lost. Certainly, classifying qualitative responses into descriptive categories allows us to simplify, summarize, compare, and aggregate data, but this kind of approach forfeits the nuance and complexity of the original text. Reporting on observations in this way is unlikely to contribute much to our understanding of the meanings and operations of social structures and processes and people's interpretations and behaviour in relation to them. It is more attuned to the thrust of qualitative research to analyze data gained from a questionnaire by sifting and sorting to identify key themes and dimensions as well as the broader concepts that might underlie them (see the discussion of analytical coding in Chapter 14). Reporting findings in these terms is much more meaningful than falling back on awkward attempts to quantify the data.

Further, in analyzing qualitative responses, we need to be aware that qualitative research makes no assumption that respondents share a common definition of the phenomenon under investigation (be that quality of neighbourhood, experience of crime, understanding of health and illness, and so on). Rather, it assumes that variable and multiple understandings coexist in a given social context. We need to incorporate this awareness into how we make sense of respondents' qualitative answers. Indeed, one of the strengths of using questionnaires in qualitative research can be their ability to identify variability in understanding and interpretation across a selected participant group, providing the groundwork for further investigation through additional complementary methods such as in-depth interviews.

Finally, keep in mind that qualitative data analysis is sometimes referred to as more of an art than a science (Babbie 2001). It is not reducible to a neat set of techniques. Although useful procedures can be followed (see Chapter 14), they may need to be customized to the unique concerns and structure of each questionnaire and the particular balance of quantitative and qualitative data it gathers. For this reason, and others, at all stages of the process of analysis we need to be mindful of engaging in critical reflexivity, especially when considering how our own frames of reference and personal positions shape the ways in which we proceed with analysis (see Chapters 2 and 17).

CONCLUSION

In seeking qualitative data, questionnaires aim not just at determining qualitative attitudes and opinions but at identifying and classifying the logic of different sets of responses, at seeking patterns or commonality and divergence in responses, and at exploring how they relate to concepts, structures, and processes that shape social life. This is no easy undertaking, and questionnaires struggle with the tensions of seeking explanation while being generally limited in their form and format to obtaining concise accounts.

Hoggart, Lees, and Davies (2002) argue that the necessarily limited complexity and length of questionnaires prevent them from being used to explain action (since this requires us to understand people's intentions), the significance of action, and the connections between acts. Compared with the depth of information developed through more intensive research methods such as in-depth interviews, focus groups, or participant observation, questionnaires may provide only superficial coverage. Nonetheless, they can help us begin the explanation in that they are useful for identifying regularities and differences and highlighting incidents and trends (see de Vaus 1995 for an extended critique). Indeed, as Beckett and Clegg's (2007) work shows, in some contexts they can enable the collection of full and frank, thoughtful and detailed accounts in ways that more intensive methods involving interviews and interviewers' presence may inhibit.

There are ways of constructing and delivering effective questionnaires that are largely qualitative in their aspirations, being mindful of the possibility of acquiring deep analytical understandings of social behaviours through careful collection of textual materials. Certainly, the interview, through its record of close dialogue between researcher and respondent, provides a particularly powerful way of uncovering narratives that reveal the motivations and meanings surrounding human interactions, and questionnaires can only ever move incompletely in this direction. However, by not requiring close and prolonged engagement with the research subject, the questionnaire offers opportunities to reach a wide range and great number of respondents, in particular through on-line applications, and to collect data on their lived experiences. This extensiveness and diversity makes questionnaires an important, contemporary research tool.

KEY TERMS

closed questions
co-constitution of knowledge
combination questions
computer-assisted telephone interviewing (CATI)
cultural safety
mixed-method research
open questions

population
pre-testing
probability sampling
purposive sampling
sample
sampling frame
satisficing behaviour
voice capture

REVIEW QUESTIONS

1. Why are open questions more suited to qualitative research than closed questions?
2. Why is the choice of the mode of questionnaire distribution specific to the nature of the sample and the nature of the research topic?
3. Why should we avoid 'closing' open question responses for the purpose of reporting findings?
4. What are the limitations of the use of questionnaires for qualitative research?
5. What are the particular benefits of administering questionnaires on-line?

USEFUL RESOURCES

Babbie, E. 2001. *The Practice of Social Research*. 9th edn. Belmont, CA: Wadsworth. See Chapters 9 and 13.

Cloke, P., et al. 2004a. *Practising Human Geography*. London: Sage. See Chapter 5.

de Vaus, D.A. 2002. *Surveys in Social Research*. 5th edn. Sydney: Allen and Unwin. See Chapters 7 and 8.

Hoggart, K., L. Lees, and A. Davies. 2002. *Researching Human Geography*. London: Arnold. See Chapter 5.

Kenyon, P. 2008. 'Skills package: How to put questionnaires on the Internet, Part 1'. Department of Psychology, University of Plymouth. http://mofetsrv.mofet.macam98.ac.il/~olzang/questionary/forms.htm . This is a two-part guide on how to put questionnaires on the Internet.

Parfitt, J. 2005. 'Questionnaire design and sampling'. In R. Flowerdew and D. Martin, eds, *Method in Human Geography: A Guide for Students Doing a Research Project*. Harlow: Pearson/Prentice Hall. See pp. 78–109.

Sarankatos, S. 2005. *Social Research*. 3rd edn. New York: Palgrave Macmillan. See Chapter 11.

———. 2008. 'Social research 3e'. Palgrave MacMillan. http://www.palgrave.com/sociology/sarantakos/workbook/questionnaires.htm . This is a companion website for Sarantakos's book *Social Research*. It offers a 'workbook' on questionnaire surveys.

Sue, V., and L. Ritter. 2007. *Conducting Online Surveys*. London: Sage.

SurveyMonkey.com. 2008. 'SurveyMonkey.com—because knowledge is everything'. http://www.surveymonkey.com. This is a commercially available web-based interface for creating and publishing custom web surveys and then viewing the results graphically in real time.

SurveyMonkey—The Monkey Team. 2008. 'Smart survey design'. s3.amazonaws.com/SurveyMonkeyFiles/SmartSurvey.pdf. SurveyMonkey's guide to effective design and question-writing for on-line questionnaires.

University of Leeds, Information Systems Services. 2008. 'Tutorial documents'. www.leeds.ac.uk/iss/documentation/top/top2.pdf. A general introduction to the design of questionnaires, including references to on-line support resources.

Doing Foucauldian Discourse Analysis— Revealing Social Realities

11

Gordon Waitt

CHAPTER OVERVIEW

Discourse analysis is now a well-established interpretive approach in geography to identify the sets of ideas, or discourses, used to make sense of the world within particular social and temporal contexts. Discourse analysis is quite different from other qualitative research methods through its use of the challenging ideas of the French philosopher Michel Foucault. Following Foucault, discourse is a mediating lens that brings the world into focus by enabling people to differentiate the validity of statements about the world(s). The goals of this chapter are twofold. The first goal is to outline why Foucauldian discourse analysis is a fundamental component of geographers' methodological repertoire. The second goal is to provide a methodological template. The chapter begins by outlining meanings of discourse. Foucault's interest in discourse was to explain how those statements accepted as 'true' are always historically variable, being the outcome of uneven social relationships, technology, and **power**. According to Foucault, to believe at face value what one hears, reads, or sees as truth would lead to the serious error of overlooking the social circumstances within which particular sets of ideas are produced, circulated, and maintained. Hence, discourse analysis offers insights into how particular knowledge becomes common sense and dominant, while simultaneously silencing different interpretations of the world. The chapter then outlines a methodological template to conduct discourse analysis. Examples are considered to illustrate why discourse analysis has many benefits for geographical research, particularly projects committed to addressing social and environmental injustice and challenging unequal power relationships.

Introducing Discourse Analysis

What is **discourse**? This chapter relies upon Foucault's concept of discourse. Thus, it is important to briefly consider what Foucault understood by this term before moving on to discuss a methodological template for conducting **discourse analysis**. From the outset it should be noted that Foucault's (1972) meaning of 'discourse' is different from conventional linguistic definitions, which understand discourse as passages of connected writing or speech. Nevertheless, providing a clear-cut definition of discourse is complicated, because Foucault employs many definitions. In his work, at least three overlain explanations of discourse can be identified:

1. all meaningful statements or **texts** that have effects on the world;
2. a group of statements that appear to have a common theme that provides them with an unified effect;
3. the rules and structures that underpin and govern the unified, coherent, and forceful statements that are produced.

Foucault's definitions of discourse cohere around the production and circulation of knowledge. He is interested in how particular knowledge systems convince people about what exists in the world (meanings) and determine what they say (attitudes) and do (practices). For Foucault, discourse is defined by rules of knowledge production that outline clear distinctions about what can be said and its degree of validity. Such an approach is labelled **constructionist**—an approach that demands asking questions about the ways in which distinct social 'realities' become naturalized. For example, why in Western societies is science widely understood as the source of factual knowledge? Alternatively, why in Western societies are tropical islands often imagined as earthly paradises or San Francisco as a lesbian and gay homeland? In addition, a constructionist approach demands remaining alert to the ways by which particular knowledge systems categorize individuals, animals, plants, and things. For example, why are certain plants spoken about in positive terms by environmental managers as 'native' while others are commonly categorized in negative terms such as 'weeds' or 'invasive'? Similarly, how do dominant gender discourses in the West constitute understandings of all women athletes as somehow less physically able than men? Simply put, a Foucauldian discourse analysis seeks to uncover the social mechanisms that maintain structures and rules of validity over statements about particular people, animals, plants, things, events, and places.

At the start, it should also be noted that there are many different types of discourse analysis. A website hosted at Sheffield Hallam University, UK, provides a helpful discussion on the range of different discourse analysis techniques and their application in various disciplines (http://extra.shu.ac.uk/daol/howto). Furthermore, it should be noted that discourse analysis is distinct from other forms of qualitative analysis including **semiology** (which considers qualitative source materials in terms of inherent meanings) and **manifest content analysis** (which quantifies the number of occurrences of particular themes or words). So how to begin conducting a Foucauldian discourse analysis?

Interestingly, Foucault struggled with writing 'how to' conduct discourse analysis. He feared that a methodological template would become too formulaic and reductionist. Perhaps it is this absence that has often led others to describe Foucault's methodological statements as 'vague' (Barret 1991, 127). Several handbooks for qualitative methods in the social sciences are equally hesitant to give formal guidelines (Phillips and Hardy 2002; Potter 1996). For a range of different reasons, some scholars argue that guidelines undermine the potential of discourse analysis. For example, it has been suggested that guidelines work in opposition to discourse analysis as a 'craft skill' (Potter 1996, 140), limit the researcher's ability 'to customise' (Phillips and Hardy 2002, 78), or inhibit demands for 'rigorous scholarship' (Gill 1996, 144) and 'human intellect' (Duncan 1987, 473). For these scholars, the maxim is 'learning by doing'. Through a combination of scholarly passion and practice, discourse analysis is typically held to become intuitive. Thus, the methodology is often left implicit rather than made explicit. For scholars who defend Foucauldian discourse analysis as an 'art', any methodological template would be understood as too systematic, mechanical, and formulaic (Burman and Parker 1993). In practical terms, such counsel is not especially helpful for those seeking advice on how to do discourse analysis! To help grapple with the methodological implications of doing a Foucauldian discourse analysis, this chapter draws on the important lessons given by the geographer Gillian Rose (1996; 2001) and the linguist Norman Fairclough (2003).

DOING FOUCAULDIAN DISCOURSE ANALYSIS

Initially, discourse analysis may not be intuitive. Hence, to help conduct a discourse analysis, Rose (2001) provides seven stages through which the technique moves (see Box 11.1). These axioms are not intended to stifle your own interpretations. Instead, they are offered as a starting point for your own thoughtful analysis.

In what follows, each stage is reviewed, using examples from tourism geographies and geographies of sexuality. This choice of examples is not accidental. They illustrate a larger effort of the application of discourse analysis in geography to challenge what can be said, what is worthy of study, and what is regarded as factual. For example, until the 1990s, research on tourism and sexuality was often shunned in the discipline as unworthy of geographical analysis (Binnie and

Valentine 1999). There is nothing intrinsic to sexuality and tourism that makes them more or less relevant for geographers to study than, say, ethnicity or car manufacturing. Instead, dominant approaches in geography made it seem self-evident that the topics of sexuality and tourism should be avoided. The very idea of spaces and bodies being gendered and sexed, let alone mutually constituted through space, was unthinkable. Equally, tourism involved people having fun and so was positioned as having little significance, irrespective of its impacts, within a scientific binary logic of work/play and masculine/feminine. One would obtain greater academic kudos, depth, and authority by quantitatively analyzing the economic geographies of men at work in the industries of steel, textiles, coal, and automobiles, spatialized by the **metaphor** of 'industrial heartlands'.

Strategies for Doing Discourse Analysis   **Box 11.1**

1. Choice of source materials or texts
2. Suspend pre-existing categories: become reflexive
3. Familiarization: absorbing yourself in and thinking critically about the social context of your texts
4. Coding: once for organization and again for interpretation
5. Power, knowledge, and persuasion: investigate your texts for effects of 'truth'
6. Rupture and resilience: take notice of inconsistencies within your texts
7. Silence: **silence** as discourse and discourses that silence

Source: Adapted from Rose 2001, 158.

Choice of Source Materials or Texts

Source materials or texts include: advertisements, brochures, maps, novels, statistics, memoranda, official reports, interview transcripts, paintings, sketches, postcards, photographs, and the spoken word. What kinds of texts are required for discourse analysis? Choosing texts for discourse analysis will in part be informed by the research goals. Here a number of examples will be helpful to illustrate the diversity of applications of discourse analysis and relevant sources materials. Geraldine Pratt (1999) applied discourse analysis to explore the ongoing marginalization of Filipinas in the Vancouver labour market. In the context of neo-liberal political rationalities, Kevin Stenson and Paul Watt (1999) employed discourse analysis to explore how the 'social' was being reconfigured by two local government authorities in southeast England. Kay Anderson (1995) demonstrated the critical insights of discourse analysis when applied to the question of how animals are fashioned for the general public within zoos. Using the case of Adelaide, South Australia, she outlined the changing relationship between animals and humans and understandings of nature by carefully considering the transformation in viewing technologies from the time of menagerie-style caging in the late nineteenth century, through the era of the fairground between the mid-1930s and the early 1960s, up to the contemporary era of naturalistic enclosures. John Urry (2002) investigated implications of the viewing technologies of the tourism industry. Drawing on the example of Niagara Falls, he argued that because of the vast number of tourism industry images instructing visitors on how to gaze at the falls, it is no longer possible for tourists to 'see' the falls themselves. Jessica Carrol and John Connell (2000) conducted a discourse analysis of the songs of the inner-city Sydney-based band, The Whitlams, to examine how lyrics help to evoke a sense of place. Finally, Peter Jackson (1994) undertook a discourse analysis of Ogilvy and Mather's 1980s soft drink advertising campaign for Lucozade involving Olympic athlete Daley Thompson to examine the associations in Britain between sport, sexuality, 'race', and gender. As these examples illustrate, discourses are expressed through a wide variety of source materials.

Different categories of source materials or texts can be further subdivided into **genres**. For example, there are many different genres of movies, including YouTube, home movies, and commercial films. The latter can be subdivided further into comedy, art-house, science fiction, horror, and romance. Similarly, there are different genres of written texts, including biographies, autobiographies, fiction, travel writing, government papers, newspapers, scientific reports, and so on. Being mindful of the genre of text is important, because the producer of each genre or category of text (e.g., journalist, traveller, academic, medical practitioner) is addressing a particular **audience** and may have to conform to a particular style of writing. Each source will be produced, circulated, and displayed by means of a particular technology (such as printing, painting, photography, handwriting, and e-mail).

Your research project may dictate the collection of one particular genre of text— say, tourist maps, music album covers, tourist brochures, holiday photographs, or picture postcards. In that case, to inform the interpretation, you will need to conduct a good deal of background research on the place, text, and author in question. For example, John Goss (1993) explored the 'official' tourism image of the Hawai'ian Islands. Goss collected 34 advertisements commissioned by the Hawai'i Visitors Bureau (HVB) published between 1972 and 1992 in various North American magazines and newspapers. Through conducting background research, he was able to identify how the elaborate Western fiction of Hawai'i as a tropical paradise was constituted for a suburban American audience. The fiction of Hawai'i as a tropical paradise is circulated in a wide range of texts targeting suburban audiences, including movies, novels, and songs. In contrast to understandings of suburbia as homogenous, sexually repressive, and mundane, Hawai'i as a tropical paradise is portrayed as diverse, sexually liberating, and exciting.

Alternatively, your project may require that you collect and interpret several different types of texts. Consider, for example, some work by Kevin Dunn (2001). He examined opposition to the construction of mosques in some Sydney suburbs in the early 1990s. His analysis drew on a diversity of texts, including semi-structured interviews, daily newspapers, the archives of local authorities (including development applications, planners' reports, campaign leaflets, and letters to local authorities from resident objectors), and Islamic guidebooks and other introductory material for converts. His analysis required working across these different texts. In this way, he identified the dominant discourse that helped to stabilize common sense ideas of being and becoming a 'local' resident in particular Sydney suburbs. **Intertextuality** is the term often employed to describe the assumption that meanings are produced as a series of relationships between texts rather than residing within the text itself. An example of intertextuality is the way in which sampling in some contemporary popular music draws on previous songs to enable artists to take advantage of their audiences' knowledge to develop fuller meanings of a song. For example, Madonna's song 'Hung Up' was influenced by the works of Abba ('Gimme, Gimme a Man after Midnight'). Similarly, tourist brochures of tropical destinations may depict semi-naked 'native' women, clad in grass skirts, set against the backdrop of a rainforest. Such gender representations are sampled from a long-standing set of elaborate Western ideas that constitute non-Western women as both exotic and erotic (Goss 1993). Intertextuality acknowledges that meanings are co-created with an active audience.

Your initial problem will most likely be having too many potential source materials. Identifying initial sources with which to begin your analysis is an essential research task. What, then, might inform your choice of texts? The following questions are designed to help you to identify your preliminary or opening texts:

- *Which texts to sample?* In making decisions about selecting or excluding texts, student researchers are often advised to select 'rich' or 'in-depth' texts. For example, Tonkiss (1998, 253) explains, 'What matters is the richness of textual detail.' Attention to richness of texts allows the researcher to interpret the effects of discourse in normalizing understandings. But what qualities make a text qualitatively rich? For example, why are texts comprised of lists of numbers not considered qualitatively rich? While such quantitative data may provide insights into the number of occurrences of particular events or practices or the strength of an attitude, it is virtually impossible to use them to find an interpretation of how and why people understand a place in a particular way. Instead, qualitatively rich texts are those that provide detailed, descriptive insights into how understandings of a place are forged. In turn, taking advantage of the insights into how an individual's understanding of a place is fashioned can then better explain that person's experiences, attitudes, and practices. Meanings may be forged through government reports, magazines, photographs, paintings, song lyrics, and so on. Clearly, for an interview to be categorized as a 'rich' text, it requires more than 'yes' or 'no' answers. A qualitatively rich interview transcript may therefore require longer-term involvement and consideration of particular techniques (see Riley and Harvey 2007 for further discussion on how to enhance the richness of oral histories).

- *How to select texts.* A purposeful sampling strategy may often be deployed initially to help identify texts that are likely to provide insights. For purposeful sampling to work most effectively requires early and ongoing secondary research, but because the research process is a continuing building process, you are likely to find that as the project unfolds, other, unexpected texts may become relevant and important to its success. As a reflexive researcher, you must remain open to such unexpected possibilities. Justifying why a particular text is meaningful to your project is an integral step in establishing **rigour**. Such justification is important to writing research that is transparent and open to scrutiny by both the participant and interpretive communities. (See Chapter 4 for additional discussion.)

- *How many texts to select?* As outlined in other chapters of this book, there are no rules for sample size in the context of qualitative research. Whereas in inferential statistics sample size can be prescribed by demands of representativeness, in discourse analysis the number of texts depends on what will be useful and what is *meaningful* in the context of your project. Background research may lead you to the view that the sample size may be as small as one or extend into hundreds of texts. Again, explaining why you have selected a particular number of texts for inclusion in your project becomes an integral step in establishing qualitative rigour (see Chapter 4).

What, then, might be a meaningful choice of texts for investigating the ways that Sydney is sometimes portrayed by the tourism industry as the 'gay capital of the South Pacific'? What might your initial starting point be? What visual or written texts socially construct Sydney as a 'gay and lesbian capital'? What sort of background research of related works might be of assistance? Who might you wish to interview? How many texts do you need to consult? (See Box 11.2.)

Suspend Pre-existing Categories: Become Reflexive

Foucault (1972) regarded the starting point of discourse analysis as reading, listening, or looking at your texts with 'fresh' eyes and ears. The imperative of

Imagine that the goal of your project is to examine the tourism industry's portrayal of Sydney, Australia, as the 'gay capital of the South Pacific'. The portrayal often relies upon Sydney hosting the annual Sydney Gay and Lesbian Mardi Gras and a general understanding among many lesbians and gay men that Oxford Street, Darlinghurst, is Sydney's lesbian and gay homeland. First: which initial texts to select? In the first instance, you may find tourism statistics for Sydney. While these statistics will tell you about visitor arrival, motivation, and spending patterns, the sexuality of visitors is not likely to be documented. Nor will the figures include detailed, descriptive insights. Hence, travel statistics alone are not particularly helpful for this project. Instead, interviews with managers at Tourism New South Wales as well as the stories of travellers and travel reports from the lesbian and gay media may be a more useful starting point. Supplementary texts may include the marketing pitches of Tourism New South Wales and the (New) Gay and Lesbian Mardi Gras organization. Second: how to select texts? Some background research may help in this process. In this case, the work of Lynda Johnston (2001), Kevin Markwell (2002), and Gordon Waitt and Kevin Markwell (2006) may be a helpful starting point. Reading this research may alert you to the paradoxical understandings of Sydney as a 'gay capital'. For example, while the Sydney Gay and Lesbian Mardi Gras is a celebration of non-normative sexualities, the audience is primarily heterosexual people. Hence, in selecting your texts, you may decide to focus on a particular target audience of the marketing campaigns of Tourism New South Wales. Alternatively, you may wish to focus on portrayals of gayness within the tourism industry. Secondary research suggests that the gay tourism industry reproduces a very mainstream, clean-cut understanding of gay. Previous tourism campaigns have tended to help make lesbians, gay men of colour, and certain subculture groups, such as 'bears' or 'leathermen', invisible. Hence, pertinent questions in the selection process include: Who is portrayed in the marketing campaigns? What versions of gayness are marketed?

shelving preconceptions is underpinned by the objective of discourse analysis—to disclose the created 'naturalness' of constructed categories, subjectivities, particularities, accountability, and responsibility. Foucault (1972, 25) pointed out that all preconceptions

> must be held in suspense. They must not be rejected definitively, of course, but the tranquility with which they are accepted must be disturbed; we must show that they do not come about by themselves, but are always the result of a construction the rules of which must be known and the justifications of which must be scrutinized.

Foucault acknowledges that this request to defer pre-existing categories is an impossible task. It is unattainable because, according to Foucault, all knowledge is socially constituted. There is no independent position from which to suspend pre-existing knowledge. Instead, he calls for researchers to become self-critically aware of the ideas that inform their understandings of a particular topic. It is possible to begin to implement his call in at least three ways. First, deploy the techniques of reflexivity outlined by Robyn Dowling (see Chapter 2). As a critically reflexive researcher, it is imperative to discuss why you selected a particular research topic (context) and your initial ideas about the topic (partiality). The term **positionality statement** refers to a researcher's explicit report 'locating' their lived experiences (**embodied knowledge**) within a project. Second, remain alert to Gillian Rose's (1997) warning that it may be impossible to ever fully locate oneself in a research project. The ideas we carry into the work may change while conducting the research. Hence, a vital part of a positionality statement is noting and reflecting upon these changes as the project unfolds. One crucial question to ask is: how has the research process made different aspects of your subject position irrelevant/relevant or visible/invisible? Third, as discussed by Matt Bradshaw and Elaine Stratford (see Chapter 4), you must endeavour to keep careful and transparent documentation of your interpretation process. In particular, illustrate how discourses identified in your project arose from your selected texts. The vital question to ask is: how are the discourses illustrative of the producer's understanding of the world, not the researcher's (Wood and Kroger 2000)? These are three particular steps you should take to ensure that your interpretation must always be considered provisional and does not impose taken-for-granted discourses from elsewhere.

Familiarization: Absorbing Yourself in and Thinking Critically about the Social Context of Your Texts

Familiarization with your texts is essential. A helpful starting point is to begin thinking critically about the 'social production' of your text, becoming alert to its authorship, technology, and intended audience. These social dimensions of a text are a good starting point for a critical interpretation because discourses operate as a process, restricting not only what can be said about the world but also who can speak with authority (Wood and Kroger 2000). Foucault understood discourses to be grounded within social networks in which groups are empowered and disempowered relative to one another. He saw discourse as subtle forms of social control and power. One effect of discourse is the privileging of relatively powerful social groups. That is, particular voices and technologies are favoured over others, often counted as sources of 'truthful' or 'factual' knowledge, while other voices may be excluded and silenced, perhaps by becoming positioned as untrustworthy, anecdotal, hearsay, or folklore. Take, for example, the distinction in Australian national parks authorities between indigenous and 'scientific' approaches to environmental management. Until relatively recently, 'objective' scientific knowledge was valued by most national park authorities at the expense of Aboriginal Australian knowledge. Scientific methods were equated with research, objectivity, and environmental management solutions. Oral Aboriginal Australian knowledge was constituted as lacking and was associated with negatively valued concepts, including the irrational and subjective (Lawrence and Adams 2005).

Foucault's ideas on how knowledge about a topic is constituted and maintained within social networks have implications for doing discourse analysis. The key point is that all texts are the outcome of a power-laden process, fashioned within a particular social context. Hence, an integral part of the familiarization process is to conduct background research to help anchor your texts within a particular historical and geographical context. Boxes 11.3, 11.4, and 11.5 provide a series of questions to help you investigate all texts as expressions of knowledge production and a subtle form of social power that constitutes particular social realities.

First, consider authorship as the outcome of a highly social process (Box 11.3). Regardless of whether your text is a transcript, photograph, painting, song, or novel, you must investigate the author's relationship with the intended audience. Hence, you must reflect carefully on what social dynamics are carried into the production of the text and may operate as a subtle form of social control. Box 11.3 provides a starting point to interpreting the relationship between discourse, knowledge, and power. *Who* produced your selected text? Give particular attention to how the author is positioned within historically and geographically specific understandings of gender, class, sexuality, and ethnicity. Remember that like sexuality, gender, class, and ethnicity are not 'natural' categories but are instead constituted through discourse. *When and where* was your selected text produced? *Why* was your selected text produced? Are the social identities of the author or audience portrayed in your selected text: that is, does the text help to naturalize belonging to a particular group in society—like a family snapshot?

Second, ask more specific questions about the social conventions linked to producing the text (Box 11.4). The key point is that you must examine *how* your selected text was produced—in particular, the implications of particular technologies of production. More subtle questions then follow, including thinking carefully about how different categories of texts have their own social histories and geographies.

Garlick (2002), Markwell (1997), and Nicholson (2002) provide helpful discussions of how the technology underpinning film and photograph production can be thought about as a way of constituting understanding of the world as factual knowledge. In Western societies, photography is commonly regarded as a science. Light reflected from the photograph object is etched into chemicals or translated into binary codes. The result possesses the objectivity of a science-

Strategies for Investigating the Social Circumstances of Authorship in Which Discourse Is Produced   Box 11.3

Social circumstances of authorship	Questions	Why is this answer important in the context of establishing or maintaining particular social realities?
Social	• Who made the text? • Who commissioned the text? • Who owns the text? • What are the relationships between the maker, the owner, and the subject of the text? • When was the text made? • Where was the text made? • Why was the text made?	
Technological	• What technologies does the author rely upon to produce the text—for example, the Internet or a camera?	
Content/ aesthetic	• Does the subject matter of the text address the social identities and relationships of its maker, owner, or intended audience?	

based technology. Hence, photographs on Facebook or picture postcards or in brochures, advertisements, or family photograph albums appear to be a 'true' or 'accurate' representation of the 'real' world, involving no subjective intervention. Forgotten are the highly social decisions of production, including choice of subject, lighting, colouring, cutting, cropping, etching, and cloning. When photographs are understood as a reflection of the 'real' world, they actually conceal the opportunities for modification by the photographer for commercial or other purposes. For Foucault, the way that photographs mask the making of particular social realities is one form of disciplinary power that may help to naturalize particular understandings of the world. Waitt (1997) and Waitt and Head (2002) discuss the deployment of different genres of photography by the Australian tourism industry, including postcards and advertisements. Photography is integral to 'circulating' and 'maintaining' particular social realities of places as holiday destinations.

Alternatively, Lisa Law (2000) provides an interesting discussion of how map-making helps to maintain particular sets of ideas. In the context of the Health Department of Cebu, the Philippines, she notes how the making of the official health map of the commercial sex industry naturalized hierarchical, value dichotomies of 'good' and 'bad' women by contributing to the appearance of the so-called 'red-light district' as an empirical 'reality'. The boundary on the map helps to make apparent the Catholic-inspired distinction between marital and commercial sex. Within the marital/commercial binary, women who participate in marital sex are valued at the expense of those who practise commercial sex. In this way, sex workers are equated with negatively valued concepts, including 'poverty', 'dirt', disease', and 'pollution'. The key point is that the technology of mapping is a power-laden process. In this case, mapping technology deployed within the Cebu Health Department resulted in constituting Catholic-inspired thinking about sexual relationships as an empirical reality. In turn, the map may help convince many health officials that all women living in Cebu could be categorized along a simplistic spatialized division of good/bad. An integral part of discourse analysis is remaining alert to the role that different technologies play in such **strategies of conviction**.

Finally, investigate audiences along the lines of the strategies outlined in Box 11.5. Audiences are not pre-given but are again thought of as the outcome of a highly social process. How a text is produced is in part dependent on its intended audience. In other words, an author will draw on particular discourses, mindful of the needs, demands, and fantasies of the intended audience. Hence, audiences can be conceptualized as co-authors of a text.

For example, both Crang (1997b) and Markwell (1997) illustrate the active role of prospective audiences as co-authors of vacation snapshots. They demonstrate how this genre of photograph is more about framing, making, and circulating social realities for a particular audience than it is about 'capturing' or 'mirroring' the real world. Vacation snapshots are the outcome of a highly selective social process influencing the choice of subjects to photograph. At one level, this choice reveals how a person is set within, or challenges, particular common sense understandings of, for example, tourism discourses. Making vacation snapshots is not value-free but reflects particular ideologies and sets of ideas contained in tourist brochures and guidebooks. At another level, each photographer frames a place in a particular way to meet both their own needs and those of their intended audience. The vital point, as Crang (1997b, 362) emphasizes, is that 'It is not a case of pictures showing what is "out there" . . . but rather how objects are made to appear for us'. How vacation snapshots are framed is an example of how individuals make sense of and then, often with a particular audience in mind, communicate particular versions as 'truths' about places.

Strategies for Investigating the Social Circumstances of *the Text* in Which Discourse Is Produced

Box 11.4

Production and circulation of a text	Questions	Why is this answer important in the context of establishing or maintaining particular social realities?
Social	• Does your selected text re-circulate texts found elsewhere? For example, can the same texts be found elsewhere—say, on the Internet, in tourist brochures, or on postcards?	
Technological	• How has technology affected the production of the text? For example, what use is made of print, colour enhancement, photography, digital technologies (airbrushing)?	
Content/ aesthetic	• What is the subject matter? • What is, or are, the genre(s) of your selected text? For example, if a photograph, is it a family snapshot or advertisement? Alternatively, if your text is in an electronic form, is it, for example, a television soap opera? • Is your text one of a series? • What are the conventional characteristics of your selected genre of text? • Is your selected text contradictory, critical, or in some way different from those circulated elsewhere? • What is the 'vantage point' of the viewer/reader in relationship to the text? That is, how is the viewer/reader positioned in relationship to the text?	

Strategies for Investigating the Social Circumstances of *the Audience* in Which Discourse is Produced		Box 11.5

The audience for the text	Questions	Why is this answer important in the context of establishing or maintaining particular social realities?
Social	• Who is/are the original audience(s)? • What are the conventions of how an audience engages with this text? Think about the norms or social conventions of going to the cinema, going to a classical music performance, or attending an art gallery. • How actively does an audience engage with your selected text?	
Technological	• How is your selected text displayed? • How is your selected text stored? • Do the technologies of display/storage affect the audiences' interpretation? • What are the social norms of reading, viewing, or listening to your selected text?	
Content/ aesthetic	• Where is the reader/viewer/listener positioned in relationship to your selected text • What sort of relationship is produced between the reader/viewer/listener and your selected text?	

Of course, Foucault did not envisage audiences as passive recipients of meanings of texts. As emphasized earlier, interpretation of texts is always a socially, spatially, and temporally contingent process. Kathleen Mee and Robyn Dowling

(2003) illustrate how film reviewers—one audience for an Australian film entitled *Idiot Box*—maintained conventional understandings of life for young men in Australian suburbia despite the alternative meanings intended by the producers. Filmed in the western suburbs of Sydney, this is a film about a fictional bank robbery. The planning and implementation of the unsuccessful robbery by the key characters, Mick and Kev, was—according to the writer-director David Caesar—intended to challenge common sense assumptions among cinema audiences about young, unemployed, suburban, working-class men as 'yobbos'. Yet as Mee and Dowling (2003) illustrate, the meanings of *Idiot Box* for film reviewers were shaped by previously circulating discourses presenting young working-class men in the western suburbs of Sydney as 'rebels' and suburbia as 'boring'. In this way, film reviewers sustained one set of understandings about the socio-spatial identities of western Sydney as 'truth'. Clearly, the key point illustrated by the *Idiot Box* example is that while the intended audiences and the social context of viewing/listening/reading play an important role in how narratives are told and in their intensity of engagement, meanings of a text are never singular or uni-directional. Each member of the audience will bring different meanings and power dynamics linked to their lived experiences and personal affiliations.

Coding: Once for Organization and Again for Interpretation

Coding is a process by which researchers structure and interpret qualitative data. When doing discourse analysis, coding serves two primary functions: organization and analysis of texts. Hence, one possible way to start coding is to draw on two different types of codes: descriptive and analytical. In Chapter 14, Meghan Cope provides a helpful discussion of the difference between these two coding structures. For a discourse analysis, descriptive codes offer one initial way of organizing your data using fairly obvious category labels. Drawing on Foucault's ideas, it is helpful to pay particular attention to four types of category labels when developing a list of descriptive codes:

1. context (for example, in an analysis of diaries kept by surfers, *where*, *when*, and *who* participated in surfing);
2. practices (the *events* [for example, what happened while surfing], *interconnections* [for example, who influenced the surfers' style of surfing and how they interact with other surfers and beach-goers], and *actions* [for example, what type of manoeuvres they performed while surfing]);
3. attitudes (for example, statements of judgment about other surfers, beach-goers, the ocean, or sharks);
4. experiences (for example, statements of feelings about surfing or interactions with other surfers or beach-goers).

For each category label, start a list of codes. However, remember that coding is an iterative process. Hence, your initial descriptive codes will change. In some cases, you will need to divide initial codes into finer detail. In other cases, you will have to amalgamate descriptive codes into broader categories. Following each iteration of coding, your coding structure will become more refined.

Alternatively, Rose (2001) suggests content analysis as the starting point for discourse analysis. Content analysis is essentially a quantitative descriptive coding technique. It proceeds with the coding of instances of key themes and then quantifying the priorities of key themes (Lutz and Collins 1993). But what is a key theme? Again, it may be helpful to start with the headings of context, practices, attitudes, and experiences. Thomas McFarlane and Iain Hay (2003) provide a helpful example that illustrates the application of content analysis. In the context of the World Trade Organization (WTO) ministerial conference in Seattle during December 1999, they coded the word clusters used to describe the protests in articles that appeared in Australia's only national newspaper, *The Australian*. They demonstrated how articles in *The Australian* effectively demonized and marginalized anti-WTO protestors. Remember, when doing discourse analysis, descriptive codes, or content analysis, are envisaged only as an initial starting point.

Coding for a Foucauldian discourse analysis also involves devising a list of analytical codes. Analytical codes involve some form of abstraction or reduction. Hence, analytical codes may be envisaged as interpretative themes rather than descriptive labels. In discourse analysis, analytical codes typically provide insights into why an individual or collective holds particular sets of ideas by which they make sense of places, themselves, and others. Take again the example of Goss's (1993) discourse analysis of the Hawai'i Visitors Bureau's (HVB) portrayal of the Hawai'ian Islands for the North American tourism market. For his analysis of the 34 advertisements commissioned by HVB and published between 1972 and 1992, he initially coded the recurring images and words employed to portray the island, including indigenous Hawai'ians, plants, volcanoes, location, and climate. This coding then suggested five analytical themes deployed by the HVB to 'invent', or pitch, the Hawai'ian Islands outside the 'normal' construction of North American geography: earthly paradise, marginality, liminality, femininity, and *aloha*. Goss went on to interpret each of these themes. He argued that the tourism industry reproduced long-standing colonial knowledge about Hawai'i. To provoke travel desire among potential tourists, the advertisers' preferred image of Hawai'i remained with the socially constructed fictional colonial knowledge of a sexually permissive black female sexuality and verdant tropical nature. Hawai'i is portrayed as a timeless location, a portal to the past, inhabited by people who are 'naturally' friendly. In this (socially constructed) timeless and luxuriant tropical island paradise, visitors are promised possibilities to become themselves, free from the repressive regulations of mainland North America.

The lesson for doing discourse analysis is that coding must be implemented as a twofold process, once for organization and again for interpretation.

Power, Knowledge, and Persuasion: Investigate Your Texts for 'Effects of Truth'

Persuasion entails establishing and maintaining sets of ideas, practices, and attitudes as both common sense and legitimate. Foucault sees persuasion as a form of disciplinary power that operates through knowledge. He positions the mutually interdependent relationship between power and knowledge as indistinguishable, arguing that 'Truth isn't outside power. . . . Truth is a thing of this world; it is produced only by virtue of multiple forms of constraint. And it induces regular effects of power' (1980, 131). Hence, questions about the 'truth' of knowledge are fruitless, for truth is unattainable. Instead, Foucault focuses on how particular knowledge is sustained as 'truth' (truth effects). According to Foucault, the mutual relationship between power and knowledge is underpinned by **discursive structures**.

Discursive structures are the relatively rule-bound sets of statements that impose limits on what gives meaning to concepts, objects, places, plants, and animals (Phillips and Jørgensen 2002). Foucault uses this term to refer to sets of ideas that typically inform dominant or common sense understandings of and interconnections between people, places, plants, animals, and things. In Western thought, for example, scientific knowledge has become understood as the most appropriate way of thinking about what exists in the world. Western rational thought is characterized by a set of binary rules underpinned by Descartes's mind/body separation. Today, this dualistic thinking is still evident in a whole series of hierarchically valued dichotomies, including rationality/irrationality, man/woman, mind/body, straight/gay, masculine/feminine, and humanity/nature. Hence, while Foucault understands discourses to be inherently unstable, discursive structures are understood to 'fix' ideas of the world within particular social groups at specific historical and spatial junctures. In sum, for both individuals and collectives, discursive structures establish limits to, or operate as constraints on, the possible ways of being and becoming in the world by establishing normative meanings, attitudes, and practices. Simply put, discursive structures are a subtle form of social power that fix, give apparent unity to, constrain, and/or naturalize as common sense particular ideas, attitudes, and practices. Foucault refers to this form of social control as the 'effects of truth'.

Take, for example, the concept of sexuality. There are many different discourses about sexuality in the world. However, within a particular time and place, a specific set of ideas will come to define socially acceptable practices of sexuality. For example, in nineteenth-century Europe, North America, and Russia, the privileged knowledge about sexuality was inspired by religious, scientific, and medical institutional narratives. Among most scientists of this time, sexual difference was understood by theories of gender inversion. Homosexuals were mistakenly constituted as effeminate men. When combined with a lethal mix of fundamental Christian morals, same-sex-attracted men were 'invented' through rhetorical strategies that grounded meaning in binary oppositions to heterosexual men as the 'natural' and 'healthy' sex. Hence, the cultural identity of the homosexual man was constituted as lacking. Homosexuality became associated with negatively valued concepts: the primitive, the irrational, the feminine, the diseased, and the sinful. In this way, it seemed self-evident that the homosexual was only worthy of study in efforts to find a medical cure. Simultaneously, the knowledge of science made same-sex-attracted women 'invisible' by constituting sex as a penetrative act. Thus, social limits were placed on understandings about same-sex-attracted women and men. Indeed, the casting of homosexuality in the nineteenth century with so many negatively valued concepts often resulted in laws criminalizing sodomy in the West.

This example illustrates a number of key implications for doing discourse analysis. First, persuasion, or the effect of truth, is a power-laden process through which particular knowledge is deployed by institutions as a mechanism of social control. In this case, social power operates through the way that male-female distinctions came to define the appropriate and dominant conception of gender. Drawing on scientific knowledge, relatively powerful groups in society were able to naturalize meanings, attitudes, and practices towards another social group constituted as diseased and mad. Hence, when doing discourse analysis, it is imperative that you remain alert to institutional dynamics and the social context of a text. Second, while discourses are always inherently unstable, multiple, and contradictory, discursive structures operate to give fixity, bringing a common sense order to the world. Particular sets of ideas become accepted and repeated by most people as 'common sense', unproblematic, unquestionable, and apparently 'natural'. Hence, when doing discourse analysis, it is essential that you be aware of the ways in which particular kinds of knowledge become understood as valid, legitimate, trustworthy, or authoritative. The way that knowledge becomes understood as appropriate is not restricted to the use of particular technologies in the production of texts (for example, computers, maps, and photographs) but also encompasses the way that sets of ideas are legitimized by the subtle deployment of different knowledge-making practices (statistics, medicine, policies, anecdotes) or categories of spokesperson (politician, scientist, academic, lawyer, judge, priest, eyewitness).

Resilience and Rupture: Take Note of Inconsistencies Within Your Sources

While one rationale for doing discourse analysis is to identify the limits on how a particular social group talks and behaves, another is to explore inconsistencies within your sources. On the one hand, taken-for-granted sets of ideas about who and what exists in the world help to impose bounds beyond which it is often very hard to reason and behave. When particular relationships become understood as common sense, they set limits to the cultural know-how of a particular social group. As we saw earlier, Foucault understood dominant or common sense understandings as discursive structures (see Phillips and Jørgensen 2002). On the other hand, while discursive structures may appear eternal, fixed, and natural, because they are embedded within different social networks they are fragile and continually ruptured. Hence, there are always possibilities for meanings, attitudes, and practices to change or be challenged. Therefore, an essential part of doing discourse analysis is to be alert to possible contradictions and ambiguities in texts.

For example, take again Law's (2000) research on Philippine sex workers. She demonstrates the resilience within the Philippine media and non-governmental agencies of neo-colonial discourses that constituted sex workers as victims of a Western-imposed 'flesh' industry. In the early 1990s, sex work was portrayed in the media as imposed on Filipina women by the West, initially by the presence of US military bases in the 1960s and later by the demands of overseas sex tourists. In sum, neo-colonial sets of ideas fashioned understandings of sex work as a social problem imposed by global capitalism and of sex workers as both victims and lacking force in uneven sets of social relationships. At the same time, the 'Brunei Beauties scandal' ruptured neo-colonial ideas. In contrast to the portrayal of all sex workers as women with little education or life choices and relying on flows of Western tourists, this incident involved a sultanate and sex workers who were also well-paid models and entertainers. An inconsistency became evident. Not all women were involved in sex work as a result of an oppressive capitalist social system. Nor were all of their clients Western tourists. To generalize, doing discourse analysis involves remaining alert to contradictions and ambiguities within your texts.

Lesley Head and Pat Muir's (2006) work on suburban gardens in Wollongong, New South Wales, also illustrates the concept of resilience and rupture. Interviews with suburban gardeners provided insights into their understandings, attitudes, and practices towards nature. Head and Muir discuss the resilience among many European-Australian gardeners of the idea of cities as places devoid of nature. According to some of the gardeners, nature is only found beyond the spatial limits of the city in, say, national parks or the Australian 'bush'. Yet others talked about cultivating nature in their backyards, thereby rupturing conventional ideas about where nature is found and repositioning humans as nurturing rather than destroying nature. Remaining alert to such ambiguities and contradictions is a fundamental challenge of discourse analysis.

Silences: Silence as Discourse and Discourses That Silence

Finally, becoming attuned to silences in your texts is as important as being aware of what is present. Gillian Rose (2001, 157) reminds us that 'silences are as productive as explicit naming; invisibility can have just as powerful effects as visibility.' Similarly, Elizabeth Edwards (2003), while discussing her approach to discourse analysis as 'dense context', draws attention to the importance of silences:

> 'Dense context' is not necessarily linked to the reality effect of the photograph in a direct way, indeed to the extent that it is not necessarily *apparent* what the photograph is 'of'. Often it is what photographs are *not* 'of' in forensic terms which is suggestive of a counterpoint. [Edwards 2003, 262–3, emphasis in original]

The key point is that identifying silences produced by texts is an integral part of discourse analysis. However, becoming alert to silences is always challenging. Clearly, to be able to interpret your texts for what is omitted requires conducting background research into the broader social context of your project and texts. Only then will you become aware of the existence of various social structures that inhibit what is present in your texts.

According to Foucault (1972), silences operate on at least two levels. First, silence as discourse is a reminder of how speakers' subjectivities are created within discourses. Who has the right to speak or is portrayed as an authority is itself constituted through discourse. Considering the social circumstances of authorship leads us to consider who is given the authority to speak. Silence may become particularly relevant when the voices within texts are considered in terms of the intersections of gender, age, class, ability, sexuality, and race. Being mindful of whose 'voices' are silenced within your texts is an integral part of discourse analysis. For example, think about an advertisement for McDonald's restaurants. These advertisements often contain the 'voices' of children. Silenced are the voices of parents who are resolving the conflicting sets of ideas about fast food and parenting. These are silences to which you can speak.

Second, Foucault's ideas alert us to how a **privileged discourse**—or dominant discourse—operates to silence different understandings of the world. Here, of vital importance are Foucault's arguments concerning the intersection between power, knowledge, and persuasion. According to Foucault, silence surrounding a particular topic is itself a mechanism of social power within established structures. For example, if you are interpreting a tourism advertising campaign targeting backpackers, ask yourself, 'what is not being said?' Maybe you will find that the backpackers portrayed never appear as not-so-young people or as having a personal preference for downtime but only as 'adventure seekers'. Another helpful example of a discourse that silences is the elaborate Western fiction of the frontier. In the nineteenth century, the frontier discourses among European politicians, scientists, priests, ministers, social scientists, and many settlers helped to naturalize and justify colonial settlement and nation-making in North America (see Goss 1993; Turner 1920) and Australia (see McGregor 1994; Schaffer 1988; and

Ward 1958). Frontier discourses relied upon maintaining Western understanding of colonized places as 'uncivilized', 'empty', 'natural', 'wild', and 'timeless'. In frontier places, European history was seen as yet to begin. As Deborah Bird Rose (1997) argues, settlers' discursive strategies towards the frontier erased the presence of indigenous people or, at best, cast them as 'primitive' people whose natural fate according to social Darwinism was extinction. In Australia, frontier discourses that portrayed Australia as a place waiting for history to begin became law, dictating that before the arrival of Europeans, the continent belonged to nobody (*terra nullius*). Frontier discourses also helped to underpin widely held scepticism towards the ability of Aboriginal Australians to make particular knowledge claims or land claims. Frontier discourses acted as a system of power, sustaining a particular set of relationships and meanings that effectively silenced indigenous people. It was only in 1993, following the Australian High Court's Mabo decision, that the 'truth' of *terra nullius* was overturned. No longer could Aboriginal Australians' knowledge be ignored in decisions over land ownership.

Nevertheless, Waitt and Head (2002) note how the frontier discourse still enjoys widespread currency within the contemporary Australian tourism industry. Indeed, segments of the tourism industry keep alive the imagined frontier geographies. For example, the Kimberley, a northwest region of Australia, is sold as the 'last frontier'. One brochure for a local airline pitches the Kimberley as 'a mysterious, ancient land. So remote, so immense, so rugged—it challenges you to discover it . . . the Last Frontier' (Slingair tourist brochure n.d.). Waitt and Head (2002) are not surprised at the selectiveness of the advertising strategies. Potential tourists are offered a portal to a timeless land. Imagined as timeless, the Kimberley can then fulfill the specific market demands of primarily metropolitan Australian and international visitors, including: to discover the 'real' Australia, imagined as the outback; to experience places imagined as wilderness; to gaze upon places portrayed as sublime 'natural' beauty; or to explore the Kimberley as an adventure setting. However, the Miriuwung-Gajerrong's (the local indigenous people) knowledge constituting this place as 'home'—named, known, and cared for over tens of thousands of years—is silenced by the colonial frontier discursive structures. Also silenced are the geographical knowledges constituted by Anglo-Celtic Australian pastoralists working on cattle stations. Discourse as silence requires appropriate background research. To interpret how particular texts omit or silence different understandings of the world necessitates the identification of the temporally and spatially contingent social structures such as class, gender, race, ethnicity, sexuality, and so on.

REFLECTING ON DOING DISCOURSE ANALYSIS

This chapter has argued that the goal of discourse analysis is to reveal how particular ideas that help to forge social realities become understood as common sense. The technique has become widely applied in geography, partly because discourse analysis can provide insights into processes of social and environmental injustice. The chapter has also argued that conducting discourse analysis requires both understanding the concept of discourse and familiarity with a methodological template. Foucault's concept of discourse alerts us to the social constitution of all knowledge. Further, Foucault warns us that within many competing knowledge systems, particular sets of ideas emerge as dominant in the definition of appropriate forms of knowledge. According to Foucault, how certain knowledge is constituted as truth is not an accidental or haphazard process. Instead, through careful tracing of ideas expressed in the authorship, technological production, and circulation of diverse texts located in specific geographical and historical social contexts, it becomes possible to unravel how and why particular forms of knowledge have become ascendant.

Foucault does not provide a formal set of guidelines for the analysis of discourse. Instead, geographers, along with other social scientists, have contributed to designing checklists to assist with analysis (Box 11.1). Are there disadvantages to relying on a methodological template? One objection, perhaps, is that the effect of writing geography that employs these criteria is inevitably both selective and prescriptive. To implement the checklist criteria uncritically would mask their potential to both silence and demand particular responses. In other words, the checklist may wrongly become equated with objectivity and rationality at the expensive of subjectivity and creativity.

Nevertheless, with such a caveat in mind, the checklist has at least four crucial advantages for the researcher new to discourse analysis. The first is to keep in mind the iterative relationship between the researcher and his or her work: the researcher shapes the analysis as much as the analysis shapes the researcher. The researcher must be conceived as integral to rather than separate from the discourse analysis. Therefore, reflexivity, or writing one's self into a project, is a vital part of discourse analysis (for a more detailed discussion of related ideas, see Chapter 17). The second is that the checklist encourages researchers to remain alert to a key aim of discourse analysis: discourse analysis is not about determining the 'truth' or 'falsity' of statements but instead seeks to understand the geographical and historical circumstances that privileged particular discourses. Thus, in addition to selecting, familiarizing, and coding, discourse analysis requires background research into a source material or text's social circumstances, including authorship, production, and circulation. Third, the checklist operates as a reminder that all knowledge production is caught up in power relationships. Certain ways of knowing the world are privileged while others are silenced. Discourse analysis involves being alert to different strategies of conviction deployed by authors to help persuade audiences that a particular form of knowledge is intrinsically better than others. Finally, it is crucial to understand that while discourses may manifest themselves in ways that bring order to social life as rules, maxims, common sense, or the norm, they are always unstable and may be ruptured. Discourse analysis requires remaining alert to such instability, ambiguity, and inconsistency. Well-conducted and thoughtful discourse analysis enables insights into the resilience and rupture of multiple and sometimes conflicting discourses that give meaning to our everyday lives.

KEY TERMS

audience	metaphor
constructionist approach	positionality statement
discourse	power
discourse analysis	privileged discourse
discursive structure	rigour
embodied knowledge	semiology
genre	silence
intertextuality	strategy of conviction
manifest content analysis	text

REVIEW QUESTIONS

1. What implications do Foucault's ideas about power and knowledge have for conducting discourse analysis?
2. Why should both background research and the researcher be regarded as an integral part of discourse analysis?
3. According to Foucault's arguments, in what ways are the terms 'real' and 'truth' misleading?
4. What are the differences between analytic and descriptive codes? How do these different codes relate to one another in conducting discourse analysis?
5. How can a researcher endeavour to produce 'trustworthy' discourse analysis through the application of a checklist?

Useful Resources

Michael Larkin's webpages linked to De Montford University provide a list of questions that may be helpful before starting your discourse analysis (http://www.psy.dmu.ac.uk/michael/qual_analysis.htm).

Linda Graham, Queensland University of Technology, provides a helpful paper about how Foucauldian discourse analysis is used in education research. It is available at http://eprints.qut.edu.au/archive/00002689/01/2689.pdf.

Helpful starting points for more comprehensive reviews of Foucault's concepts include Cousins and Houssain (1984), Hall (1997), Hook (2001), McNay (1994), Mills (1997), and Thiesmeyer (2003).

Alternative discussions of 'doing' discourse analysis are provided by Dryzek (2005), Phillips and Jørgensen (2002), Rose (2001), Shurmer-Smith (2002), and Tonkiss (1998).

Seeing with Clarity: Undertaking Observational Research

12

Robin A. Kearns

CHAPTER OVERVIEW

> . . . for a discipline often preoccupied with the visual, it seems . . . [geography] has not studied the practices of seeing rigorously enough. [Crang 1997b, 371]

'Seeing is believing', as the saying goes. While visual observation is a key to many types of research, there is more to observation than simply seeing: it also involves touching, smelling, and hearing the environment and making implicit or explicit comparisons with previous experience (Rodaway 1994). Further, seeing implies a vantage point, a place—both social and geographical—at which we position ourselves to observe and be part of the world (Jackson 1993). What we observe from this literal or metaphorical place is influenced by whether we are regarded by others as an 'insider' (i.e., one who belongs), an 'outsider' (i.e., one who does not belong and is 'out of place'), or someone in between. The goal of this chapter is to reflect on the reasons why observation is fundamental to geographical research and to consider critically what it means to observe. It explores various positions the researcher can adopt vis-à-vis the observed—from the viewing of secondary materials such as photographs or video footage to participating in the life of a community. Two key contentions are:

1. that observation has tended to be an assumed, and consequently undervalued, practice in geographic research; and
2. that ultimately all observation is a form of participant observation.

In the chapter, I argue that observation has been taken for granted as something that occurs 'naturally'. Indeed, observation is commonly—and unfairly—regarded as 'inherently easy' and 'of limited value' (Fyfe 1992, 128). As a research approach, it has not been regarded as requiring the degree of attention granted to more technical aspects of our methodological repertoire as human geographers (e.g., questionnaire design, survey sampling). With critical reflection, however, observation can be transformed into a self-conscious, effective, and ethically sound practice.

PURPOSES OF OBSERVATION

Among the definitions of **observation** in the *Oxford English Dictionary* is 'accurate watching and noting of phenomena as they occur', implying that observation has an unconstrained quality. To regard observation as random or haphazard would be a mistake, however, for we never observe everything there is to be seen. Observation is the outcome of active choice rather than mere exposure. Our choice—whether conscious or unconscious—of first *what* to see and second *how* to see it means that we always have an active role in the observation process. Following Mike Crang, I wish to argue for observation as a way of 'taking part in the world, not just representing it' (Crang 1997b, 360).

There is a range of purposes for observation in social scientific research, and they can be summarized in three words, which conveniently all begin with 'c': counting, complementing, and contextualizing. The first purpose—*counting*—refers to an enumerative function for observation. For example, we might accumulate observations of pedestrians passing various points in a shopping mall or airport in order to establish daily rhythms of activity within these places. Research in the time-geographic tradition has used this approach to chart the ebb and flow of spatio-temporal activity (Shapcott and Steadman 1978; Walmsley and Lewis 1984). Under this observational rationale, other elements of the immediate setting are (at least temporarily) ignored while the focal activity occurs. The resulting numerical data is then easily displayed graphically or analyzed statistically. This is an approach to observation that may be useful for establishing trends but is ultimately too reductionist for developing any comprehensive understanding of place.

A second purpose of observation is providing *complementary* evidence. The rationale here is to gather additional descriptive information before, during, or after other more structured forms of data collection. The intent is to gain added value from time 'in the field' and to provide a descriptive complement to more controlled and formalized methods such as interviewing. Complementary observation might involve 'hanging out' in a neighbourhood after completing a household survey and taking notes on the appearance of houses, the types of cars, and the upkeep of gardens. I used this approach in research on housing problems in Auckland and Christchurch. Interviewers were instructed to observe the dwellings of those they interviewed. Their **field notes** added to (and often contrasted with) what people said about their own dwellings (Kearns, Smith, and Abbott 1991). Such information complements the aggregated data gathered by more structured means and assists in interpreting the experience of place.

The third purpose of observation might be called *contextual* understanding. Here the goal is to construct an in-depth interpretation of a particular time and place through direct experience. To achieve this understanding, the researcher immerses herself or himself in the socio-temporal context of interest and uses first-hand observations as the prime source of data. In this situation, the observer is very much a participant.

It should be noted at the outset that these purposes are not mutually exclusive. As I will later show by way of examples from research, one can approach observation with mixed purposes, seeking, for instance, to both enumerate and understand context during a period in the field (Kearns 1991a).

TYPES OF OBSERVATION

Some social scientists identify two types of observation: *controlled* and *uncontrolled*. **Controlled observation** 'is typified by clear and explicit decisions on what, how, and when to observe' (Frankfort-Nachmaias and Nachmaias 1992, 206). This style of observation is associated with natural science and its experimental approach to research, which has been imported into physical geography. Thus, a geographer might set up an electronic stream gauge to gather data at periodic intervals throughout a flood, resulting in a series of so-called 'observations'. However, such data are collected remotely without the aid of human senses. Indeed, the human eye may have only been involved in the secondary sense of observing the gauge subsequent to the actual data collection. This recognition that controlled observations can be made through mechanical means adds weight to the common perception that such research is rigorous and easily replicated. However, controlled observation is also limiting in terms of the sensory and experiential input that is admissible as 'findings'. Human geographers are unlikely to employ such controlled methods of observation, except perhaps in the earlier example of pedestrian counts.

In two respects, controlled observation is necessarily limiting. First, there is an imposed focus on particular elements of the known world, and second, it is only *directly* observable aspects that are of interest. Thus, imputed characteristics of place and the feelings of residents may be out of range. Most of the observation conducted by contemporary social and cultural geographers could be described as **uncontrolled observation**. Such observation is certainly directed by goals and ethical considerations but is not controlled in the sense of being restricted to noting prescribed phenomena. Therefore, although observation refers literally to that which is seen, in social science it may involve more than just seeing. Most obviously, observation also includes listening, a critical aspect of several approaches in human geography, including interviewing and participant observation. Effective listening can assist visual observation by both confirming the place of the researcher as a participant (Kearns 1991b) and by attuning oneself to **soundscapes** and the aural aspects of social settings (S.J. Smith 1994).

One might argue that all research involves observation or at least comprises a series of observations. Thus, in a social setting, we can 'observe' the population by employing questionnaires through which the researcher establishes the frequency of certain variables (for example, occupation, sex). The activity of conducting a questionnaire survey invariably places the researcher in the position of an 'outsider', marked as 'other' by purpose, if not by appearance and demeanour (for example, clothing, age, ethnicity, or type of language used). Identifying a sample and evaluating the responses to questions involve concerns related to the goal of generalizability to broader populations. But the cost of these concerns is that only a restricted subset of social phenomena is deemed to be of interest. These phenomena can easily be isolated from their context, unaccompanied by less directly observable values, intentions, and feelings. Qualitative approaches allow the consideration of human experience, potentially suspending traditional concerns about researcher bias and recognizing instead the relationship between the researcher and the people and places he or she seeks to study.

A further distinction can be drawn between **primary observation** and **secondary observation**. The former activity would have us adopt the position of participants in and interpreters of human activity, whereas as secondary observers we are interpreters of the observations of others. Examples of secondary observation are analyses of picture postcards available to tourists (Crang 1997b) or the photography of one 'racialized' group by a representative of another (Jackson 1992). In this chapter, I will not dwell on this use of secondary materials but rather note the all-pervasive nature of participant observation. This 'all-pervasiveness' of participation can be illustrated first through the point that a solitary researcher observing images of place is also an active participant in the process. This is because the observer is co-creating meaning through bringing their own perspectives and life experiences to their analysis and interpretation. Further, **secondary data** is increasingly being generated *through* participatory methods. **Photovoice**, for instance, is an approach formulated within community health development work in the US in which participants are given disposable cameras to record sites of significance to them. Once printed, the photographs serve as catalysts for discussion, which in turn is recorded and interpreted (Wang and Burris 1997). Recently, variations on this approach have been adopted by geographers (e.g., Bijoux and Myers 2006; Mitchell, Kearns, and Collins 2007). The relevant point for our discussion is the importance of engaging with *participants* (not 'respondents') and at least partially participating in their worlds. Only then can we meaningfully observe their photographs and engage in interpretive discussion. Similarly, but with more ethical complexity, video has been used as an observational complement to direct engagement in social settings such as cafés (Laurier and Philo 2006).

PARTICIPANT OBSERVATION

Participant observation is most closely associated with social anthropology (Sanjek 1990; Srinivas, Shaw, and Ramaswamy 1979). Significantly, its profile among the repertoire of approaches used by human geographers was raised by Peter Jackson (1983), who had studied both geography and social anthropology. The approach has been adopted and adapted by geographers seeking to understand more fully the meanings of place and the contexts of everyday life. Examples include a number of now classic studies within the humanistic tradition, such as David Ley's (1974) work on Monroe, a neighbourhood in inner-city Philadelphia, John Western's (1981) *Outcast Cape Town*, and Graham Rowles's (1978) research on the ways older people experience place. Like many social geographers, these writers talked to 'locals' in the course of their research, but it was the depth of their involvement in a community, their recurrent contact with people, and their relatively unstructured social interactions that stood out in their work. While many of their contemporaries were solely interested in perception or behaviour, these geographers were concerned with experience.

Developing a geography of everyday experience requires us to move beyond reliance on formalized interactions such as those occurring in interviews. As Mel Evans (1988, 203) remarks, 'although an interview situation is still a social situation . . . it is a world apart from everyday life.' Evans is suggesting that no matter how much we are able to put people at ease before and during an interview, its structured format often removes the researcher from the 'flow' of everyday life in both time and space. In other words, an interview ordinarily has an anticipated length and occurs in a mutually agreeable place often set apart from other social interactions. In contrast, the goal of participant observation is to develop understanding through being part of the spontaneity of everyday interactions.

There is a consensus that although definable, participant observation is difficult to describe systematically. Commentators have remarked on its 'elusive nature' (Alder and Alder 1994), its breadth (Bryman 1984), and the fact that it remains 'ill-defined and . . . tainted with mysticism' (Evans 1988, 197). Part of the 'mysticism' Evans refers to stems from the fact that there are few systematic outlines of the method, and students who have been curious about the approach have commonly been referred to classic studies such as William Whyte's *Street Corner Society* (1943) for models. One explanation for this tendency to refer to examples rather than to offer step-by-step guidelines is that every participant observation situation is unique. Another reason offered by Evans (1988, 197) is that the success of the approach depends less on the strict application of rules and more upon 'introspection on the part of the researcher with respect to his or her relationship to what is to be (and is being) researched'.

While introspection or reflection on what we see and experience is important, it is surely guidance in the act of observing that is needed. Jackson draws on Kluckhohn (1940) for a concise description of participant observation, defining it as 'conscious and systematic sharing, in so far as circumstances permit, in the life activities and, on occasion, in the interests . . . of a group of persons' (1983, 39). It is therefore the intentional character of observations that contrasts the activities of a participant observer with those of routine participants in daily life (Spradley 1980). To generalize, participant observation for a geographer involves strategically placing oneself in situations in which systematic understandings of place are most likely to arise.

One rationale for participant observation is recognition that in a more structured setting, the mere presence of a researcher potentially alters the behaviour or the dispositions of those being observed. This point is illustrated in a 'Far side' cartoon by Gary Larson. In the drawing, two men wearing pith helmets are approaching a thatched hut. The 'primitive' occupants, stereotypically adorned with bones through their noses and wearing head-dresses, are exclaiming 'Anthropologists! Anthropologists!' as they rush to hide their television set. The cartoon ironically suggests that scholarly explorers do not expect 'primitives' to have technology (and that primitives might want to live up to this stereotype). In terms of research methods, the cartoon's message is that the undisguised entry of others into a social situation is bound to alter behaviour. The point for us as qualitative researchers is that conscious participation in the social processes being observed increases the potential for more 'natural' interactions and responses to occur.

Use of the term 'observation' in highly controlled scientific research might tempt us to think too easily in terms of a simple dichotomy between participant and non-participant. According to Atkinson and Hammersley, this is an unhelpful distinction 'not least because it seems to imply that the non-participant observer plays no recognised role at all' (1984, 248). Indeed, as the 'Far side' cartoon succinctly illustrates, there is really no such thing as a non-participant in a social situation: even those who believe that they are present but not participating in a research context often unwittingly alter the research setting. To move beyond this false binary construct of participant/non-participant, Gold (1958) suggests that there is a range of four possible research roles:

1. **complete observer** (for example, a psychologist watching a child through a one-way mirror or a prisoner being observed through closed-circuit TV cameras)
2. **observer-as-participant** (for example, a newcomer to a sport being part of the crowd—see Latimer 1998);
3. **participant-as-observer** (for example, seeking to understand social change in one's own locality—see Ponga 1998)
4. **complete participation** (for example, living in a rural settlement to understand meanings of sustainability—see Scott, Park, and Cocklin 1997).

While it is difficult to imagine how being a complete observer might be incorporated into geographical research, a potential example could be remote surveillance of a shopping mall or public square, perhaps with access to footage from closed-circuit television (CCTV). This technology allows the 'remote' observation of activity. Behaviour may be less influenced by the observer 'being seen' than it would be if the research were conducted in the presence of an embodied and visible observer. Frequently, however, the mere presence of cameras influences and moderates behaviour (Fyfe and Banister 1996; Williams and Johnstone 2000). Further, the use of such footage for research purposes raises various ethical issues around consent and privacy. As the references above indicate, work by geographers can be easily categorized according to the remaining three roles (i.e., not a 'complete observer'). Each category represents a form of participant observation, and the difference is essentially a matter of the degree of participation involved. This division implies a continuum of involvement for the observing researcher from detachment to engagement. However, whatever one's positioning vis-à-vis 'the observed', it is important to acknowledge that the act of observation is imbued with power dynamics.

POWER, KNOWLEDGE, AND OBSERVATION

Before considering the stages of observation in the field in greater detail, it is important to reflect on aspects of the process itself. As emphasized earlier, observation involves participating, both socially and spatially. We cannot usually observe directly without being present, and our bodily presence brings with it personal characteristics that mark our identity such as 'race', sex, and age. Belonging to a dominant group in society can mean that we carry with us the power dynamics linked to such an affiliation (Kearns and Dyck 2005). Being a white, adult male, for instance, will invariably create challenges for being a participant in a group whose members do not share those characteristics, such as a new mothers' support group. In other words, our difference in terms of key markers of societal power (or lack thereof) contributes to our (in)ability to be 'insiders' and participants in the quest to understand place (matters of 'insider' and 'outsider' status are also discussed in Chapters 2, 3, and 13).

More subtle challenges may be generated by our level of education and affiliation with a university. In undertaking university-based research, we can carry institutional dynamics into our acts of observing through subtle forms of social control. Social control can occur through the ways in which members of one group (the relatively powerful) are able to maintain watch over members of another (the relatively disempowered). Perhaps the most memorable image of this dynamic is Bentham's **panopticon**, a circular prison designed to maximize the ability of guards to see into every cell and to watch prisoners (Foucault 1977a). Prisoners do not know exactly when they are being observed but learn to act as if they were always being watched. Foucault sees this surveillance as a form of disciplinary power that is enforced through the layout of the built environment rather than through the exertion of force per se. The key to the resulting institutional dynamic is the knowledge on the part of prisoners that they could be observed at any time. This knowledge, according to Foucault, results in self-surveillance and self-discipline. The use of CCTV in prisons to maintain an ever-present eye on detainees reflects an ongoing manifestation of the principles of the panopticon.

Foucault's ideas have implications for observation as a research approach. The **surveillance** to which he refers is a very visual and disembodied form of observation. Gillian Rose (1993) has linked it to the traditional geographical activity of fieldwork. To Rose and other feminists, geography has been an excessively observational discipline characterized by an implicit **masculine gaze**. The key point is that observation can be a power-laden process deployed within institutional practices. As we are based in and representative of academic institutions, it is imperative that we be aware of the ways in which others' behaviour may be modified by our presence (Dyck and Kearns 1995).

A challenge posed by feminist geographers and anthropologists is to see fieldwork as a gendered activity (Bell et al. 1994; Nast 1994; Rose 1993). Their argument is that in being participant observers, we unavoidably incorporate our gendered selves into the arena of observation. An extract from the field notes of Louisa Kivell (1995) illustrates how observation in human geography is far from a simple matter of uni-directional watching; rather, it involves interactions that are potentially laden with sexual energy. The resulting situations may shed light on the gendered constitution of the field site itself (Box 12.1).

'And Black Underwear Too': The Gendered Constitution of a Field Site   Box 12.1

Me. Walking onto an orchard looking for the manager with whom I had an appointment to discuss employment relations in his operation. Me. Dressed in my 'orchard clothes', a reflection of my expectation that he would have been out working in his 'orchard clothes' too. My orchard clothes include: dark shorts suitable for ladders, dirt and other eventualities; a plain green T-shirt without any brand labels or other logos; a jumper tied around my waist because I know it can get cold down the apple rows; socks and cross-trainers. . . . Anything else? Oh yes, a band tying my hair out of the way, a watch and a medic alert bracelet, mascara, suntan lotion and deodorant, lip balm, a pen and paper in my pocket and maybe a linger of this morning's perfume. Stop.

A white male wearing his 'orchard clothes' approaches me on a tractor and stops. He smiles, then laughs out a question-come-statement: 'You aren't here looking for a picking job are you?'

For a moment I am incapable.

> I reply. 'As it happens, no. I have an appointment to see Mr Sky but I'm interested to hear what makes you think I wouldn't be picker material.'
> 'Oh well . . . nothing really . . . it's just that um . . . in two years I've never seen a girl like you on an orchard before.'
> What is a girl like me?
> White, 21, pretty, on my own.
> My education status, my socio-economic background, my place of origin, most of the other 'girl like me' things are not immediately visible.
> What pale concept of a girl like me did he dispatch with his laugh? (Kivell 1995, 49).

STAGES OF PARTICIPANT OBSERVATION

Participant observation is far from haphazard. There are some commonly recognized stages through which the process moves, from choice of research site through to presentation of results. I will review these stages using case examples from my research in the Hokianga district of Northland, New Zealand. In this work, my objective was to understand the influence that sense of place had in shaping the social meanings of the local health care system. In the course of this research, I adopted three of the four roles identified by Gold (1958): observer-as-participant, participant-as-observer, and complete participant (see Kearns 1991b; 1997).

Choice of Setting

Choosing a setting for participant observation might be dictated by the goals of a larger research project, in which case the setting may be unfamiliar to the researcher (for example, Scott, Park, and Cocklin 1997). In such a case, there will be a need for a good deal of background research on the community in question as well as reconnaissance visits before a period of immersion into community life. However, it is likely that student researchers will choose to study settings at least partially familiar to them. But familiarity can bring pitfalls. There is a danger that the researcher may be over-familiar with the community, with the result that there is 'too much participation at the expense of observation' (Evans 1988, 205). Indeed, some argue that doing fieldwork in one's own society is just as challenging as it is in an 'alien' one (Srinivas, Shah, and Ramaswamy 1979).

What, then, might be the ideal? Possibly the best balance to strive for in choosing a research setting is a mid-point between the 'insider' and 'outsider' statuses discussed earlier. The conventionally recommended stance is that of stranger, but this is not necessarily a position in which one simply does not belong. Rather, it is one in which the researcher's status is what Evans (1988) terms 'marginal' in relation to the community. By 'marginal', I take Evans to mean socially, and possibly spatially, on the edge of a community or group. Thus, although cheering on the terraces at Eden Park (a sports ground in Auckland), Bill Latimer (1998) was marginal because he had never been to a high-profile rugby game before. He was from northern England, and the game of his culture and class was soccer. He was thus able to be a critical, only partially involved observer of the place of rugby in Auckland (see Box 12.2).

At the Game: An Observer-as-Participant   **Box 12.2**

Saturday March 21st, walking alone to Eden Park I was struck by the sea of decorative clothing and painted faces—blue and white for Auckland; red and black for Canterbury. One man had made himself a hat out of DB Export beer coasters. Once inside the ground I steered away from the loud 'yobs' who had positioned themselves behind the goal posts. I didn't choose a good spot. I felt uneasy as I took my seat between two sets of monotonic droning young men (wailing 'AAAAUUUUUAUCKLAAAAAAND!!!' like Tarzans of the terrace).

An eerie silence accompanied the match, interrupted only by the mo(ro)notonic drone of those seated around me. As I was about to discover, this was the calm before the storm. The Blues scored. Before I was able to offer a congratulatory clap of hands Eden Park erupted. Music blasted onto the terraces and was taken as the cue to scream, dance and throw arms, and their contents, into the air. As the music continued to pump out, cheerleaders thrusted and gyrated suggestively to the very deep, sexy, sensuous bass beat. I was soaked with beer and hit about the body with plastic cups and other missiles. I was also very unimpressed. Each time Auckland scored, a young guy in front of me would stand up and say to his friend, 'did you feeeel that?' He would then embrace his mate in a hand wrestling fashion which meant that their forearm muscles bulged, and both would look each other in the eye, faces inches apart, and grunt 'AAAAUUUUUUCKLAAAAND!!!' (Latimer 1998, 91).

Access

Gaining entry to social settings and places is potentially a fundamental challenge. A crucial issue is identifying key individuals who can act as gatekeepers, facilitating opportunities to interact with others at the chosen research site. There are some settings into which one can simply walk and take on the role of participant. Investigating the way in which a public shopping mall is used is a good example. Because a mall is commonly used and perceived as a public place, fewer permissions need to be sought than in more 'private' settings such as health clinics. Nor is there a problem with 'blending in', given the diversity of users. One can observe through simply participating in the mall's functions: shopping, resting on the seats provided, or using the food court. But gaining access is potentially more challenging when the place is smaller in scale or less public in character.

Gaining access may well be more straightforward if one has a known role. Even being 'the visiting student' in a workplace may give one a role in a way that just being an anonymous visitor or stranger would not. However, there are often no convenient roles in hospitals or factories, so once ethical approval is gained (see the discussion in Chapter 2), perhaps there is good reason to resist being typecast into any role except that of outsider (see Kearns 1997). This ambiguous position worked for me in the health clinics of the Hokianga district. I could not legitimately pass as either a health professional or a patient, since local residents knew each other too well, so acting as a visitor reading the newspaper helped me to remain reasonably inconspicuous for a short period (see Box 12.3).

Gaining Access to a Field Site	Box 12.3

My research in, on, and with Hokianga people began with a period of fieldwork during October 1988. My wife Pat had an opportunity to become a medical *locum tenens* in the district, and I was also keen to go north. Once in Hokianga, having a spouse with a temporary part in the health care system gave me some legitimacy in seeking approval to undertake some research at the hospital and community clinics. The medical director as gatekeeper to the health system knew who I was, and as I was already there, the formality of letters and telephone conversations could largely be circumvented. I hastily devised a research plan that involved spending time in the waiting areas of each clinic, where I observed social dynamics with the goal of understanding the social function of these places.

Field Relations

The role that the researcher adopts within an observed setting (for example, complete participant or participant-as-observer) will define the character of the relations she or he generates. The idea of impression management is critical here, for the impact you make on those encountered will determine, to a large extent, the ease with which they will interact with you and incorporate you into their place. Embedded within the word *incorporate* is 'corpus', the Latin word for body. My purpose in discussing incorporation is to stress the idea of the researcher's embodiment and to recognize that as researchers, we take more than our intentions and notebooks into any situation: we also take our bodies. The way we clothe ourselves, for instance, can be a key marker of who we are or who we wish to be seen as in the field. By way of example, while researching the inner-city experiences of psychiatric patients, I chose to wear older clothes to drop-in centres in order to minimize being regarded as yet another health professional or social worker intruding on patients' lives (Kearns 1987). Since I was a student at the time, it was easy for me to find place-appropriate clothes for a drop-in centre! But a troubling question is whether I could have 'blended in' if I had been attempting to study the more elite social relations of place (for example, the dynamics of a lawyers' convention or a restaurant frequented by politicians). At the most fundamental level, I would not have had the right clothes to wear. It is generally easier to dress 'down' than 'up', and this perhaps explains in small part why participant observation is more often used in studies of people less powerful than researchers themselves.

Our ability to relate to others in the field depends not just on appearance but also on the level and type of activity undertaken. To extend the above example, passively 'hanging out' in a psychiatric drop-in centre is one thing, but feigning ability in more expert pursuits is another. This need to pass as an *active* participant is reflected in Phil Crang's (1994) research on the workplace geographies within a Mexican-theme restaurant in England. Here the researcher sought to understand the dynamics of performance in the restaurant by being employed as a member of the serving staff and becoming part of the daily routines of the place.

A further point is that although concern for appearance and clothing is appropriate, it potentially reinforces geographers' fixations with the visual (see Rose 1993). Hester Parr (1998) describes how, while researching the geographical experiences of people with mental illness in Nottingham, she became conscious of non-visual aspects of her 'otherness' such as smell: her perfumed deodorant served to set her apart from those she was attempting conversation with and inhibit free interaction. Parr's reflection reminds us that senses such as smell add to the character of place (Porteous 1985) and merit consideration for the way they can mark as 'other' the 'bodiliness' of the researcher (see Rodaway 1994).

A related point influencing **field relations** is that codes of behaviour are attached to different settings. Parr (1998) discusses how she suddenly established a rapport with Bob, an acutely schizophrenic man with whom communication was difficult. The setting for this 'breakthrough' was a city park, neutral ground that neither the researcher nor the informant regarded as home. This site at which social boundaries were blurred contrasted with the drop-in centre where roles were more established and routine. The lesson is that research relations may be enabled or constrained by the (often unspoken) ways in which social space is codified and regulated. Successful participant observation thus involves not only the (temporary) occupation of unfamiliar places but also the adoption of alternative ways of using time. Box 12.4 continues an account of my Hokianga research, describing my use of time in community clinics.

Field Relations in an Observational Study	Box 12.4

At some of the 10 hospital and community clinics, I blended into the small crowd of clinic attendees, inconspicuously observing events under the guise of reading the newspaper. At others, however, I was clearly a Pakeha ('white') visitor, since all the others present were both local and Maori and seemed to know each other well. It was immediately evident that going to the doctor involved more than medical interactions. Rather, the occasion frequently provided an opportunity for locals to tell stories and reflect not only on their own well-being but also on that of their families and friends. From this participant observation in the clinic waiting areas, I noted that the most frequent conversation category was community concerns. Comments on the deleterious impacts of the restructuring of public services were frequently expressed. Residents adopted a relaxed approach to clinic attendance and their use of waiting areas. Patients were observed arriving well before their appointment time (sometimes in the company of others who had no intention of consulting the doctor or nurse) and lingering afterwards 'having a yarn'. These observations led me to interpret the clinics as de facto community centres, analogous to the village market in other countries. The difference, however, was that the place (the clinics) and the time (clinic day) prompted a considerable amount of conversation that explicitly centred on health concerns (adapted from Kearns 1991b, 525).

Talking and Listening

We cannot blend in as researchers unless we participate in the social relations we are seeking to understand. Listening ought to precede talking so that we become attuned to what matters in a particular time, place, and social setting (Kearns 1991a) (see Box 12.5). Asking questions can be manipulative, and simply conversing about what otherwise might seem obvious may result in a less threatening entry into the social relations of place. The 'how' and 'where' of talking and listening are also crucially important. For research with children, for instance, adopting their level—both physically and in terms of style of language—may be the key to successful observation. Sarah Gregory (1998) found that playing with children on the floor was a useful precursor to asking them about their experiences using consumption goods. Similarly, Hannah Mitchell engaged with children within their 'workplace' (the classroom) and 'walkplace' (the space between home and school) (Mitchell, Kearns, and Collins 2007). In other words, observation is least conspicuous when one is interacting most naturally with the research subjects. The lesson is that as researchers, we are our own most crucial tool and 'it is the fact of participation, of being part of a collective contract, which creates the data' (Evans 1988, 209).

Talking and Listening: The Embodied Observer   Box 12.5

My embodiment as researcher was central to the construction of this knowledge. Within the waiting area I had to position myself in such a way as to neither be threatening (and thus inhibit conversation by gazing at others) nor be overly welcoming of conversational engagement (hence my 'hiding' behind a newspaper). Such choreography was bound to break down, and did. On one occasion two locals offered to help with the crossword puzzle I was half-heartedly completing, and on another, a *kuia* (female elder) entered, kissed, and welcomed all present to 'her' clinic, including myself. My corporeality within the observed arena of social interaction thus rendered binary constructs of researcher/researched and subject/object thoroughly permeable. In this *kuia*'s clinic, my conceptions of being an 'autonomous self' were dissected and (re)embodied within their rightful web of socio-cultural relations (Kearns 1997, 5).

Recording Data

A clipboard or audio recorder are the standard means of recording information in other qualitative approaches involving face-to-face communication such as interviews. However, because these tools would be disruptive to the flow of conversation or interaction, a participant observer can rarely use them. Rather, there is a greater reliance on recollection and a need to work on detailed note-taking after a period of field encounters. At the end of any day or session of observation, one is likely to feel tired and not inclined to take out a pen or go to a computer to record reflections. However, developing a discipline for such 'homework' is a key part of field observation: notes are invaluable sources of data and prompts to further reflection (Scott, Park, and Cocklin 1997). Field notes become a personal text for the researcher to refer to and analyze. They represent the process of transforming observed interaction into written communication (see the discussion in Chapter 17 on relationships between writing and research). Jean Jackson (1990, 7) describes field notes creatively as 'ideas that are marinating'. Preliminary field notes may be taken on any materials at hand, such as the margins of a newspaper (Kearns 1991b). However, back at 'home base', annotations are now almost invariably entered into a computer, and just as earlier researchers feared the destruction of paper notes by fire (Sanjek 1990), a contemporary concern is loss of electronic data. Keeping backup files and/or print-outs of your field data is thus a crucial precaution.

Analysis and Presentation

Analyzing the results of any period of observation will vary according to the purposes for which it was undertaken. Observations that have involved counting, or carefully recording each instance of some phenomenon, lend themselves to tables that enumerate occurrences and express data as frequencies or percentages. Observation that has complemented a more explicitly structured research design commonly leads to the presentation of quotations or descriptions that assist in our interpretation of findings derived from other sources. In research into the impact of logging trucks on Hokianga roads and people, for instance, we complemented interviews with community members and analyses of secondary data, such as log-harvest projections, with a first-hand narrative of the experience of driving along selected stretches of road (Collins and Kearns 1998). Elsewhere, an assessment of the impact of a 'walking school bus' scheme on children's and parents' commuting patterns involved not only a calculation of car trips saved but also verbatim narratives from children gathered while the lead author walked with them to school (Kearns, Collins, and Neuwelt 2003).

When observation is embedded in an attempt to reach contextual understanding, a considerable volume of text will typically be accumulated, and strategies for the storage, classification, and analysis of information will need to be carefully considered (for a discussion of some of these issues, see Chapter 14). A number of useful texts are also available—for example, Bryman and Burgess (1994). This scale of observational engagement also suggests the merits of using computing software, such as NUD*IST and NVivo, now employed by human geographers (see Chapter 15).

Ethical Obligations

What are our obligations to those whom we have observed? Clearly, if observation has been fleeting and devoid of personal contact (for example, watching pedestrian behaviour), any return of research results to the individuals observed would be impractical and perhaps unnecessary. Indeed the 'observed' in this example are only nominally 'participants' in the research. However, observations such as 'reading' the presence of graffiti in the built environment (see Lindsey and Kearns 1994) have the potential to be a little more challenging. Observing the 'what' and 'where' of graffiti is unproblematic, since 'tagging' is part of the publicly observed landscape. But because the act of inscribing graffiti is invariably illegal, what are our obligations if we witness 'taggers' at work: report their breach of the law or preserve their anonymity? Arguably, our role as citizens takes precedence, and the former stance should hold sway.

When observation entails involvement in a geographical community or social group, it might be argued that there is an ethical imperative to maintain contact after the formal research period. This imperative may be formalized by the requirements of university ethics committees, but the stronger influence should surely come from the researcher, especially if pre-existing relationships have been developed (or reactivated) through the research process. As Maria Ponga noted in her study of the restoration of a *marae* (meeting house) in her home community in New Zealand, 'there is a social responsibility to carry on the ties. I have tried to keep in touch and send letters thanking people for the information they have given me and keep them updated on my progress. As we are all *whanau* (extended family), this has also had an added dimension of maintaining the family ties' (1998, 54). In short, moral as well as ethical obligations can make the process of 'exit' problematic. (See also Chapters 3 and 13.)

Such obligations may be taken for granted when pre-existing ties are involved but require careful consideration when social situations are entered into, or generated, for the sake of research. It is generally agreed that cross-cultural fieldwork is particularly problematic (see Chapter 3 for a full discussion). This is the case for two reasons, both of which may be linked to the 'field' metaphor: first, the researcher is potentially venturing onto another's turf, and second, because fieldwork involves a researcher working in a field of knowledge, there is the risk of overlooking local understandings and priorities and of (possibly unintended) one-way traffic of knowledge from the field (periphery) to the academy (centre) (Kearns 1997). Jody Lawrence's (2008) work with the Auckland Somali community had very porous boundaries that saw her attending weddings and sewing groups and completing grant applications on the community's behalf. Why did she do this? The answer has as much to do with relationship as it has with research, even though it was the latter that initially took her into that community. Box 12.6 draws on these ideas to complete the series of examples from my Hokianga research. In such situations, the development of 'culturally safe' research practice (Dyck and Kearns 1995) is important. Such practice recognizes the ways in which collective histories of power relations may affect individual research encounters. It also stresses the need for appropriate translation of materials into everyday language and the return of knowledge to the communities that provided or generated it (Kearns 1997). (See also Chapters 2, 3, and 13.)

Completing the Circle: Cultures, Theory, and Practice   Box 12.6

While on overseas sabbatical leave in 1995, I had an opportunity to recount the story of place and health in Hokianga and to connect its plot lines to the coordinates of other struggles for identity and turf. Drawing on Feinsilver's (1993) Cuban experience, for instance, I could identify the community action as a 'narrative of struggle' in which there is a greater symbolism to health politics than just a defence of local services. However, I was left searching for a rationale for the place of the researcher in the narrative of struggle.

To impose theory (upon observation) without reference to the community would be as foolish as the unfettered importation of exotic wildlife or viruses into New Zealand. Clearly, the use of theory must be regulated. My own form of self-regulation has been to return draft papers to people involved with Hokianga's health trust for comment. In one such draft, I had interpreted their struggle as a 'postmodern politics of resistance' (Kearns 1997). The returned manuscript was annotated with the comment from a health trust worker that my words sounded like 'undigested theory'. On reflection, this was a fair assessment. I had, perhaps unwittingly, used theory imported from recent visits to conferences in Los Angeles and Chicago. It was time to return, to digest theory, and to reconnect with the source of the story. Such returns are made easier through adoptive *whanau* (extended family) relationships in which research is, at times, indistinguishable from the *aroha* (love, affection) and *korero* (purposeful talk) among friends. Hokianga has taught me about being a bicultural geographer, a role that requires respecting the rituals we can never fully enter while reforming the rituals of our own research (Kearns 1997, 6).

REFLECTING ON THE METHOD

As observers, our goal should be to achieve seeing that is, as the saying goes, believing. However, this chapter has argued that believable observation is the outcome of more than simply seeing; rather, it requires cognizance of the full sensory experience of being in place. If the questionnaire is the tool for survey researchers and the audio recorder for key informant interviewers, researchers themselves are the tool for participant observation. If we are to observe, we must take the time to do it properly. Indeed, as Laurier and Philo (2006, 194) remark, 'If being academic researchers allows us certain privileges, one of them is to take the time to attend carefully and patiently to spatial phenomena.' The question remaining is: how do we know whether the fruits of participant observation are valid? Evans (1988) reminds us of the very important point that any method is, to a degree, valid when the knowledge that it constructs is considered by stakeholders to be an adequate interpretation of the social phenomena that it seeks to understand and explain (see Chapter 4).

Are there disadvantages to reliance on observation? One danger, perhaps, is privileging face-to-face interactions over less localized relations that remain beyond the view of the researcher in the field (Gupta and Ferguson 1997). To believe only what we see would be to make the serious mistake of denying the existence of structures such as social class or communicative processes such as Internet relationships that occur 'off-stage'.

Whether we seek to count, to gather complementary information, or to understand the context of place more deeply, the key to taking observation seriously is being attentive to detail as well as acknowledging our positions as researchers. Those recognitions imply that we are aware of both our place within the social relations we are attempting to study and the reasons that we have the research agendas we do (White and Jackson 1995).

KEY TERMS

complete observer

complete participation

controlled observation

field notes

field relations

masculine gaze

observation

observer-as-participant

panopticon

participant-as-observer

participant observation

photovoice

primary observation

secondary data

secondary observation

soundscape

surveillance

uncontrolled observation

REVIEW QUESTIONS

1. What are some of the ethical considerations that arise from observation?
2. In what ways might access to a specific social setting (for example, a sports club, an industrial workplace) be achieved for research purposes?
3. In what ways does observation involve more than seeing?
4. Suggest some ways in which our presence might influence interactions in the research setting.
5. Why, and in what circumstances, might one opt for participant observation as a research method?
6. In what types of circumstances might observational research be usefully aided by still or video photography?

USEFUL RESOURCES

A helpful guide for students using participation in dissertation research is provided by Ian Cook's (2005) chapter in *Methods in Human Geography* (by R. Flowerdew and D. Martin, eds, Pearson). Another accessible resource, with concepts interleaved with research examples, is Mel Evans's chapter in Eyles and Smith's (1988) book *Qualitative Methods in Human Geography* (Polity Press, Cambridge).

Participatory Action Research

13

Sara Kindon

CHAPTER OVERVIEW[1]

Participatory action research (PAR) is one of the fastest emerging approaches in human geography. It involves academic researchers in research, education, and socio-political action with members of community groups as co-researchers and decision-makers in their own right (Thomas-Slayter 1995). As such, it is quite different from many other research methods and demands different types of attitudes and behaviours from a researcher. In this chapter, I discuss the overall process of PAR and its cycles of action and reflection. I also discuss different types of relationships and various strategies and techniques that can be used to enable the involvement of research participants in all stages of the process. These strategies and techniques aim at establishing a more democratic research process, which respects and builds co-researchers' capacity and generates more rich, diverse, and appropriate knowledge for community change. If done well, PAR has many benefits for human geographers, particularly for those committed to challenging unequal power relationships and increasing **social justice**. PAR can also be challenging to carry out: the emphasis on power, relationships, and change is a potent mix. Finally, I consider how we can present PAR-generated research information to a range of audiences in effective and ethical ways.

WHAT IS PARTICIPATORY ACTION RESEARCH?

Participatory action research has been defined as:

> [A]n experiential methodology for the acquisition of serious and reliable knowledge upon which to construct power, or countervailing power, for the poor, oppressed and exploited groups and social classes—the grassroots—and for their . . . organizations and movements. Its

> purpose is to enable oppressed groups and classes to acquire sufficient creative and transforming leverage as expressed in specific projects, acts and struggles to achieve goals of social transformation. [Fals-Borda and Rahman 1991, 4]

PAR has been evolving since at least the 1970s and is fundamentally different from many other social science methods because its goal is not just to describe or analyze social reality but to help change it (Pratt 2000). This change occurs through the active involvement of research participants in the focus and direction of the research itself (Kindon, Pain, and Kesby 2007a; Kindon 2009). To clarify, as an academic researcher thinking about embarking on PAR, you would not usually determine a research agenda independently. Rather, you would work with a social group (usually regarded or self-defined as marginalized or disenfranchised, as Fals-Borda and Rahman note above) to define the issues facing them (see Chapter 3). Together, you would then generate and analyze information, which would hopefully lead to action and, ultimately, positive change for those involved (Bowes 1996; Cameron and Gibson 2004; Cooke 2001; Thomas-Slayter 1995). In short, a PAR researcher does not conduct research *on* a group but works *with* them to achieve change that *they* desire.

PAR has its origins in **action research** (Lewin 1946) in the United States, which sought to inform change by testing theory through practical interventions and action, and **participatory research** (Hall 2005), which emerged out of Africa, India, and Latin America where educators and others involved in community development devised a new **epistemology** grounded in people's struggles and local knowledges. When combined, participatory action research sought to 'develop new alternative institutions and procedures for research that could be **emancipatory** and foster radical social change' (Kindon, Pain, and Kesby 2007b, 10).

Today, diverse PAR generally involves a number of stages through which academic and social group members work together to define, address, and reconsider the issues facing them (Kindon, Pain, and Kesby 2007a; Parkes and Panelli 2001). Such issues commonly include the lack of access to information or resources, the threat of removal of services or subsidies, and the need to respond to and mitigate further unanticipated events. The emphasis on this iterative cycle of **action–reflection** is one of the key distinguishing features of PAR. It can also enable multiple perspectives of different **stakeholders** to be taken into account throughout the research, which can lead to more informed decision-making and more equitable and potentially sustainable outcomes.

PAR attends to power relations (Kesby, Kindon, and Pain 2007) and as such can be challenging, particularly in the context of an undergraduate research project. It often involves the researcher in a facilitative rather than 'extractive' role and demands that he or she pay considerable attention to ethics and issues of representation. That said, PAR can be very rewarding, and even if it is not possible to involve research participants deeply in every step of a research project, it may be possible to make your research more participatory by adopting some of the ideas discussed in this chapter.

CONDUCTING 'GOOD' PARTICIPATORY ACTION RESEARCH

PAR is an approach that ideally grows out of the needs of a specific context, research question, or problem and the relationships between researcher and research participants. It is more about the value orientation of the work and its approach (epistemology) than about the specific techniques used, although participatory techniques are certainly important (see Kesby, Kindon, and Pain 2005 for discussion of deep versus other forms of **participation**). It is also an approach that values the process as much as the product so that the 'success' of a PAR project rests not only on the quality of information generated but also on the extent to which skills, knowledge, and participants' capacities are developed (Chatterton, Fuller, and Routledge 2007; Cornwall and Jewkes 1995; Kesby, Kindon, and Pain 2005; Maguire 1987).

According to some practitioners (see Chambers 1994), the most important aspects of participatory work are the attitudes and behaviours of 'outside' researchers (usually academics or practitioners). These attitudes and behaviours affect the relationships formed with research participants and the outcomes achieved. Whether you, as a researcher, respect people's knowledge or perpetuate unequal power relations and extract information largely for your own benefit depends to an extent on your attitudes and behaviours and the nature of the research relationships you establish. To illustrate this point further, Box 13.1 shows common connections between the attitudes of the researcher (illustrated here by things a researcher may say to a researched group), the kind of research relationships they form, the resultant mode of participation possible, and the relationship between the researched group and the research itself.

Researchers using PAR generally strive to adopt and practise the attitudes and behaviours that result in people's **co-learning** and **collective action**. They also generally follow an iterative process of action–reflection (see Box 13.2), although the specifics of what actually happens, how, and when vary depending on the particular context and circumstances of those involved.

The cycles of action–reflection outlined in Box 13.2 ideally involve us and the people with whom we are working in each stage. However, this may not be possible within the confines of an undergraduate research project. Do not be put off, since this 'ideal' process can and should be adapted collectively to suit the particular needs and constraints of the situation. For example, in a four-month-long PAR project conducted in 1990 with a Costa Rican women's cooperative,

Box 13.1

The Importance of Attitudes to Relationships within PAR

Attitude of researcher and example of attitude reflected in what researcher might say to researched group (RG)	Relationship between researcher and researched group (RG)	Mode of participation	Relationship between research and researched group (RG)
Elitist *'Trust and leave it to me. I know best.'*	Researcher designs and carries out research; RG representatives chosen but largely uninvolved; no real power-sharing.	Co-option	ON
Patronizing *'Work with me. I know how to help.'* (i.e., I know best.)	Researcher decides on agenda and directs the research; tasks are assigned to RG representatives with incentives; no real power-sharing.	Compliance	ON/FOR
Well-meaning *'Tell me what you think, then I'll analyze the information and give you recommendations.'* (i.e., I know best.)	Researcher seeks RG opinions but then analyzes and decides on best course of action independently; limited power-sharing.	Consultation	FOR/WITH
Respectful *'What is important to you in the research? How about we do it together? Here's my suggestion about how we might go about this.'*	Researcher and RG determine priorities, but responsibility rests with researcher to direct the process; some power-sharing.	Cooperation	WITH
Facilitative *'What does this mean for you? How might we do the research together? How can I support you to change your situation?'*	Researcher and RG share knowledge, create new understandings, and work together to form action plans; power-sharing.	Co-learning	WITH/BY
Hands-off *'Let me know if and how you need me.'*	RG sets their own agenda and carries it out with or without researcher; some power-sharing.	Collective action	BY

Source: Adapted from Parkes and Panelli 2001.

		Box 13.2
Key Stages in a Typical PAR Process		

Phase	Activities
Getting started	• Assess information sources. • Scope problems and issues. • Initiate contact with researched group (RG) and other stakeholders. • Seek common understanding about perceived problems and issues. • Establish a mutually agreeable and realistic time frame. • Establish a **memorandum of understanding** (MoU) if appropriate.
Reflection	On problem formulation, power relations, knowledge construction process.
Building partnerships	• Build relationships and negotiate ethics, roles, and representation with RG and other stakeholders. • Establish team of co-researchers from members of RG. • Gain access to relevant data and information using appropriate techniques (see Box 13.3). • Develop shared understanding about problems and issues. • Design shared plans for research and action.
Reflection	Reformulation, reassessment of problems, issues, information requirements.
Working together	• Implement specific **collaborative research** projects, • Establish ways of involving others and disseminating information (see Box 13.3).
Reflection	Evaluation, feedback, re-participation, re-planning for future iterations.
Looking ahead	Options for further cycles of participation, research, and action with or without researcher involvement.

Source: Adapted from Parkes and Panelli 2001, 98.

members decided that it would be best if the primary research into their ongoing economic development was undertaken by me and another student rather than by any of them. They felt that we had more time than they had and would be able to involve everyone equally. In this case, we worked with a core advisory group and external development worker to reflect on our actions as we undertook interviews and analysis. We then ran a participatory workshop and produced a report in cartoon format to invite members' reflection upon and analysis of our findings and to develop action plans for the future (Kindon and Odell 1990).

What is most important in PAR is that the design and process are negotiated with the researched group and carried out in ways appropriate to the context, the time available, and the people involved, including yourself (Manzo and Brightbill 2007). Adopting even some of the strategies above will enable 'partial' participation to occur. This will bring benefits to your project and some of your participants (see Chapter 3), particularly if you combine them with some of the participatory techniques discussed below (also see Kesby, Kindon, and Pain 2005).

The strategies and techniques used within a PAR process can involve and adapt some of the other methods discussed in other chapters of this book if they occur in the context of reciprocal relationships. Common methods and techniques are interviewing and visualization (sometimes also referred to as **participatory diagramming** or **participatory mapping** in which participants create diagrams, pictures, and maps to explore issues and relationships (see Box 13.3; also see Alexander et al. 2007 for a discussion of participatory diagramming and Sanderson et al. 2007 for a discussion of participatory cartographies, including mapping). These methods and techniques emphasize shared learning (researcher and researched group), shared knowledge, and flexible yet structured collaborative analysis (*pla notes* 2003). They embody the process of **transformative reflexivity** in which both researcher and researched group reflect on their (mis)understandings and negotiate the meanings of information generated together (see Crang 2003, 497).

Some Strategies and Techniques Used within PAR   **Box 13.3**

Establish a support base and platform for your research

- *Find and critically review secondary data.* Secondary data can help to establish the direction of the research and identify where gaps or contradictions in understanding exist.
- *Involve those who are experts about specific issues and processes.* Local experts always exist and can help facilitate the participation of others and inclusion of their knowledges.
- *Negotiate and establish a memorandum of understanding (MoU) for the research team[2].* MoUs and the process that leads to them clarify research expectations, agreed-upon norms of behaviour, modes of interaction, and 'ownership' of information generated and are important for sustaining participatory partnerships and realistic expectations.

Get involved with people and their lives

- *Observe directly (see for yourself)*. Visiting people and places is essential. Taking a walk through an area with members of the researched group enables you to observe firsthand and question things directly.
- *Do-it-yourself.* By living like the people you are working with, you can learn something—although never all—of their realities, needs, and priorities. You can also learn how they communicate with each other and get things done.
- *Work with groups.* Groups can be casual or encountered 'randomly', **focus groups** that are representative or structured for diversity, or community, neighbourhood, or specific social groups. Group interviews are usually a powerful and efficient way of generating and analyzing information.

Use interviewing and story-telling approaches

- *Collect case studies and stories.* Focusing on specific events or cases, such as a household profile or history or how a group coped with a crisis, is a helpful way of teasing out issues. A variety of cases can reveal and illustrate common themes and important differences.
- *Use open-ended questions and key probes.* Asking questions that start with 'what', 'when', 'where', who', or 'how' can generate specific information without leading respondents to particular answers. Probing (i.e., 'what happens when . . . ?' and 'why is that?') can identify key issues, local rationales, and current activities and procedures (see Chapter 6 on interviewing).

Use types of participatory mapping and diagramming such as:

- *Maps.* Mapping, drawing, and colouring using locally appropriate materials can represent resources, issues, and relationships in ways that enable more democratic participation than verbal discussions alone. Maps might focus on tangible resources such as land, forests, houses, or services. Diagrams can tease out relationships between people, institutions, and resources (see below). Do remember—as Howitt and Stevens note in Chapter 3—that participatory mapping and diagramming may sometimes be suspect if used in the wrong context.
- *Timelines and trend/change analysis.* Locally defined chronologies of events showing approximate dates, people's accounts of how customs, practices, and things have changed, ethno-biographies or local histories of particular crops, animals, or trees, and changes in land use, population, migration, fuel uses, education, health, credit, and so forth may enable analysis of cause and effect factors over time.
- *Seasonal calendars.* Focusing on seasonal variation of particular factors (for example, rain, crop yields, workload, travel) can enable insights into

matters such as climatic variation, labour patterns, migration, diet, and local decision-making processes.

- *Daily time-use analysis.* Indicating the relative amount of time, degree of drudgery, and level of status associated with various activities may reveal local power relations and identify the best times for research activities.
- *Institutional or Venn diagramming.* Drawing out the relationships between individuals and institutions using overlapping circles signifying the importance or closeness of the relationships enhances understanding of power relations, local and surrounding contexts, and the identification of where there are blocks to or possibilities for change.
- *Well-being grouping (or wealth ranking).* Grouping or ranking households according to local criteria, including those considered poorest and worst off, can be a helpful lead into discussions about the livelihoods of the poor and how they cope depending on the particular cultural context.
- *Matrix scoring and ranking.* Drawing matrices of resources, such as different types of trees, soils, or methods of health provision, then using seeds to score or rank how they compare according to different criteria (such as productivity, fertility, or accessibility) can reveal local preferences (what scores highly) and the aspects that inform decision-making strategies.

Use the Internet and electronic technologies to engage people and disseminate information

- Use cellular phones as a helpful tool for involving people, particularly if this is their usual mode of communication and/or they are highly mobile. Phones can be used to text or call co-researchers or participants with invitations to meetings, to invite responses to questions to inform process and analysis, and to share results or outcomes.
- Establish Internet chat sites, wiki, or blogs to create spaces in which co-researchers and participants can share ideas and responses to emerging analysis. As with all group spaces, these need to be carefully managed and moderated.
- Design and promote websites as a productive means of engaging participation and establishing a core identity for a group, as well as disseminating the process, findings, and outcomes of participatory action research.
- Use sites like YouTube to disseminate information widely, but take care because once posted on the site, images and videos can be used and appropriated by others for their own means, with little regard for the original ethical agreements or MoUs negotiated between collaborating parties.

Engage people in joint analysis, reflection, and future planning

- *Shared presentations and analysis.* Involving people in the presentation and analysis of the maps, diagrams, and information generated through-

out the research shares power and allows information to be checked, corrected, and discussed.
- *Contrast comparisons.* Asking group A to analyze the findings of group B and vice versa can be a useful strategy for raising awareness and establishing dialogue, particularly between different groups. This technique has been used for gender awareness, asking men, for instance, to analyze how women spend their time.

Further information on these and other strategies and techniques, with examples of their use, can be found in Pretty et al. 1995.

While the above list of techniques may seem exhaustive, it is not prescriptive. Integrating one or several of them, where appropriate, will enhance your research. However, as Richie Howitt and Stan Stevens suggest in Chapter 3, a few participatory techniques will not, in and of themselves, make your project PAR. For this to occur, the open negotiation of the research design and methodology with the people with whom you are working is critical, as is an emphasis on supporting people's capacity to do their own research and analysis.

For more ideas about what this might mean in practice, the work of Caitlin Cahill and the Fed Up Honeys is instructive. Over several weeks, a US academic researcher (Cahill) met with six young women (also known as the Fed Up Honeys) to carry out a PAR project, Makes Me Mad: Stereotypes of Young Urban Women of Color, in New York City's Lower East Side. The process involved a period of time deciding first on the research focus and then on the approach they would adopt. They decided that each of them would engage in reflective journal writing and analysis of their own thoughts, plus the reading and discussion of each others' writing, to explore how they had internalized racist and sexist stereotypes and applied the same representations to others. It was an intense and at times frustrating process for participants because of the project's close attention to detail and the challenge it presented to their personal beliefs. The project required dedicated and reflexive facilitation by Cahill and commitment from each participant to enable them to work through their differences. It was worthwhile, however, for the depth of friendships it produced, the degree of increased awareness it provoked, and the level of political action that the participants engaged in through their development of a website and their implementation of a sticker campaign (see Cahill 2004 and the Fed Up Honeys website for more details). Box 13.4 summarizes some ideas about how to approach doing PAR. The points integrate and reinforce ideas about the attitudes, behaviours, relationships, research design, process, and techniques discussed above.

Some Ways to Promote Participation in Geographic Research	Box 13.4

- Involve/be involved with the group with whom you are working as equal decision-makers to define the research questions, goals, and methods and as co-researchers and analysts of information generated.
- Show awareness that you are an outsider to the group you are research-ing, even if you are working together as co-researchers.
- Be clear about the potential impacts people's involvement may have and what will happen to the information generated (ideally through an MoU).
- Take care not to promise too much or inflate people's expectations of what might happen as a result of the research.
- Develop facilitation skills, which can stimulate initiative and sensitively challenge the status quo without imposing your own agenda.
- Work at fostering participatory processes and research techniques, which will release creative ideas and enthusiasm but not take too much time for those involved (see Box 13.3).
- Seek out the perspectives and participation of the most vulnerable and marginal people.
- Find ways of limiting the dominance of interest groups and powerful people (including yourself, where appropriate).
- Acknowledge that process is as important as product (and sometimes more important), and factor in enough time to involve people appropri-ately at various stages of the research, including time for reflection.
- Support the group with whom you are working to share the benefits of their involvement with others and to take initiative to address their concerns.
- Involve the group with whom you are working in the writing and dis-semination of relevant information; at the very least, acknowledge their contributions to any sole-authored work.
- Practice honesty, integrity, compassion, and respect at all times.
- Keep a sense of humour!

Source: Adapted from Botes and van Rensburg 2000, 53–4; Kesby, Kindon, and Pain 2005; and the author's own experiences.

THE VALUE AND REWARDS OF PARTICIPATORY ACTION RESEARCH

Through attention to attitudes and behaviours, as well as the use of appropri-ate strategies and techniques to support people's participation in collaborative research and action, it is possible to examine and challenge forms of oppression and inequality. PAR is used most frequently by geographers with an activist agenda to work for social change, because it offers a tangible way of putting the aims and principles of **critical geography** into practice (Kesby 2000). Often, this means specifically addressing issues of racism, ableism, sexism, heterosexism, and imperialism (Ruddick 2004, 239) and how they are manifested through people's unequal access to and control over resources or their positions within inequitable social relationships (See Box 13.5).

Selected Geographers and PAR	Box 13.5

Geographer	*PAR work with . . .*
Caitlin Cahill	Young women of colour—the Fed Up Honeys—in New York City
Jenny Cameron and Katherine Gibson	Economically 'depressed' communities, Australia
Sarah Elwood	Urban community groups using participatory GIS, Chicago
Duncan Fuller	Graffiti artists, Newcastle-upon-Tyne, UK
J. K. Gibson-Graham	Women in mining communities, Australia
Robin Kearns	Primary schools and parents, Auckland, Aotearoa/New Zealand
Mike Kesby	People with HIV/AIDS, Zimbabwe
Sara Kindon	Indigenous women and men in Indonesia and Aotearoa/New Zealand
Rob Kitchin	People with disabilities, Ireland
Fran Klodawsky	Homeless people and women in local government, Ottawa
Audrey Kobayashi	Immigrant and ethnic minority groups, Canada
Jan Monk	Women's and non-governmental organizations, US and Mexico
Carolyn Moser and Cathy McIlwaine	Communities coping with violence, Colombia and Guatemala
Karen Nairn	Young people, Dunedin, Aotearoa/New Zealand
Rachel Pain	Locally born and asylum-seeking young people, County Durham and Newcastle, UK
Ruth Panelli	Rural communities, Australia and Aotearoa/New Zealand

Linda Peake	Women's handicraft cooperative, Guyana
Geraldine Pratt	Philippine Migrant Workers Collective, Vancouver
Maureen Reed	Women in logging communities, Vancouver Island
Diane Rocheleau	Rural and farming communities, Africa and South America
Kevin St Martin and Madeleine Hall-Arbor	New England fishing communities, US
David Slater	Third-World non-governmental organizations, India
Janet Townsend	Female rainforest settlers, Colombia and Mexico

Sources: Kindon, Pain, and Kesby 2007a; Pain 2003; Ruddick 2004.

Because of this activist orientation, PAR can build capacity and alliances within a community. For example, in Canada, geographer Geraldine Pratt has worked for more than 12 years with the Kalayaan Centre in Vancouver using a PAR approach. Together, they have conducted four main projects exploring different dimensions of the lived experiences of Filipina migrants. The first project involved collecting women's stories of being live-in carers to Canadian nationals and mobilized them to change their working conditions. The second used storytelling to explore how the same women's lives had changed over time as a means of understanding long-term migrant integration. The third project adopted a focus group approach with first- and second-generation young people to record their stories of racism and to begin to foster a greater sense of collective identity and belonging. The most recent project has involved women and young people to collect the stories of transnational families in an effort to develop and share collective community stories, which can challenge individual feelings of isolation and despair (Pratt et al. 2007).

In my own work as part of an ongoing PAR project with *Te Iwi o Ngaati Hauiti* in the central North Island of Aotearoa/New Zealand, several *iwi* (tribe) members established themselves as a community video research team to explore the relationships between place, cultural identity, and social cohesion (Hume-Cook et al. 2007). They undertook training in video production and community research with an academic colleague and me and then carried out video interviews with other members of the *iwi*. Sometimes I was a co-interviewer and my colleague a co-videographer; at other times they worked independently and later shared their tapes and analysis with us (Kindon 2003; Kindon and Latham 2002). We have also been involved in the collaborative editing of a documentary of a key event in the *iwi*'s efforts to foster social cohesion and cultural identity—a five-day rafting journey down their most significant river. A DVD of the documentary will be given to all participants in the event and is likely to be shared more widely with other *iwi* members, further strengthening their relationships and connections.

PAR enables rich and varied information embedded within specific 'communities' to be shared, analyzed, and evaluated collectively (Cooke 2001). This information may be more accurate and relevant for other uses than had a researcher worked alone. In PAR work with rural communities in Bali, Indonesia, in the early 1990s (Kindon 1995; 1998), men associated with one village used information generated through our regular participatory research meetings and focus groups (see Chapter 8) to develop an action plan for their community. The plan addressed the need for better roads to open up access to markets. In the four years that followed, these men established a savings/credit fund and liaised with government agencies to raise enough money to seal a remote road and enable more efficient transport of produce to market. Elsewhere in the village, government planners acted on PAR information generated by women's groups about the need for more accessible health care facilities and established a local health post. On my return to the village in 1998, these two key development needs had been met by the collective actions of these men and women and the support received from government agencies.

In summary, people's participation in their own research may challenge prevailing biases and preconceptions about their knowledge on the part of others in positions of power (Sanderson and Kindon 2004), such as government officials and policy-makers. In addition, PAR can bring about desired change more successfully than 'normal' social science research methods (Brockington and Sullivan 2003; Kesby 2000) and often results in improvements in living or working conditions for those involved (Kesby, Kindon, and Pain 2005; Kindon, Pain, and Kesby 2007a; Pain 2003, 2004; Parkes and Panelli 2001).

CHALLENGES AND STRATEGIES

There are challenges associated with collaborative research endeavours as well (Monk, Manning, and Denman 2003), especially when participatory techniques are involved (Pain and Francis 2003). While participation is becoming increasingly popular, not all researchers are doing it well (Parnwell 2003).

Most PAR takes place in a group setting, which is both a strength and a weakness. Particular techniques (like the participatory mapping and diagramming techniques listed in Box 13.3) require group participation, often in public spaces. This spatial aspect of PAR shapes the construction of knowledge (Brockington and Sullivan 2003; Cooke 2001; Kesby 2007; Mosse 1994), and it usually tends to generate knowledge that reflects dominant power relations in wider society (Kothari 2001). It is therefore important to pay attention not only to *who* participates but *where* they participate and *how*. Keeping a field diary with this information can be helpful when you come to analyze the products of group work. If funds permit, involving a colleague or friend to keep notes on the process or make video recordings can provide a detailed and more dispassionate record of where and how people participated (yourself included) for later analysis (Kindon 2003).

In terms of the participatory techniques themselves, because they often appear to be quick and easy to use (Leurs 1997), it can be tempting to use them repeatedly in the same ways rather than adapt and modify them to the particular contexts involved (Chambers 1994). Certainly, many current participatory processes associated with development projects involve a sequence of participatory techniques such as community mapping, social well-being ranking, agricultural land-use transects, and institutional diagramming, regardless of the particular context or issues being assessed. This formulaic and researcher-led use of participatory techniques can result in what some academics have called the 'tyranny of participation' (Cooke and Kothari 2001) in which any inequalities, particularly between the researcher and researched group, are reinforced (Parkes and Panelli 2001; Wadsworth 1998). A way around these difficulties is to consider who defines participation or who initiates what activities at each stage of the PAR process. Being open to sharing facilitation and innovating techniques in response to specific contexts can play a vital role in helping to monitor who is framing participation and how (Hickey and Mohan 2004; Williams et al. 2003). Discussing these aspects early on is a particularly useful way of clarifying expectations, establishing greater collaboration, and specifying roles and responsibilities (Kindon and Latham 2002; Manzo and Brightbill 2007).

Participatory techniques can generate information quickly, but they are not a substitute for more in-depth social research methods (Kesby 2000) such as those discussed elsewhere in this book. Understanding the contexts within which information is generated is critical to our ability to rigorously analyze it. For this reason, you might wish to consider first undertaking PAR within a community or location already familiar to you. Within an undergraduate dissertation, this could provide you with some of the necessary contextual information, freeing you to focus more energy on the process.

Sometimes our desire to avoid exploitation or extractive research relationships can mean that we become so involved with our co-researchers that we are unable to work effectively for change. Establishing outside support networks (see Bingley 2002) can help to prevent this situation and sustain our endeavours.

A related point is that long-term relationships, even friendships, with participants and co-researchers commonly develop through PAR, and while some studies may become lifetime projects, we typically have to leave the group with whom we have been working. Investing time into a sensitive and appropriate leaving strategy at the beginning of the research can help to avoid raising expectations and assist in navigating the changing status of relationships (Kindon and Cupples 2003). Formal meetings, celebrations, feedback sessions, visits, and the exchange of gifts may all be appropriate mechanisms to assist with closure.

In other cases, it may be academically and professionally important to maintain a sustained engagement long after the official research project is over—particularly to ensure the spread and impacts of any empowering change (Kesby 2005) and/or to practice what Paul Chatterton, Duncan Fuller, and Paul Routledge (2007) have called 'solidarity action research'. Your engagement may be in the capacity of support person, community board member, fund writer, publicist, or campaigner. There may be ethical challenges if your status changes from co-researcher to 'friend' or 'resisting other' (Chatterton, Fuller, and Routledge 2007), and being clear about what you can commit to in any of these relationships is vital. Overall, a key way to manage this and other challenges associated with PAR is to be realistic with yourself, your co-researchers, and other stakeholders about what is possible within the time and resources available to you.

PRESENTING RESULTS

In Chapter 3, Richie Howitt and Stan Stevens observed that representing people with whom we work is no easy undertaking, even if we involve them in the process. In their work with urban communities in Latin America, Cathy McIlwaine and Caroline Moser (2003) propose that a balance is needed with respect to the presentation of information. Information should influence policy-makers (to affect change) and should be meaningful to those involved in the initial research. In practice, this requires culturally appropriate ways of sharing knowledge and may, for example, require the production of several reports or presentations for different audiences by different members of the research teams using different media (see Cameron and Gibson 2005).

Project or policy reports are powerful advocacy tools to advance action plans developed during the research. Clear, simple presentations work best, providing policy-makers with a sense of the process and how it generated reliable, meaningful 'data' upon which practicable policy can be developed. Presenting the results of PAR at public meetings, conferences, or other gatherings with co-researchers can be appropriate and often enjoyable. If their direct participation is not possible, then discussing what they would like you to emphasize in a presentation can go some way towards addressing the power imbalance and lend you some authority to speak on their behalf. Taking the findings 'to the streets' in accessible media (for example, newsletters, magazine articles, plays, posters, websites, art) should ideally be part of the iterative process of PAR and can have some of the greatest impact at the local level (see Chapters 17 and 18 and Cahill and Torre 2007 for more discussion).

For you as a student, the results of your work also need to meet the requirements of the academy. It can be challenging to present the 'results' of an iterative and participatory research process within the context of a typical thesis or dissertation but certainly not impossible.[3] Engaging in PAR provides you with the opportunity to negotiate explicitly how you will use information and how you will represent others' experiences and/or views (Kindon and Latham 2002). Although sometimes

---

An Example of Participatory Publishing   Box 13.6

The following text is from US geographer Sarah Elwood, who has been working for many years with urban-based community groups and municipal councils in Chicago to support research, education, and community capacity-building within processes of neighbourhood redevelopment. It is a collaboratively authored chapter and represents one way in which multiple research partners' voices can be respected and shared.

In this chapter, we provide a collectively authored discussion of our use of GIS as a negotiating tool, and the lessons we have learned about sustaining community-university PGIS (**participatory geographical information systems**) projects. Our author group includes a university-based researcher (Sarah), the executive directors of the Near Northwest Neighbourhood Network (NNNN) and the West Humboldt Park Family and Community Development Council (the 'Development Council') (Bill and Eliud), a university-based research assistant (Nandhini), and past and present staff members of NNNN and the Development Council (Kate, Reid, Niuris, Lily and Ruben). . . .

From the perspective of the community organisation staff, PGIS partnerships are likely to be sustainable and effective if they produce immediately-applicable results, and if university partners do not dictate results. One staff participant writes:

> There has to be a clear understanding that the action and the research produced are usable for the community, not just something to publish.

Both directors argue that leaders of organisations have a special role to play in sustaining GIS resources and skills. One writes:

> Part of the job of executive director is recognizing the importance of the project, but also finding a way to keep it for the organisation, and not let it go.

In sum, for community organisations a great deal of the power and impact of PGIS stems from the diversity of ways it can serve as a process for community change.

Source: Elwood et al. 2007, 170–8.

time-consuming, such steps can temper your powerful position as the sole author in what has been, until now, a collaborative process. Sharing your choices and discussing how you intend to construct your argument continues the participatory process and goes some way towards ensuring that your final product respects the people and diversity of issues involved. An MoU item about this at the beginning of the research can prevent misunderstandings when you later want to quote people or include maps or diagrams produced through the research process.

Within the dissertation itself, including direct quotations from a range of people, which tease out common or disparate threads, can illustrate the multiple perspectives in circulation. Citing a disagreement or exchange between people can highlight where there are tensions or differences of perspective. However, take care to contextualize and analyze them adequately or they could be overwhelming to the reader. In addition, discussing aspects of methodology—so important to the participatory process—can honour people's involvement, acknowledge that process is as important as product, and enrich the analysis of the 'results' produced.

It may be appropriate and courteous to include co-researchers as co-authors on any papers that may emerge from PAR (see for example, Box 13.6; Hume-Cook et al. 2007; Peake 2000; Pratt et al. 1999; Townsend et al. 1995). However, we do not have to have our names on publications at all—we can work behind the scenes to enable our co-researchers to publish or disseminate their understandings independently of us. Or we can establish a collective name under which we write and publish. This avoids the problem of whose name comes first or who gets cited. It also demonstrates to others a collaborative and collective approach to research and writing (for example, see mrs kinpaisby 2008 and mrs c kinpaisby-hill 2008). Overall, having multiple research products written by different combinations of people can enrich the knowledge produced and be critically important if ongoing action is to be sustained (see Chapter 17).

REFLECTING ON PARTICIPATORY ACTION RESEARCH

A key question of PAR is often 'to whom is the research relevant?' (Pain 2003, 651). For us as researchers, if we accept that we have an opportunity and an obligation to co-construct responsible geographies (McLean et al. 1997; Williams et al. 2003), then PAR offers us an exciting means of undertaking relevant, change-oriented research. While academe does not usually reward such **activism**, the central role of space in many people's oppression (Ruddick 2004) means that human geographers are uniquely positioned, and morally beholden, to adopt ways of researching that build collaborative communities of inquiry (Reason 1998, cited in Hiebert and Swan 1999, 239) and challenge oppression.

Fortunately, certain parts of human geography, such as social geography, have a rich tradition of activism (Kindon 2009; Panelli 2004). In addition, PAR is becoming more common within geographic research, providing a growing body of work and experience from which to draw. PAR is not without its challenges, particularly within the confines of student research projects, but it is possible to adopt many of the principles discussed in this chapter to enable a rigorous research approach, which also results in tangible benefits for those involved. Perhaps the greatest challenge of all is for academics, including undergraduate researchers, to 'cross boundaries of privilege and confront their personal stake in an issue, and the ways they are positioned differently from members of the [groups] they work with' (Ruddick 2004, 239). I hope that this chapter has given you some ideas with which to begin this journey within your own work and some methodological resources for respectfully and ethically facilitating others' participation throughout the process.

KEY TERMS

action–reflection

action research

activism

co-learning

collaborative research

collective action

critical geography

emancipatory

epistemology

focus groups

memorandum of understanding (MoU)

participation

participatory action research (PAR)

participatory diagramming

participatory geographical information systems (PGIS)

participatory mapping

participatory research

social justice

stakeholder

transformative reflexivity

REVIEW QUESTIONS

1. Why is participation important in qualitative research?
2. Find an example of a participatory approach to geographic research in a recent book or journal. How is rigour established and maintained?
3. Given the importance of facilitation in PAR, make a note of the skills and attributes needed to be an effective PAR researcher. How might you develop and/or strengthen these skills and attributes in yourself?

4. What are some of the major challenges associated with doing PAR in geography? Make a list, and then devise strategies to manage these challenges productively.
5. Devise a list of ways to make your current research project more participatory. What are some of the implications for how you design and carry out each phase of the research? (You might like to focus on the implications for who is involved and in what capacity, what kinds of methods will be used, how long each phase might take, and how information will be used and presented.)

USEFUL RESOURCES

Action Research, Action Learning Association. 2008. 'Welcome to ALARA'. http://www. alara.net.au.

Cahill, C. 2004. 'Defying gravity: Raising consciousness through collective research'. *Children's Geographies* 2 (2): 273–86.

Cameron, J., and K. Gibson. 2005. 'Participatory action research in a poststructuralist vein'. *Geoforum* 36 (3): 315–31.

Fed Up Honeys. 2008. 'Welcome'. http://www.fed-up-honeys.org/mainpage.htm.

Kesby, M. 2000. 'Participatory diagramming: Deploying qualitative methods through an action research epistemology'. *Area* 32 (4): 423–35.

Kesby, M., S. Kindon, and R. Pain. 2005. '"Participatory" diagramming and approaches'. In R. Flowerdew and D. Martin, eds, *Methods in Human Geography*, 2nd edn, 144–66. London: Pearson.

Kindon, S. 2009. 'Participation'. In S. Smith, R. Pain, S. Marston, and J.P. Jones III, eds, *The Handbook of Social Geography*. London: Sage.

Manzo, L., and N. Brightbill. 2007. 'Towards a participatory ethics'. In S. Kindon, R. Pain, and M. Kesby, eds, *Participatory Action Research Approaches and Methods: Connecting People, Participation and Place*, 33–40. London: Routledge.

mrs c kinpaisby-hill. 2008. 'Publishing from participatory research'. In A. Blunt, ed., *Publishing in Geography: A Guide for New Researchers*, 45–7. London: Wiley-Blackwell.

Pain, R., and P. Francis. 2003. 'Reflections on participatory research'. *Area* 35 (1): 46–54.

Participatory Geographies Research Group. 2008. 'Participatory Geographies Research Group'. http://www.pygywg.org.

Young, L., and H. Barratt. 2001. 'Adapting visual methods: Action research with Kampala street children'. *Area* 33 (2): 141–52.

Notes

1. Readers of this chapter are strongly encouraged to also read Chapter 3, which provides a useful, complementary overview of cross-cultural research ethics, methods, and relationships.

2. The research team may consist of researchers only, researched people only, or both researched people and researchers.

3. If you are using PAR in a thesis, for example, and because universities typically expect a thesis or dissertation to be the 'original work' of the student alone, you should discuss matters of authorship with your supervisor when preparing your project.

Part III **'Interpreting and Communicating' Qualitative Research**

Coding Qualitative Data

14

Meghan Cope

CHAPTER OVERVIEW

This chapter defines coding, which is a process of identifying and organizing themes in qualitative data, reviews different types of codes and their uses, and discusses several ways to get started with coding in a qualitative project. Specifically, a distinction is drawn between descriptive codes, which are category labels, and analytic codes, which are thematic, theoretical, or in some way emerge from the analysis. Borrowing from the work of grounded theory's Anselm Strauss, a basic four-point plan is reviewed as a strategy to begin coding focused on looking for conditions, interactions, strategies/tactics, and consequences. The building of a 'codebook' is also discussed, stressing the importance of looking critically at the codes themselves, identifying ways in which they relate, minimizing overlap between codes, and strengthening the analytical potential of the coding structure. Finally, several related issues are covered, such as coding with others, integrating coding and mapping, and viewing the world from the perspective of coding.

INTRODUCTION

Geographers are increasingly engaged not only in doing qualitative research but also in thinking and writing critically about methodologies, including the ways that we evaluate, organize, and 'make sense' of our data through the **coding** process (Cope 2003; Jackson 2001). Coding social data (for example, text, images, talk, interactions) is sometimes derided as tedious, but if you think of it as a kind of detective work, it can be intriguing, exciting, and very valuable to the research process.

The purposes of coding are partly **data reduction** (to help the researcher get a handle on large amounts of data by distilling along key **themes**), partly organization (to act as a 'finding aid' for researchers sorting through data), and partly a substantive process of data exploration, analysis, and theory-building. Further, different researchers use coding for different reasons depending on their goals and epistemologies; sometimes coding is used in an exploratory, inductive way such as in **grounded theory** in which the purpose is to generate theories from empirical data, while other times coding is used to support a theory or hypothesis in a more deductive manner. Several approaches are discussed here, with pointers on how to organize and begin the coding aspect of a research project.

TYPES OF CODES AND CODING

One common type of coding is **content analysis,** which is essentially a *quantitative* technique and by no means represents the full extent of coding for qualitative research. Content analysis can be done by 'hand' or by computer (see Chapter 15 for a discussion), but either way it is a system of identifying terms, phrases, or actions that appear in a document, audio recording, or video and then counting how many times they appear and in what context. For example, a researcher might be interested in how many times the word 'democracy' is used in newspaper articles from a particular country or in how many and what type of places are portrayed in a television program. Frequently in content analysis, sampling is used in similar ways to quantitative analysis of populations; perhaps only front-page newspaper stories are included in the analysis, or a television program is sampled for five minutes out of each hour. Similarly, researchers using content analysis typically subject their coded findings to standard statistical analysis to determine frequencies, correlations, variations, and so on. There are many good guidebooks and instructions for conducting content analysis, including some available on the Internet (see, for example, Krippendorff 2003; Neuendorf 2001).

While content analysis is a frequently used type of coding, the primary focus of this chapter is on qualitative approaches to coding. However, one of the basic principles of content analysis has broad implications for all coding activity and is thus worth exploring further. It is the notion that there are both 'manifest' and 'latent' messages contained in the material (for example, text, images, video). **Manifest messages** are those that are blatant and obvious—they then generate manifest *codes.* For example, if I (as a feminist geographer) were performing content analysis on a set of international newspapers and the term 'sex worker' appeared with some level of frequency, I would take that as a code and proceed to scan subsequent materials for it. However, because places with high levels of prostitution tend to be places where women have few other economic opportunities or political rights, I might also code instances of the term 'sex worker' for the **latent message** of 'the status of women' as well (see also Chapter 11).

In much of ethnographic work in which researchers use coding qualitatively, the ideas of manifest and latent codes have parallels in 'descriptive' and 'analytic' codes. **Descriptive codes** are similar to manifest codes: they reflect themes or patterns that are obvious on the surface or are stated directly by research subjects. Descriptive codes can be thought of as category labels because they often answer 'who, what, where, when, and how' types of question. Examples of descriptive codes that might interest geographic researchers include demographic categories (male, female, young, elderly), site categories (home, school, work, public space), or even scale identifiers (local, regional, national, global).

One special type of descriptive code is called *in vivo* **codes**; they are descriptive codes that come directly from the statements of subjects or are common phrases found in the texts being examined (Strauss and Corbin 1990). For example, if interviews were done with elderly women and they repeatedly mentioned concern with crime in their neighbourhoods, 'crime' would become an *in vivo* descriptive code—the term is used by and describes something important to the subjects. *In vivo* codes are a good way to get started in coding, particularly in projects that are designed to be inductive or exploratory.

Ethnographers also develop **analytic codes** to code text (or other forms of data) that reflect a *theme* the researcher is interested in or one that has already become important in the project. Analytic codes typically dig deeper into the processes and context of phrases or actions. For example, it might become apparent that the elderly women mentioned above were especially afraid of young men and boys whom they perceived as threatening while walking down the street, and therefore an analytic code called 'fear of young males in public space' could be developed. This code might then be applied to the rest of the data to identify other instances of fear, perceptions of young men, and experiences in public spaces.

Often, descriptive codes bring about analytic codes by revealing some important theme or pattern in the data or by allowing a connection to be made (for example, crime, fear of youth in public), while other times analytic codes are in place from the beginning of the coding process because they are embedded in the research questions. For instance, if we were interested from the start in how elderly women navigate urban spaces, their personal mobility and impediments to mobility would be themes reflected in the analytic codes from the project's very beginning. The recursive strength of coding lies in its being open to new and unexpected connections, which can sometimes generate the most important insights.

THE PURPOSES OF CODING

There are three main purposes for coding qualitative material: data reduction, organization and the creation of searching aids, and analysis. As the prolific French theorist Henri Lefebvre noted, 'Reduction is a scientific procedure designed to deal with the complexity and chaos of brute observations' (Lefebvre 1991, 105). Qualitative research usually produces masses of data in forms that are difficult to interpret or digest all at once, whether the data are in the form of interview transcripts, hours of video, or pages of observation notes. Therefore, some form of reduction, or **abstracting**, is desirable to facilitate familiarity, understanding, and analysis. Coding helps to reduce data by putting them into smaller 'packages'. These packages could be arranged by topic, such as 'instances in which environmental degradation was mentioned', or by characteristics of the participants such as 'interviews with women working part-time', or by some other feature of the research context or subjects such as 'observations in public spaces'. By reducing the 'chaos of brute observations', data reduction helps us get a handle on what we have and allows us to start paying special attention to the contents of our data.

The second purpose of coding is to create an organizational structure and finding aid that will help us make the most of qualitative data. Similar to data reduction, the organizational process mitigates the overwhelming aspects of minutiae and allows analysis to proceed by arranging the data along lines of similarity or relationship. An important background step is constructing and maintaining a complete record of sources, dates (of participant observation, interviews, or focus groups, for example), subject contact information, and other relevant information. While this database is not part of the coding process per se, it is an important step in organizing qualitative material for coding and analysis and also allows the researcher to find specific data more easily. For example, interview transcripts might be coded not only for their content but also by their circumstances—was the interview conducted in the participant's home? were others present? did the subject seem nervous?—which can help organize information. With better computer-assisted qualitative data analysis software (CAQDAS) available now (see Chapter 15), organizing and searching within electronic documents is greatly simplified. Additionally, coding itself is also an important aspect of organizing and searching because it is essentially a process of categorizing and qualifying data. 'The organizing part will entail some system for categorizing the various chunks [words, phrases, paragraphs], so the researcher can quickly find, pull out, and cluster the segments relating to a particular research question, hypothesis, construct, or theme' (Miles and Huberman 1994, 57). While the development of the **coding structure** is by no means a simple process, it is one that—if done well—enables the data to be organized in such a way that patterns, commonalities, relationships, correspondences, and even disjunctures are identified and brought out for scrutiny.

The final, and principal, purpose of coding is analysis. While strategies for analytical coding will be examined in greater detail below, at this point it is sufficient to note that the *process* of coding is an integral part of analysis. Rather than imagining that analysis of the data is something that begins after the coding is finished, we should recognize that coding *is* analysis (and is probably never truly 'finished'!). Coding is in many ways a *recursive* juggling act of starting with **initial codes** that come from the research questions, background literature, and categories inherent in the project and progressing through codes that are more interpretive as patterns, relationships, and differences arise.

Coding also opens the opportunity for reflexivity, a critical self-evaluation of the research process (see Chapter 2). By recursively reviewing data and the connections between codes, researchers can also come to see elements of their own research practice, subjects' representations, and broader strategies of knowledge construction that had not previously been apparent. While it is sometimes difficult to be self-critical in the midst of fieldwork or data collection, the process of coding is inherently more contemplative and analytical and thus offers a suitable moment for reflection.

How to Get Started with Coding

The above discussion of types of codes addressed two main approaches to coding, which may be seen as descriptive and analytic codes, although other terms are also used (for example, initial codes and **interpretive codes**). The key distinction is that one type of code is fairly obvious and superficial and is often what the researcher begins with, such as simple category labels. The other type of code is interpretive, analytic, and has more connections to the theoretical framework of the study; it tends to come later in the coding process after some initial patterns have been identified. When coding was (and sometimes still is) done manually, researchers developed a **codebook**—a long list of codes that were categorized and organized repeatedly. Although current qualitative software packages typically do not use the term 'codebook', it is a useful concept that has relevance whether the codebook is actually a tangible item in manual coding or merely an abstraction in electronic coding.

To start a codebook, it is easiest to begin with the most obvious qualities, conditions, actions, and categories seen in your data and use them as initial codes. These elements will emerge quite rapidly from background literature, your own proposal or other research-planning documents, and the themes that stick out for you from gathering qualitative data (for example, memorable statements in interviews, notable actions seen while doing participant observation, key words that jump out in first readings of historical documents). For example,[1] in my work on how urban children conceptualize city spaces in Buffalo, New York, one of my original interests that was heavily present in my grant proposal was how children in the 8- to 12-year-old age group define 'neighbourhood' and 'community' (see Box 14.1). These terms are obvious starting points for my codebook.

29 November 2003. After-School Program Observation Notes, Children's Urban Geographies Research

Box 14.1

Text: Field Notes from the Quilt Project	Descriptive and Category Codes	Analytic Codes and Themes	Notes
As I was setting up, Jakob,* Mariana, and Ari came over and then Izzy and Salomé (a new girl I hadn't met before). We set up at a round table in between the bench and the 'café', near the pool table. The noise level was very high and I had a hard time hearing the children at my table. Next to us, three or four younger boys (Stefan et al.) were playing a war board game and making lots of terrible noises (at one point I asked them to be quieter). After I explained what I wanted (to use the materials to show your house or apartment building and family), I asked the children what a 'neighbourhood' is. There were varying answers immediately, mostly around the idea of 'a bunch of houses next to	Jakob Mariana Ari Izzy Salomé *Relations:* Izzy and Salomé are friends	Relationship between gender and violent play?	*Early release day from school—kids were wild and bored.*
			Tape recording would not have worked here!
	Stefan		
	Research setting conditions: loud		*Gave very loose instructions to allow children freedom within the project's scope.*
	'Neighbourhood'	Difference between 'neighbourhood' and 'community'	

each other'. Izzy said, 'it's when you have one house and then another one and you all get together to play.' Mariana said, 'I don't live in a neigh- bourhood, I'm part of the West Side Com- munity'(!) I couldn't hear very well, so I got out my notebook and went around the group to write down answers. Ari's answer was very long and complicated with something to do with your 'home friend'. I'd like to revisit the question of what is a neighbourhood in video interviews. Then I got out my digital camera and took pictures of the group (all five gave full permission for this). Nate came up and wanted to 'see' the camera, which I didn't want to let him do because he is so volatile and unpre- dictable. Reluctantly, I let him take a picture of me with the chil- dren working on the quilt and retrieved the camera from him immediately. [Ironically, the photo Nate took is one of my favourites of this project!]	Play 'Community' *Tactic:* attention Technology Nate *Interactions:* Nate's bullying	Children's identification with a commu- nity or neigh- bourhood Our relation- ships with specific children	*Mariana seems proud of her West Side iden- tification.* *Ari (age 5) seems to crave attention.* *Future work—video interviews* *All the children love technology and the gizmos we bring in get a lot of attention.* *Review and code photos.*

* All children's names have been changed

However, codes can also be too general and become cumbersome. Because much of my children's urban geography research is centred around issues of neighbourhood and community, I found that I needed to break each of them into more specific codes, such as codes for the particular neighbourhoods the children refer to, the use of both 'neighbourhood' and 'community' to mean 'local' (such as in reports from the city newspaper), and the way that school curriculum materials define 'community'. This is a frequent characteristic of coding: an initial category becomes overly broad and must be refined and partitioned into multiple codes.

Bear in mind that the opposite also occurs—some codes die a natural death through lack of use. For instance, in my project I had expected the children, who are for the most part in low-income families, to talk about a lack of money or not being able to afford something they wanted. However, after two years in the project, I have found little evidence of children discussing their own poverty (though that absence is itself an interesting research question). While I will probably keep a 'low-income' code for other purposes, its prevalence is much less than I anticipated in the materials generated by the children. As Miles and Huberman said, 'some codes do not work; others decay. No field material fits them, or the way they slice up the phenomenon is not the way the phenomenon appears empirically. This issue calls for doing away with the code or changing its level' (1994, 61).

So the first step is to make a list of what you think are the most important themes upfront, with the understanding that some of them will be split into finer specifics while others will remain largely unused. But how do you know what is important? Anselm Strauss, one of the founders of grounded theory, had a helpful system for beginning this awesome task (best represented in Strauss and Corbin 1990). He suggested paying attention to four types of themes:

- conditions
- interactions among actors
- strategies and tactics
- consequences

'Conditions' might include geographical context (both social and physical), the circumstances of individual participants, or specific life situations that are mentioned or observed (for example, losing a job, becoming a parent, a child changing schools). By thinking along the lines of 'conditions' and coding only for them, the coding process is easily started, and you may learn a lot about your data in a short time.

The same is true for limiting your scope to 'interactions among actors'—if you focus on relationships, encounters, conflicts, accords, and other types of interactions, a series of powerful codes will emerge that will be helpful throughout the research. For example, in her research on adolescent girls in the southern United States, Mary Thomas (2004) found that young (14-year-old) African-American girls' interactions with peers were strongly implicated in the type and level of their sexual activity. Thus, Thomas might have coded her interview transcripts regarding peer factors by *whether, how*, and *where* girls engaged in sexual activity, as well as *whom* they were influenced by or interacted with.

'Strategies and tactics'[2] is a little more complicated than Strauss's first two types of themes because it requires a deeper understanding of the things (events, actions, statements) you observe and how they relate to broader phenomena and it suggests a certain level of purposeful intent among the research subjects that may demand additional inquiry on your part. For example, feminist geographers are often interested in women's survival or 'livelihood' strategies in different areas of the world (see, for example, the special issue of *Gender, Place and Culture*, 2004, vol. 11, no. 2). Noting that women in certain economic contexts tend to use particular types of financial survival tactics (for example, growing food products for sale in a local market) can begin to illuminate broader economic, social, and political processes that shape women's options and actions, which is a valuable insight for geographic research. Other types of strategies or tactics might involve career decisions, political activism, housing choices, family negotiations, or even subversion.

Coding for strategies and tactics can be straightforward (and descriptive) in instances when respondents say something like 'I moved in with my mother so that she could care for my baby while I finished job training' or 'I got involved with a local group of residents to raise awareness of environmental contamination in our neighbourhood because I was concerned about property values.' Note the words 'so' and 'because' in these statements, which are good tip-offs that a strategy or tactic is embedded in the text.

Other times, coding for strategies and tactics may be more subtle—and more analytical—as when respondents do not explicitly state their reasons for certain actions but a connection emerges through observation, review of interview text, or other data. For instance, many geographers (for example, Blumen 2002; Cresswell 1999; Flint 2001; Nagar 2000; Secor 2004) have paid attention to ways that people engage in *resistance* against diverse forms of oppression, which may be seen as strategies for empowerment, rights, or merely survival. Orna Blumen (2002, 133) took 'dissatisfaction articulated in subtle terms' by ultra-orthodox Jewish women as small but significant indicators of the women's resistance to their families' economic circumstances and, more broadly, to the status and roles of women in that community. For the women in Blumen's study, then, referring to fatigue, hoping their husbands would soon find paying work, and 'minor, personal, nonconformist remarks suggestive of ambivalence' (2002, 140) could all be coded as tactics of resistance, in part because Blumen—through careful qualitative work—had sufficiently analyzed the broader context of the women's lives and goals.

Similar to the above, 'consequences' is a slightly more complicated code. On the surface, there are descriptive indicators for consequences, including terms such as 'then', 'because', 'as a result of', and 'due to' that may be used in subjects' statements and can be good clues to consequences and as a first-run could certainly be used in this way. Again, however, there are also more analytically sophisticated ways of discovering and coding consequences that are dependent on the unique empirical settings and events of each study. Some consequences will be matters of time passing and actions taking place that result in a particular outcome—the passage of a law, a change in rules or practices, and so on. However, other consequences are more subtle and personal, or they are not the result of changes over time and therefore may be trickier to identify and code as such. For example, when Anna Secor (2004) hears from young Kurdish women living in Istanbul that they feel uncomfortable in some areas of the city, she might code her focus group transcripts for the consequences of feeling out of place due to the women's identity as an oppressed minority in Turkey. Coding for 'consequences' of this kind requires sensitivity to both the subjects and their community context but is potentially a rich source of analysis and insight if done with care.

As an example of what a sample of coded material looks like with both descriptive and analytic codes, Box 14.1 demonstrates a small selection of field notes from my project on children's urban geographies along with codes, themes, and notes. Even this fairly short piece of text reveals several relationships (friends, bullies), tactics (ways of getting attention), and conditions (chaos, noise level) that stimulated further examination in other project analysis. Additionally, several analytic themes or questions are seen emerging here: the possible relationship between gender and violent play, some children's pride in perceived community membership (despite living in a blighted physical environment), and the importance of play in defining what a neighbourhood is among the children. Subsequent to the quilt project represented here and in combination with other Children's Urban Geographies Project data, I generated a theory of how children define and ascribe meaning to the idea of 'neighbourhood' (Cope 2008); theory-building, after all, is an important goal of most qualitative research.

DEVELOPING THE CODING STRUCTURE

Using the four types of themes reviewed here will take you a long way towards constructing a codebook, and you may find other types of themes that are helpful to you, such as 'meanings', 'processes', or 'definitions'. Using the combination of descriptive and analytic codes, you may well have more than 100 codes by this point, which is unwieldy at best and counterproductive at worst. Lists of codes that have not been categorized, grouped, and connected will be hard to remember, have too much overlap and/or leave too many gaps, and will not enable productive analysis. Therefore, the next step is to develop a coding structure whereby codes themselves are grouped together according to their similarities, substantive relationships, and conceptual links. This process requires some amount of work but is well worth the effort, both for ease of coding your material and for discerning significant results from your findings.

Developing the coding structure can proceed in various ways, and there are many resources available that demonstrate different approaches (see Denzin and Lincoln 2000; Miles and Huberman 1994), but the main purpose is to organize the codes—and therefore the data and the analysis process. Some codes will automatically cluster; for example, codes relating to the *setting* of interviews (for example, home, office, public space, clinic), *characteristics* of subjects (for example, age, gender), or other *categories* (for example, occupations, leisure activities, life events). Other codes seem to fit together because of their *common issues*; for example, you might have a group of codes related to people's goals or intentions or a group of codes related to people's experiences of oppression. Finally, codes based on the *substantive content* of text or actions—and most likely related to the analytic themes you are developing—will create another cluster of codes: for example, perceptions, meanings, places, identities, memories, difference, representations, and associations.

Once the codebook is relatively comfortable (I hesitate to say 'complete') and the coding structure is devised, you will want to go through much of your data again to capture connections that may have been missed the first (or second or third) time around. Remember that coding is an iterative process that feeds back on itself—only you can decide when it is time to move on. As Miles and Huberman (1994) point out, it is sometimes simpler when time or money pressures put a finality on projects that otherwise could always benefit from 'one more case study' or endless additional tweaking of the coding structure!

CODING WITH OTHERS

Depending on the size and resources of the research project, there may be a case for using multiple coders for the data, which complicates the process considerably. There is an inherent tension in using multiple coders on a project: is the goal to make everyone code as consistently as possible, or is the goal to allow each coder to interpret data in her or his way within the bounds of the coding structure in order to capture many diverse meanings? The answer will depend on the project and the epistemological leanings of the lead researcher, but in fact both of these goals are important. In the first instance, reliability of the data is undoubtedly enhanced when several coders independently code a piece of data the same way—a common interpretation of data means that there is agreement on its meaning. For the sake of time and data reduction, having multiple coders can certainly be helpful, but only if they are truly consistent in their coding, which is rare but could be accomplished by achieving conformity on the meanings of codes and providing thorough definitions for each code. On the other hand, text and video—as social data sources—are inherently subject to multiple interpretations and understandings, all of which may be correct or 'true'. While there may be some interpretations that are farfetched or extreme, in general we as social researchers will be interested in capturing diverse understandings, and having multiple coders can be a great benefit for the project to make deeper and broader connections from the data.

CODING AND MAPPING

New work is emerging that pushes the boundaries of coding and leads to creative analyses and representations of qualitative geographic data. With the rise of participatory research in geography (Kesby, Kindon, and Pain 2005; see also Chapter 13 in this volume), qualitative geographical information systems (GIS) (Cope and Elwood 2009; Kwan and Knigge 2006), participatory mapping, mixed methods, and other integrative practices, we need to stay attuned to how coding can keep up with new research processes and technologies. One example of this is Jin-Kyu Jung's (2009) experiments with using codes as a bridge between qualitative analysis using CAQDAS (computer-assisted qualitative data analysis) and spatial data analysis using GIS. In his work, the code literally serves as a software-level link between databases, allowing analysis programs to 'speak' to each other in a platform he calls **CAQ-GIS (computer-aided qualitative geographical information systems)**. At a conceptual level in Jung's work, the code also serves as an analytical connection between social contextual data and spatially referenced data, allowing researchers to develop new understandings of social–spatial relations. These and other innovations will continue to challenge and enrich geographic inquiry.

BEING IN THE WORLD, CODING THE WORLD

By way of conclusion, let me point out that coding is not a mysterious process that must be learned from scratch but rather is one that we are all already actively practising in our everyday lives. The recognition that we are all constantly interpreting and 'coding' the world around us may be a helpful realization for getting started in a research project and can also assist us in critiquing our own practices of data reduction, organization, and analysis. As Silverman (1991, 293) points out, there are many ways of 'seeing' and interpreting the world, and—as social beings—we never really shut those lenses off, so why not embrace diverse interpretations and turn our gaze to the process of interpretation?

> How we code or transcribe our data is a crucial matter for qualitative researchers. Often, however, such researchers simply replicate the positivist model routinely used in quantitative research. According to this model, coders of data are usually trained in procedures with the aim of ensuring a uniform approach. . . . However, ethnomethodology reminds us that 'coding' is not the [sole] preserve of research scientists. In some sense, researchers, like all of us, 'code' what they hear and see in the world around them [all the time]. . . . The ethnomethodological response is to make this everyday 'coding' (or 'interpretive practice') the object of inquiry. [Silverman 1991, 293]

Being in the world requires us to categorize, sort, prioritize, and interpret social data in all of our interactions. Coding qualitative data is merely a formalization of this process in order to apply it to research and to provide some structure as a way of conveying our interpretations to others.

KEY TERMS

abstracting

analytic code

CAQ-GIS (computer-aided qualitative geographical information systems)

codebook

coding

coding structure

content analysis

data reduction

descriptive code

grounded theory

initial codes

interpretive codes

in vivo code

latent message

manifest message

theme

REVIEW QUESTIONS

1. What is the difference between descriptive and analytic codes, how do they relate to one another, and what are their respective uses in coding qualitative data?
2. Why does the author state that coding is analysis?
3. What are some potential benefits and potential problems with having multiple people coding in a project?
4. In what ways do we 'code' events, processes, and other phenomena in everyday life? How might thinking about these ways help us to become better qualitative researchers?

Useful Resources

For an excellent step-by-step guide to coding from a grounded theory perspective, see Strauss and Corbin 1990 and the chapter by Kathy Charmaz (2000) in the venerable volume edited by Denzin and Lincoln, *Handbook of Qualitative Research* (which itself is worth a look, although it may require a trip to the library because of its high cost). Alternatively, Miles and Huberman (1994) offer a thorough discussion of several different approaches to coding, although their own coding examples are somewhat arcane and confusing. Finally, there are several examples of coding and 'making sense' of data by geographers, including collections by Clifford and Valentine (2003); Flowerdew and Martin (2005); Limb and Dwyer (2001); and Moss (2002).

Notes

1. While it is always difficult to convey examples of coding without recounting the entire scope of the research, it is hoped that these examples from a real research project are sufficiently illustrative to demonstrate different coding approaches.

2. Despite the similar pairing of these two words, I am not referring here to Michel de Certeau's (1984) notion of 'strategy' (a technique of spatial organization employed by 'the powerful') and 'tactic' (an everyday means of 'making do', typically used by those with few options), although there are certainly potential connections. Rather, I am using the terms in their most literal sense as they are employed in Strauss and Corbin (1990) to convey ideas about how people conceptualize what they want and what they do to try to arrive at those goals.

Computers, Qualitative Data, and Geographic Research

15

Robin Peace and Bettina van Hoven

CHAPTER OVERVIEW

This chapter provides a brief introduction to the use of computers in qualitative human geography research. It focuses on a description of some of the basic aspects of computer-assisted qualitative data analysis software (CAQDAS) and suggests several reasons for developing CAQDAS skills as part of a researcher's toolkit. A summary of advantages, concerns, and future directions for computer use in human geography research concludes the chapter.

INTRODUCTION

Qualitative data, in general, 'tell it like it is' and provide the researcher with stories that detail human lives. As rich, interesting, and exciting as the respondents' stories are, when turned into 'data' they can also become 'unstructured': a source of chaos and headache. In this chapter, we offer insight into the question of whether or not computers can help us deal with these rich but unstructured data. Qualitative data may be coded, sorted, retrieved, and manipulated by using coloured pencils, self-stick notes, scissor-based cut and paste techniques, or index cards (see Ritchie and Lewis 2003 or Miles and Huberman 1994 for discussion of data reduction and display by hand). Using a computer to assist with qualitative data analysis (QDA), however, is not simply a matter of replacing these techniques with a software program. Using computers also opens up new ways of thinking about data analysis (Staller 2002). This chapter illustrates the role that computers and software play in the research process as well as identifying the opportunities and limitations such technologies produce.

Researchers who are new to **computer-assisted qualitative data analysis software (CAQDAS)** may have a number of questions: How can computers help with data? How many kinds of software are there? Which one should I choose? How long does it take to learn how to use the software? Will the computer do my analysis for me, and how will I know if it has got it right? Factual queries aside, many questions are personally motivated and determined, and you will not necessarily find all the answers you are looking for in this chapter. However, at the end of this chapter, you will find a list of additional references and sources to provide further guidance on where to find answers and help with making decisions, as well as discussion lists for communicating the joys and troubles of CAQDAS with fellow researchers. Before you start, however, we wish to firmly emphasize that computers *do not do analysis*. Even the most sophisticated software is merely one of many tools in a researcher's toolkit.

WHO IS USING CAQDAS?

Qualitative researchers in disciplines as disparate as anthropology, economics, psychology, and sociology, as well as in more interdisciplinary areas such as management studies, social work, health, and education, are all engaged in exploring qualitative methods that rely on computer assistance (see, for example, Eliott and Olver 2002; Eustace 1998; Scribner 2003). The CAQDAS Networking Project website, the QSR website, and the University of Huddersfield on-line QDA website each has extensive lists of literature discussing either techniques and/or the software or specific case studies. Increasingly, articles and books are appearing that provide comparative information about particular software packages (see in particular Lewins and Silver 2007; Lewis 2004).

Human geography has been slower to respond to the opportunities suggested by the availability of CAQDAS, despite having a well-developed reputation for computational work involving statistics and geographic information systems (GIS). Among the few pieces of geography research cited in the literature on the use of and problems with CAQDAS are Baxter 1998; Crang et al. 1997; Hinchliffe et al. 1997; and more recently, Bhowmick 2006. Examples of the use of a specific program in geography (QSR N4) can be found in van Hoven's (2003) study of rural women in eastern Germany (see also van Hoven-Iganski 2000) and van Hoven and Poelman's (2003) work on sense of place of Brighton, England, as experienced by lesbian, gay, bisexual, and transgendered (LGBT) people. Meijering's (2006) research on rural intentional communities is an example of data analysis using QSR NVivo 2.0. Despite the small amount of published material, there is a growing awareness in geography that computer software can have an important role in qualitative work.

TOOLKIT TECHNOLOGY

In general, you are probably familiar with the use of computer software for your studies. You are likely to have used it for making and editing notes, storing them, searching for and retrieving text, and preparing essays and reports. More specific types of geographical activity that can be effectively handled by computers, and that you are likely to have encountered as well, include statistical analysis, modelling, graphics, cartography, **image processing**, visual imagery, and remote sensing. Before the 1990s (but since the late 1960s), computing in geography was largely associated with **geographic information systems** (GIS), with SPSS (Statistical Programs for the Social Sciences), and with SAS (Statistical Analysis System) software (Earickson and Harlin 1994; Griffith, Desloges, and Amrhein 1990; Rogerson 2001; Shaw and Wheeler 1994). These are software systems with the capacity to deal with the statistical calculations and manipulations of numeric or digitized 'real-world' data—migration statistics, demographic data, census information, or satellite readings. In geographical information systems such as ARCVIEW, the software facilitates links between graphic files (digitized pictures of some kind, such as maps) and **attribute databases** (such as information on the location or movement of some factor) in order to identify patterns or changes over time. SPSS and SAS are specialized statistical programs that are used to perform complex calculations, to test research hypotheses, and to convert structured data from numeric to graphical display forms.

Relatively recently, human geographers, along with other social scientists, have looked for computer support for analyzing unstructured, qualitative data. Such support has been forthcoming in the form of specialized software that has been 'custombuilt' for social science qualitative research. Since the 1990s, the range of geographical research activities has been extended to include **word searching**, data coding, **data storage** and retrieval, **memoing**, **graphic mapping**, **hierarchical tree-building**, **concept building**, and reflexive report-writing (see the glossary at the end of this volume for explanations of some of these terms if they are unfamiliar to you).

The emergence of visual data in geographical research also implies the development of ways to interpret and manage such data (Johnsen, May, and Cloke 2008; Radley, Hodgetts, and Cullen 2005; Rose 2007; Stanczak 2007). One emerging field links qualitative data analysis, GIS, and visual data in innovative ways and can be seen in the varied work of Jiang (2003), Knigge and Cope (2006), and Kwan (2002), in which GIS-based analysis has been integrated with ethnographic research, grounded theory and visualization, and feminist geographic research respectively. This new field of 'qualitative geographic information systems' is summarized by Cope and Jung in a new 12-volume *International Encyclopedia of Human Geography* (2009) edited by Rob Kitchin and Nigel Thrift (see also Kwan and Knigge 2006).

Although some of these activities are more closely linked to the quantitative capacities of GIS software, they also represent the blurring of the boundaries between quantitative and qualitative approaches to data collection and analysis and provide a stepping-off point for considering computer-assisted qualitative data analysis systems.

WHAT IS CAQDAS?

CAQDAS is a generic title for a range of software that is specifically designed to handle unstructured, qualitative data. They are different from the data handling and statistical systems mentioned above that have been familiar to geographers in the past.

Computer software that is designed to assist researchers involves a three-way relationship between the 'researcher', the 'research process', and the 'hardware and software'. The individual skills, attributes, and desires of the researcher inform the kind of research that is undertaken, govern the researcher's actions, and influence the kind of analysis that is performed. These two elements (i.e., 'researcher' and 'research processes') are in turn influenced by technology. The researcher uses technology to support or enhance the processes in which he or she is engaged. CAQDAS, no less than other software, relies on this three-way relationship. Electronic technology can now facilitate the manipulation and management of qualitative data. Two important technological facets of the data management processes are: (1) being able to locate and capture particular words or phrases that form part of the research corpus (word searching) and (2) being able to develop themes and identify concepts (concept building).

In 1995, Weitzman and Miles put forward a typology to help researchers distinguish the different capabilities of individual software programs. We have used this typology to create a framework for the discussion that follows.

Word searching, a relatively simple task that most word processors can perform with fairly rudimentary instruction from the researcher (and therefore a relatively low level of skill), is at one end of the range. The researcher's aim may be, for example, to find out how respondents talk about certain terms and the frequency of references to this term (content analysis). In the example below, a word search was conducted to find out how correctional officers in a New Mexico prison talked about 'control' (see Figure 15.1).[1] In this case, the 'find' or 'search' tool in the word processor can be used to look for key words and to return a complete record of all occurrences. Using this technique, large amounts of text may be searched quickly. In addition, the copy and paste functions can be used to transfer material electronically to another document without retyping.

Concept building, which is a more sophisticated task requiring specialized software and considerable researcher training, exists at the other end of the range. Box 15.1 outlines the specific features of this process.

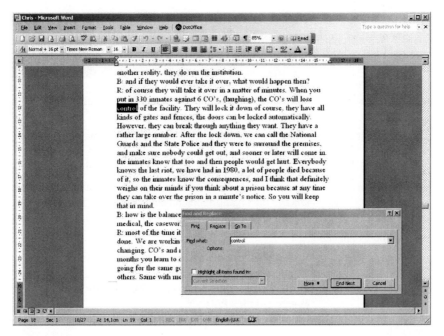

Figure 15.1: Text retrieving in Microsoft Word. Searching for 'control'.

It should be noted, however, that in the highly competitive IT (information technology) marketing environment, software capabilities are increasing all the time and many of their distinguishing characteristics are now blurred. For instance, in their clear descriptive account of different software packages Lewins and Silver (2006, 2) suggest that the distinctions 'between software formerly labeled Code and Retrieve and the more extensive functionality in Code-based Theory Builders have become blurred' and text-retrievers and **text-based managers** are 'beginning to enable tasks originally only possible in Code-based Theory Building software' (see also Weitzman 2000; Digital Research Tools n.d.).

DIFFERENT TYPES OF SOFTWARE

There is no recipe for choosing the software that will be best for your project. You need to bear in mind the kind of computer you are using, your budget, what access you have to software in your institution, your current skill level and the amount of time you have to learn new skills, and which kind of program is likely to be most useful to you. Rather than give any product specifications or recommendations, we summarize some of the key information below. Lewins and Silver (2007) provide an excellent, workbook-based study of seven key software programs, and Weitzman and Miles's (1995) earlier text canvasses 24 (see also

Some Characteristics of Research Using Concept Building by Means of Qualitative Data Analysis Software	**Box 15.1**

The researcher	Is likely to be experienced in qualitative research, possibly working at post-graduate level.
The research process	Involves qualitative analysis of transcribed in-depth interviews with numbers of respondents, intensive specialized analysis of a small (focus) group data or documents.
The researcher's question	For example, to examine the roles of looking and surveillance in the organization and control of space in a New Mexico prison (van Hoven and Sibley 2008).
Technology needed	
Hardware:	For a sophisticated program like NVivo8, minimum requirements are: Processor: 1.2GHz Pentium III-compatible processor Memory: 512 MB RAM Hard disk: approx. 1 GB of available hard-disk space.
Software:	QDA program (for example, the Ethnograph5, ATLAS/ti5, NVivo8).
The researcher's needs	Access to an appropriate computer, software, institutional and user support, ability and time to learn new skills, and an ability to understand the function and limitations of the software. The researcher also needs to be data-driven rather than technology-driven and willing to reflect on the ways in which the technology influences the analysis.

Ways in which the research process is enhanced through use of the computer:

- Transcribed text can be imported and exported electronically from a word processing program to a CAQDAS program and back again.
- More sophisticated programs can be used for audio and video data.
- The 'find' or 'search' tools in the CAQDAS program are very sophisticated and can be used to identify a range of relationships and ways in which key words or significant data occur.
- Categories and indexing systems for codes and texts allow for comprehensive retrieval.
- The CAQDAS program allows for additional data to be associated with the text, such as memos, notations, diagrams of key relationships, hierarchical trees, and graphic relationships.
- Some CAQDAS programs allow for interfacing with quantitative databases such as Excel or SPSS.

- Very large amounts of text may be searched quickly.
- Data can be effectively coded and records of all the codes kept and used for **data retrieval**.
- Concepts can be developed out of the coding processes because the coding can be flexible and fluid through the early stages of the analysis.
- Data can be coded repeatedly so that ideas and concepts can be examined from a number of different angles.
- The ability to retain records of coding patterns and coded material makes it possible for the researcher to rethink and recode—to work recursively.
- Sub-coding patterns can be developed from major coding axes so that the detail of concepts can be examined.
- Text can be searched selectively in order to discover links and relations between codes.
- Some CAQDAS allows for graphical data display.
- Theories can be built up, tested, dismantled, rebuilt, and re-examined while the software retains records of the versions, trials, and hunches.

Fielding 1994; Fielding and Lee 1998; Grbich 1999; Lee and Fielding 2004; Kelle 1995; Tesch 1990). Some programs have since been updated and upgraded (for example, NUD*IST to NVivo), while others have been taken out of production (for example, WinMAX, QSR N6). Product homepages and mailing lists are usually the best sources for the most up-to-date information (see Useful Resources at the end of this chapter). Many sites include screenshots and demonstrations. Some have program reviews and information about training.

Basic Word Search and Transcription Tools

For some researchers, complicated and expensive software packages are too difficult to manage. Inventive computer use and on-line discussion forums have led to more do-it-yourself approaches to data management and analysis that work directly with the capabilities built into your desktop or laptop. Rettie (2005) discusses ways to use desktop Microsoft™ search engines as a basic word search tool and also describes the 'autocorrect' tool as personalized shorthand that can speed up transcription. Similarly, Ryan (2004) provides ideas about using word processor capability to tag and retrieve blocks of text. In a more detailed description, La Pelle (2004, 86) suggests ways of using Microsoft Word™ to 'analyze text from key informant interviews, focus groups, document reviews, and open-ended survey questions' and 'Word functions such as *Table, Table Sort, Insert File, Find/Replace*, and *Insert Comment* to do this work'. La Pelle has used these functions successfully at a range of scales. Developing a taste for how these basic tools can facilitate research tasks may pave the way for exploring some of the stand-alone software packages.

Text Retrievers/Textbase Managers

These kinds of software are great for doing text analysis when you need to hunt words, strings of words, or characters (such as 'control' in Figure 15.1) and retrieve and batch them into categories and groups. They include software such as Textquest3.1 and WordStat5.1. Some programs' additional capacity facilitates the systematic organization of words or characters into records based on fields and subsets (i.e., a database). This software includes askSam7, Cardfile, and Readware. Some programs have specific capacity to deal with audio and video-based material, such as Transana and Qualrus. These programs, which have been developed recently and are technically sophisticated, can also be used for **code and retrieve** and code-based **theory building**.

Code and Retrieve

These types of software take the retrieval and management aspects a step further by incorporating capacity for attaching coding. Key words or codes can be applied to data chunks so that relevant sections of texts (rather than individual words or characters) can be retrieved according to a coding formula (see Figure 15.2: note the difference between Figure 15.1 and Figure 15.2, which are both concerned with 'control'). Programs in this range include Kwalitan5, Martin, and The Ethnograph5.

Code-Based Theory Building

Code-based theory builders are capable of doing all of the previous tasks because they are designed with capacity for retrieval, coding, annotating, memo making, and cross-questioning (see Figure 15.3). In particular, however, they allow researchers to interact with their data in ways that facilitate conceptual thinking about the data. They usually have some capacity for graphic representation of coding structures and patterns and increasingly are compatible with quantitative software programs such as SPSS so that you can use multi-method approaches. Included in this category are programs such as AQUAD6, ATLAS/ti5, HyperRESEARCH2.6, MaxQDA2, and NVivo8. Figure 15.3, for example, shows information about the respondents' professions and gender. Thus, the researcher can be thinking conceptually at an early stage in the research process about the respondents' attributes and how these attributes may contribute to building an understanding about their responses.

Graphic Display

Graphic display and **conceptual mapping** software programs have been developed to interface with other CAQDA software programs so that coding patterns can be exported from one program and displayed in another. Other graphical display programs include Decision Explorer and IHMC Cmap Tools.

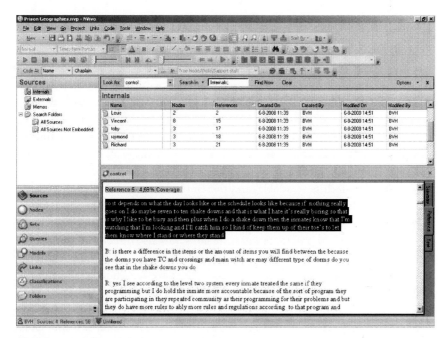

Figure 15.2: Code and retrieve in NVivo8. Coding 'Control' as concept rather than word search.

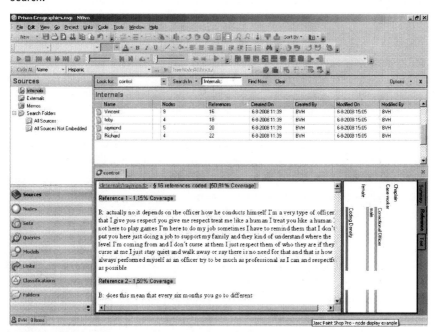

Figure 15.3: Example of a search for 'control'. What do correctional officers in a US prison say about 'control'? The figure also shows the respondents' role in prison and their gender.

CONSIDERATIONS FOR CHOOSING CAQDAS: ADVANTAGES AND CONCERNS

In sum, computer technology can be used to support a range of tasks that a qualitative researcher might wish to perform. Box 15.1 outlined some ways in which a computer program can enhance your research. There are more advantages worth noting: the speed with which examples of text can be retrieved is a significant time-saver for the researcher, but so is the ability to manage otherwise unwieldy databases. Some software enables you to keep tabs on data that are off-line—that is, not directly 'machine-readable' and therefore cannot be handled directly by the software. Off-line data might include bibliographies, untranscribed interview data on audio tape, movie or video clips, static images, or material on compact disk (CD). Several programs' latest versions now also directly accommodate audio, photo, and video materials as on-line data. Some programs are beginning to include mapping options, such as MaxQDA2. Figure 15.4 shows an example of adding a geo-reference to chunks of data (i.e., information from a prisoner about life in a prison dormitory)—the blue 'bullet' next to the underlined blue text indicates the presence of such a link. Figure 15.5 illustrates what this link looks like in Google Earth once the user clicks the underlined blue text. Being able to make memos and annotations relating to off-line data can extend the reach of your data without costly scanning or transcribing processes being involved.

More Than 'Cut and Paste'

Research, at its best, is a recursive process—one in which researchers interact intimately with their data. As Michael Agar (1986, in Crang and Cook 2007, 145) suggests, the process of coding and subdividing codes can be 'maddeningly recursive' because the categories that seem stable and appropriate one minute tend to break down as the research progresses so that new categories must be invented (see Chapter 14). This recursivity, this bending back on itself, is one of the great and exciting strengths of qualitative research, and from our experience, using CAQDAS can make the process more efficient and more enjoyable.

In the introduction, we claimed that using a computer in qualitative data analysis is not simply a matter of replacing manual techniques with a program but also opens up new ways of thinking about data analysis to the researcher. Computer software can make data coding and categorizing more flexible. Not only can you experiment freely with various categories, you can also code data more fluidly in different ways. You do not have to make up one coding system for your research project and then allocate data to one set of either/or categories. Programs such as NVivo8 allow you to put the same segments of data under a number of different headings or categories and retrieve them through a range of

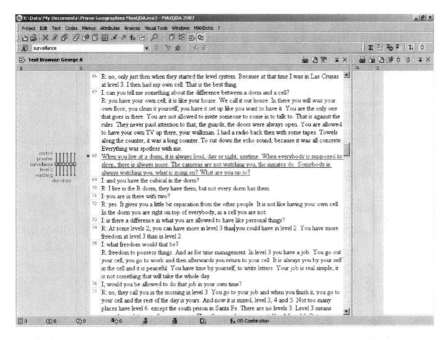

Figure 15.4: An example of 'geo-referencing' in MAXQDA2. The example also shows codes assigned to the chunk of text.

Figure 15.5: An example of jumping from a geo-reference in MAXQDA2 to its visualization in Google Earth. Source: Google Earth 2008.

indexing and searching tools that are activated by the codes the segments of data have been given. You can do multiple coding and then initiate complex inquiries of those coding patterns. This versatility can also enhance 'interactivity' between the researcher and the data.

The Vocational Argument

Apart from these broad methodological arguments in favour of using computers for qualitative research, there is a vocational argument. If you wish to use a purpose-built program, such as The Ethnograph5, ATLAS/ti5, or NVivo8, there is a specific learning curve associated with each program that cannot be circumvented. But being able to do 'fast' research, being able to demonstrate a capacity to analyze more than a dozen in-depth interviews, and having a sense of the versatility and interactive skills involved in qualitative research also makes for a very marketable researcher. Having a wide range of skills as a software user enhances your qualifications. You should consider this seriously when making a decision about whether to invest time in learning to use a particular software program or not.

Issues of Concern

Broadly, there are two sets of concerns associated with CAQDAS that we wish to raise here. One addresses the use of computers in research in general; the other targets the use of specialized software in particular. Computer technology is not universally accessible. Neither is the ability to link up to the Internet, or the ability to load up software to facilitate research. At a basic level, computer technology requires a power supply, appropriate hardware and software, and a method of gaining access to whatever system support is offered as part of the program. In many countries, for example, access to broadband is limited to larger urban centres, but broadband capability is required to download much of the Internet support for software programs. Use of the software entails a decision to commit to capital costs, and it requires that all of these elements be in place before the research gets underway. Word processing programs are usually available for use in most research institutions, but not all institutions provide or support more specialized software programs. Trying to finance the purchase of software and software training and support out of your own budget can be very costly, even when choosing no-cost alternatives (see freeware listed in Useful Resources at the end of the chapter). Trying to learn how to use new software on your own can also be very demoralizing.

Several other concerns have been raised by researchers discussing and using CAQDAS (not all critiques are by users), and they too should be taken into consideration. Richards (1997), for example, cautions against opting for large-scale projects under the illusion that good research involves collecting vast amounts of data. She suggests that although computers have the capacity for prodigious memory work, they should not be used lightly. CAQDAS are as effective working with small databases as they are with larger ones.

König (n.d.) suggests some very specific technical limitations on many CAQDAS programs—although NVivo8 is something of an exception in this regard. While the following problems are not necessarily universal issues, care should be taken with all new programs until you are familiar with their limitations. First, much of the software, such as Microsoft™, commonly used every day allows you to 'undo' files and automatically back them up. These 'safety nets' seldom exist in CAQDAS programs (NVivo8 is an exception). This means that you need to do frequent, manual backups. Second, CAQDAS programs are usually designed to handle plain text and/or rich text formats, so you need to adjust the format of your data to the specification of the program you choose (again, NVivo8 is less choosy). A file format converter program can be very useful. Third, often it is not possible to share CAQDAS data from one software package to another (importing SPSS data into SAS and vice versa is an exception, and this is a particular capacity of NVivo).

When Bhowmick (2006, 9) makes the point that 'different programs are built with some methodology or particular analysis in mind', she refers to the specific functionality of each software set: some are designed to manage only word-based text for content analysis, while others have the flexibility to work with statistical data as well. She also criticizes existing CAQDAS systems for failing to effectively handle spatial or temporal data that is often both available and important in qualitative studies (Kwan 2002; Matthews, Burton, and Detwiler 2001). Bhowmick (2006, 8) also suggests that 'CAQDAS has limited or almost no exploratory capabilities. Typical exploration elements such as linking, brushing and exploring multiple views of data are also not available in CAQDAS.'

A further, thoughtful critique has been developed by Mangabeira, Lee, and Fielding (2004), who argue that users should be alerted to three trends: that software is (1) increasingly sophisticated, (2) commercialized, and (3) generationally dependent. In relation to the second trend, they note that alongside the 'growing commercialization' and availability of CAQDA software is 'a concern that the commercial success of particular packages might translate into the closure and stabilization of particular approaches to or styles of analysis' (Mangabeira, Lee, and Fielding 2004, 167).

Box 15.2 outlines further concerns as well as some advantages mentioned above. A relatively serious concern is the overemphasis of programs on grounded theory (discussed further in Chapter 14). The research process induced by NVivo8, for example, generally follows the analytical steps of grounded theory (coding, categorizing, memoing, constant comparison). However, Barry (1998, 2.6) helpfully suggests that:

researchers will be more likely to take what they can from the software and use supplementary non-computerised methods, than to confine themselves to the limitations of computer methods. Perhaps first time qualitative researchers might be tempted to start with 'grounded theory' as a method and with computerized data analysis as a tool.... However once qualitative researchers find their feet they will soon be happy to reject methods and tools that will not serve their type of data and their type of problem.

CAQDAS: Advantages and Concerns   **Box 15.2**

Advantages	*Concerns*
Managing large quantities of data	Obsession with volume
Convenient coding and retrieving	Mechanistic data analysis/taken-for-granted mode of data handling
Ability to transcribe and analyze audio and video data	
Ability to tag text to maps (e.g., Google Earth)	
Comprehensive and accurate text searches	Overburdening a solo researcher with 'too much' data
Quick identification of deviant cases	Over-emphasis on 'grounded theory'
More time to explore 'thick data' as clerical tasks become easier	Loss of overview
Playful relationship with data—enhanced creativity	The machine takes over; alienation from data; makes qualitative research look more 'scientific'; limitations for connecting with geographical data such as GIS-type systems (van Hoven 2003)

Our message, therefore, is that undergraduates and graduates, as well as established researchers, need to remember that there are complex configurations of practical/technical and theoretical/political issues to bear in mind when assessing the role of computing in their own explorations of qualitative research. As Fielding and Lee suggest, 'With technology we can do more, but we also have more to do' (2002, 202).

Computer software does offer exciting new developments and capacities, and also offers computer users the chance to acquire marketable skills and a sense of conceptual confidence and flexibility that may be harder to achieve with other methods. However, use of software is contestable and politicized. As Thrift and French (2002, 331) caution, 'Software challenges us to understand new forms of technological politics and new practices of political invention, legibility and intervention that we are only beginning to comprehend as political at all.' Qualitative research that uses computer software demands above all else a high degree of researcher reflexivity. Any full-scale move to using computers to help make sense of qualitative research will entail ongoing reflection and critique.

CONCLUSION—FUTURE DIRECTIONS

We hope that more geographers who experiment with CAQDA software will publish accounts of their experiences and methods and the field will be opened up for widespread discussion and debate. An exciting new development might see the production of purpose-built software for interfacing between GIS programs and CAQDA software in much the same way that NVivo8 and SAS and SPSS programs are now compatible. Increasingly, there will be a trend towards working across qualitative/quantitative boundaries to produce multi-method work that draws research insights from the widest possible background. Such work will depend on the capacities of computers to manage databases—to store, organize, retrieve, and display symbolic information—in ways that make it possible to deal with data more efficiently and with greater versatility, to reduce drudgery and increase flexibility, and to provide researchers with interactive, portable, and valued skills.

KEY TERMS

attribute database

code and retrieve software

Computer-Assisted Qualitative
Data Analysis Software (CAQDAS)

concept building

conceptual mapping

data retrieval

data storage

geographic information systems
(GIS)

graphic mapping

hierarchical tree-building

image processing

memoing

NUD*IST

text-based managers

text retriever software

theory building software

word searching

REVIEW QUESTIONS

1. Qualitative data have been described on the one hand as 'rich' and as 'chaotic' on the other. Discuss the ways in which these attributes are appropriate descriptors for qualitative data.
2. Discuss the ways in which qualitative and quantitative data are different and require different styles and technologies for their management.
3. Computers are frequently described as 'tools'. Discuss some of the specific characteristics of computers that are indicative of this.
4. The letters 'CA' in the acronym 'CAQDAS' refer to the particular relationship that computers have to the analytic processes involved in research. The idea that computers 'assist' the analyst is a far cry from the notion that computers 'do research'. Discuss the relationships between data analysis and computer technologies in qualitative research.
5. Qualitative research is often described as a 'recursive' process. What is meant by recursivity? How do recursive approaches assist qualitative analysis? What might be an example of a recursive process in qualitative research?
6. Choose any one of the four generic types of CAQDA software (described in this chapter), and use the Internet to explore the capacities of the software in that category. Compare your findings with someone who chose a different category. Discuss which kinds of software would be appropriate for any research project you are engaged in or know about.

USEFUL RESOURCES

The best place to start with wider reading on the use of computer-assisted qualitative approaches is with a standard text such as Denzin and Lincoln's (2005) *The Sage Handbook of Qualitative Research* or Gilbert's (2001) *Researching Social Life*, which give you the advantage of reading the work of many different authors and seeing CAQDAS discussed in a wider social science frame. Some single-authored texts, such as that of Bryman (2004), deal with CAQDAS in the context of a robust, sociological discussion of social science methods.

More detailed information on using computer technology can be found in Hahn's (2008) *Doing Qualitative Research Using Your Computer: A Practical Guide*, Lewins and Silver's (2007) *Using Qualitative Software: A Step by Step Guide*, and Richards's (2005) *Handling Qualitative Data: A Practical Guide* or in older but still highly informative texts such as Weitzman and Miles's (1995) *Computer Programs for Qualitative Data Analysis: Software Sourcebook* or Kelle's (1995) *Computer-Aided Qualitative Data Analysis: Theory, Methods and Practice*.

Some texts focus more specifically on analysis or specific methodologies. They include Popping's (2000) *Computer-Assisted Text Analysis*, Lee and Fielding's (2004) *Tools for Qualitative Data Analysis*, Seale's (2002) *Computer Assisted Analysis of Qualitative Data*, and Corbin and Strauss's (2007) *Basics of Qualitative Research Techniques and Procedures for Developing Grounded Theory*.

Most recently, increasingly technologically focused, specialist material is emerging. QSR International, the company responsible for NVivo, produces workshop manuals such as the *NVivo8 Workbook*, and a number of publications on NVivo have appeared, such as Gibbs's (2002) *Qualitative Data Analysis: Explorations with NVivo*, Edhlund's (2007) *NVivo Essentials*, and Bazeley's (2007) *Qualitative Data Analysis with NVivo* (see also Bazeley 2002).

Not surprisingly, there is also a large volume of Internet resources on this topic, which we have subdivided here into seven categories:

- finding out more about CAQDAS
- finding out more about qualitative research
- finding out more about computers
- product pages for specific software packages
- electronic journals
- e-mail discussion lists
- Internet sites for geographers

Finding Out More About CAQDAS

CAQDAS (Computer Assisted Qualitative Data Analysis Software) Networking Project: http://caqdas.soc.surrey.ac.uk
Digital Research Tools (DiRT), digitalresearchtools Perform Qualitative Data Analysis: http://digitalresearchtools.pbwiki.com/Perform-Qualitative-Data-Analysis
Text analysis software sources: http://www.textanalysis.info/qualitative.htm
The UK Economic and Social Research Council's National Centre for Research Methods working paper series: http://www.ncrm.ac.uk/research/outputs/publications
QDA website: http://onlineqda.hud.ac.uk/Introduction/index.php
QSR website: http://www. qsrinternational.com

Finding Out More About Qualitative Research

Association for Qualitative Research: http://www.aqr.org.uk
Qualitative Research in Geography Specialty Group (AAG): http://www.uvm.edu/~mcope/QRSG/index.html
Qualitative Research Web Ring: http://kerlins.net/bobbi/research/qualresearch
QualPage resources for qualitative researchers: http://www.qualitativeresearch.uga.edu/QualPage
Sociological Research Online: http://www.socresonline.org.uk/home.html.
Text analysis software sources: http://www.textanalysis.info/qualitative.htm

Finding Out More About Computers

Exploring the Internet, Nicky Ferguson, *Social Research Update*: http://sru.soc.surrey.ac.uk/SRU4.html
Glossary of Internet Terms, Enzer Matisse, 1994–2008: http://www.matisse.net/files/glossary.html
PCWebopaedia: http://www.pcwebopedia.com/index.html

Product Pages for Specific Software Packages

American Evaluation Association: http://www.eval.org/Resources/QDA.htm. This site provides a summary, including pricing and contact information, of 31 different software packages designed to support analysis of qualitative data of all kinds, including text, audio, and video.

AnSWR: http://www.cdc.gov/hiv/topics/surveillance/resources/software/answr/index. htm. Freeware.

ATLAS/ti 5: http://www.atlasti.de/index.html. Video tutorials.

AQUAD 6: http://www.aquad.de. Free demonstration version.

AskSam7: http://www.asksam.com. 30-day free trial.

Cardfile 1.7: http://www.topshareware.com/Cardfile-software-download-6232.htm. Freeware.

CDC EZ-text: http://www.cdc.gov/hiv/topics/surveillance/resources/software/ez-text/ index.htm. Freeware.

Decision Explorer: http://www.banxia.com/dexplore/index.html. Free demonstration version.

ELAN: http://www.lat-mpi.eu/tools/tools/elan. Freeware.

ESA: http://www.indiana.edu/~socpsy/ESA/Introduction.html. Freeware.

Ethnograph 5.0: http://www.QualisResearch.com. Free demonstration version.

FolioViews: http://www.thefiengroup.com/np-views.html

HyperQual: http://home.satx.rr.com/hyperqual

Hyper RESEARCH 2.6 Software for Qualitative Data Analysis: http://www. researchware.com/hr/index.html. Free limited demonstration version.

IHMC Cmap Tools: http://cmap.ihmc.us. Freeware.

Inspiration Software, Inc.: http://www.inspiration.com/productinfo/inspiration/index. cfm. 30-day free trial.

Kwalitan 5: http://www.kwalitan.net/engels. Free demonstration version.

MaxQDA2: http://www.maxqda.com

Qualrus: http://www.ideaworks.com/qualrus/index.html. Free demonstration version.

QSR N6/NVivo: http://www.qsrinternational.com//default.aspx

SAS Institute: http://www.sas.com/technologies/analytics/datamining/index.html

SPSS Statistical Product and Service Solutions: http://www.spss.com

TextQuest 3.1t: http://www.textquest.de. Free trial version.

Transana: http://www.transana.org. Freeware.

WordStat 5.1: http://www.provalisresearch.com/wordstat/Wordstat.html. Free trial version.

Electronic Journals

The following journals provide on-line access to articles dealing with issues of computer use in qualitative research or in human geography:

Field Methods http://fieldmethods.org

Forum Qualitative Social Research http://www.qualitative-research.net/index.php/fqs

International Journal of Qualitative Methods http://ejournals.library.ualberta.ca/index. php/IJQM/index

International Journal of Social Research Methodology http://www.tandf.co.uk/journals/ tf/13645579.html

Progress in Human Geography http://phg.sagepub.com

Social Research Update http://sru.soc.surrey.ac.uk

Sociological Research Online http://www.socresonline.org.uk/home.html

Social Science Computer Review http://hcl.chass.ncsu.edu/sscore/sscore.htm

The Qualitative Report: An Online Journal Dedicated to Qualitative Research and Critical Inquiry http://www.nova.edu/ssss/QR

E-mail Discussion Lists

Qual Research: http://www.jiscmail.ac.uk/cgi-bin/webadmin?A0=QUAL-RESEARCH
Qualitative Research: http://www.jiscmail.ac.uk/cgi-bin/
  webadmin?A0=QUALITATIVE-RESEARCH
Qual-software: http://www.jiscmail.ac.uk/cgi-bin/webadmin?A0=qual-software
QSR Forum: http://forums.qsrinternational.com

Internet Sites for Geographers

Intute: Social Sciences—The Human Geography Gateway: http://www.intute.ac.uk/
  socialsciences/geography
Internet Resources for Geographers: http://www.colorado.edu/geography/virtdept/
  resources/contents.htm
The Virtual Geography Department: http://www.colorado.edu/geography/virtdept/
  contents.html

Note

1. The examples in Figures 15.1 to 15.3 draw on a study undertaken by Bettina van Hoven and David Sibley in a prison in New Mexico (van Hoven and Sibley 2008; Sibley and van Hoven 2009).

Writing a Compelling Research Proposal

16

Janice Monk and Richard Bedford

CHAPTER OVERVIEW

This chapter aims at helping you with the challenging task of initiating research by writing a proposal that will focus your thoughts, plan your approach, and convince others that your project is important. We address how ideas for research originate, how to specify research questions, how to demonstrate your ideas' significance, and how to define your research methods. We discuss the process of writing research proposals and comment briefly on how reviewers evaluate them. Our concluding comment sums up why we think writing compelling research proposals is so challenging while at the same time being one of the most enjoyable parts of a geographer's post-secondary training. Learning to write such a proposal is also one of the most valuable contributions you can make to the development of your skills as a researcher, especially if you choose a career that involves defining research priorities or bidding for funding to support research activity.

THE RESEARCH CHALLENGE

Writing a compelling research proposal is a real challenge. In our view, it is probably the most demanding task that any undergraduate or graduate student experiences. It is also one of the most exciting, because it is a chance for you to define your research questions, frame them in the context of existing knowledge, identify appropriate methods of inquiry to address the questions, negotiate ethical issues, and establish data collection and analysis procedures.

Writing a research proposal requires you to bring into focus all of the essential elements of your university education. It demonstrates your ability to synthesize and question existing knowledge on a topic and your understanding of how ideas are used to formulate theoretical frameworks. It draws on your knowledge of and experience with different ways of approaching problems and asking questions, as well as your ability to construct a coherent design for new research. It indicates the extent to which you are stimulated by the 'cutting edges' of knowledge and inquiry in the geographical questions that really interest you.

As we show in this chapter, a proposal offers a roadmap for the research journey, but as on many journeys, you are likely to encounter crossroads, roadblocks, and many twists and turns that you need to negotiate. You need to be flexible enough to accommodate these changes. Your proposal is not a static, definitive statement; it is a vibrant, living component of your research. It is subject to revision, extension, and amendment as the research unfolds.

Writing research proposals is stimulating and challenging. Carrying out the research to address the questions in the proposal is even more of an adventure. Research is to be enjoyed, not endured, even if at times it is difficult to see how you will ever address all the research questions you started with. No worthwhile journey is easy; research is not easy, but research makes university-based study really rewarding.

WHERE DO RESEARCH IDEAS COME FROM?

The first thing one has to establish when developing a research proposal is a topic. In our experience, ideas for research come from at least three sources: personal experience, reading, or conversations with other geographers and scholars. In most cases, some combination of these sources is involved. It is important to appreciate that you need to have a strong personal interest in the research you are planning to do. The work has to be something that you care about and have the skills to carry out (or know where you can get these skills).

If you are writing a thesis, you will be spending a lot of time on the research, so you want to be sure you have a topic that really interests you. Do not be surprised if it takes time to refine the topic; the initial idea may come from some experiences you had a long time before you actually became involved in research. In the course of writing a proposal, the ideas will be clarified, and you may find that by the time the proposal is written, the original idea you started with is substantially altered and refined.

To illustrate how research ideas can evolve, we draw on our own experiences as graduate and post-graduate students embarking on our first major research exercises (see Boxes 16.1 and 16.2).

Something we both found highly motivating about our research was that the ideas that contributed to the development of our research proposals were of interest to participants, local government officials, and other academics studying

The Origins of a Dissertation in Social Geography   Box 16.1

Jan Monk grew up in a fairly low-income family in Australia at a time when non-English-speaking immigrants were arriving in substantial numbers. She was conscious of social inequalities, ethnic differences, and stereotyping of those we would now label 'Other'. Shortly after graduating with her BA, she participated in a work camp that built a house for an Aboriginal family so that they could move from a reserve into a small town in New South Wales. The project reflected the state's policy of assimilating Aboriginal people into the white community. The government paid for the building materials, and a church group brought together young professional men and women who donated the labour.

The project raised geographic questions about who lives where and who has the power to shape those residential patterns. The experience, and some of the project's ethical and political implications, made a strong impression on Jan. At that time, Australian geographers had not been writing about Aboriginal affairs,[1] and anthropologists were mainly interested in 'traditional' cultural patterns or in psychological questions about life on reserves. There was really no geographic precedent for Jan to link her personal interests and potential research at this time. The ideas for possible research were stimulated by personal experience; further reading in the social sciences while she undertook post-graduate studies was needed before the research proposal could be developed.

Jan moved on to doctoral study in geography in the United States. Attention to relations between groups was rare in geography, but she took courses in sociology and anthropology that spoke to her interests. The anthropology professor's ecological framework took into account demography, economics, and politics. With this new perspective, Jan saw a way to address what had been on her mind in Australia. It so happened that Australian policies towards Aboriginal people were also coming under scrutiny at the time she embarked on doctoral research, so her ideas happened to coincide with an emerging area of policy concern. This was a bonus—government officials were also interested in the research, and this made the whole project seem more useful and relevant.

She now had context that identified her question as one of policy importance, she was more familiar with relevant interdisciplinary literature, and she had a set of research tools to apply to questions grounded in personal experience. Imbued with the geographical perspective that place matters, Jan designed a comparative analysis of the social and economic lives of Aboriginal communities in six small towns in the state of New South Wales and went on to explore how the lives of these Aboriginal people were influenced by white social, political, and economic history.

contemporary social, cultural, and economic transformation (for an elaboration, see Chapters 3 and 13). Personal as well as broader interest in what you are researching can be critically important, both for sustaining your own engagement with the project as well as for stimulating new research ideas and opportunities.

Beginning a Career in Research on Migration   Box 16.2

Serendipity can play a major role in the selection of research topics. Richard Bedford's undergraduate training in geography, history, geology, and Pacific studies at the University of Auckland led him to consider two quite different directions for a graduate research degree: fluvial geomorphology and population geography. In the end, it was an invitation to visit a central Pacific atoll country (Kiribati) for a holiday that swung the balance towards research about people rather than rivers. There are no rivers on atolls, but there are some very interesting people–environment relationship issues, especially apparent to someone from a much larger, mountainous country who is visiting coral atolls for the first time.

A combination of reading for a graduate paper on the geography of the Pacific Islands, discussions with senior government officials in Kiribati, and several months of fieldwork in the islands provided the ideas that were to become the basis for a Master's thesis on migration in an atoll environment. As was the case with Jan Monk's research on Aboriginal social geography, Richard Bedford's research on migration as a response to population pressure on coral atolls was of interest to the colonial government of the day. The concerns the government had about population change thus fed into informing the ideas that underpinned the research proposal that was developed to guide the research over the subsequent 12 months. This interest in the migration of Pacific peoples then took Richard to the Australian National University's Research School of Pacific Studies (now the Research School of Pacific and Asian Studies) where he completed his doctorate.

Doctoral research in Vanuatu and post-doctoral research in Papua New Guinea and Fiji during the 1970s and 1980s, as these countries moved from colonial rule to independence, was exciting and very challenging for a 'white' male New Zealander. A mix of research methods and sources of information was always required: archival research, analysis of census data, questionnaire surveys, in-depth interviews, group discussions. It was not a question of 'either quantitative or qualitative research methods and approaches'; it was always necessary to use a mix of both, especially if the analysis of population movement, as described in the reports of government officials and the statistics collected in surveys and censuses, was ever to be informed by the personal experiences of those who moved.

One of the great benefits of much research done by geographers is that it seems 'relevant' to others: that is often very important for researchers when they are trying to define a topic on which they will work for a substantial period. However, it should be noted that research does not have to be 'relevant' to particular interest groups to be worthwhile; much theoretical inquiry is driven by curiosity, not by a concern with addressing a question that is of interest to a particular group. As we have already said, the key thing about doing research is to be genuinely interested in the topic you choose to work on.

GETTING STARTED

When and how you start work on a research proposal will depend on the stage of your post-secondary education, what your department and supervisor expect, the nature of your project, and whether or not you need funding for the research. If you are an undergraduate, your department may offer a course that includes some training in writing a research proposal. At this level, it is likely that you will be expected to develop a full research proposal, but you may not be required to actually carry out the proposed research or to seek outside funding to support your research. If you are an honours, Master's, or doctoral student, however, you will usually be expected to write a proposal and present it in a colloquium prior to having it approved. Once it is approved, you will then undertake the proposed research. You may also need to prepare a proposal to a granting body to obtain financial support for the research.

Even if you are not required to write a formal proposal along the lines of what is suggested in Box 16.3, we suggest that you do—it will clarify your ideas, taking them from a broad theme to specific questions. The proposal will provide you with a roadmap for the research journey. It will help you to say why your project is important, how you will carry out the work, what resources (financial and other) you will need, and what timetable you will follow for collecting data, analyzing the materials, and writing up your results.

In Box 16.3, we summarize some of the key components of a research proposal. Several of the items that are bulleted in this summary are discussed in greater detail in subsequent sections.

In the next three sections, we use examples from a research proposal prepared by Ranjana Chakrabarti, a doctoral student in the United States who has given us approval to quote from her work, to illustrate the major parts of the proposal. These parts are: specifying the research questions, framing the research in terms of its theoretical and empirical contexts, and developing the methodology for the project. We then provide some suggestions about the way successful research proposals are written and return briefly to the issue of funding.

Components of a Research Proposal   Box 16.3

A good proposal will enable the reader to appreciate immediately:

- the *question/problem* that is being addressed in the research;
- the current state of *knowledge* on the topic that is the subject of inquiry;
- the *methods* that will be used to collect the information required;
- the *ethical issues* that have to be addressed before undertaking any fieldwork;

- the *resources* that will be required to collect the required information and complete the research report;
- the *intended outcomes* of the research.

The proposal will have several sections in which the points raised above will be covered. The key sections are:

- introduction to and background of the proposed research, including a statement as to why the topic is worth studying;
- the key research questions/problems (which might be specified in the form of hypotheses, but this is not always appropriate);
- the wider theoretical and empirical context for the research;
- research methods and associated ethical issues;
- research plan (timetable outlining the various stages of the research);
- budget and sources of funding for the research;
- references cited.

SPECIFYING YOUR RESEARCH QUESTION

One of the most challenging aspects of writing a proposal is articulating your major research question and the sub-questions that flow from it. Everything else follows. You have to keep the 'big picture' in mind, but think through how to identify its component parts so that you have a manageable project. Suppose you are interested in policy aspects of housing and homelessness in Australia. You know, stereotypically, that homelessness is associated with urban areas and 'vagrant' men; remedies are couched in terms of providing shelters, mostly in inner-city neighbourhoods. Then you see a report that young people make up almost one-third of the homeless.[2] You grew up in a rural area and know of students who dropped out of your high school and ended up homeless within the local community. This knowledge prompts your first decision: you will conduct research on aspects of youth homelessness in country towns.

Now you have more decisions to make. Will you focus on the causes of the problem, the experiences of homeless youth themselves, the types of services provided or needed, or on some combination of these aspects? Will you undertake a case study of a single community or of several? What criteria will you use to select the study area(s)? Will you focus on homeless young men, young women, or both? How will you connect your research back to policy concerns? There is no 'right' answer to these questions. Which questions you address will depend partly on the time and resources you have but also on what you learn about existing related research so that you can show that yours will make a new contribution.

Your choice of research question will also be heavily influenced by your knowledge of and skills with different research methods. Some geography students have a strong interest in statistical analysis, especially if they work with geographic information systems (GIS) or have subjects like psychology and economics in their degrees. Other students have a strong preference for qualitative modes of inquiry, especially if they have studied cultural geography, critical perspectives, gender and ethnicity, and so on. There is no 'right' set of skills that everyone doing research must have, aside from an ability to think and write clearly. The particular research and analytical skills needed will depend very much on the nature of the research questions being addressed.

Research offers the best opportunity there is to use the skills you have already developed in the course of getting your degree, while at the same time enabling you to gain new skills. You should not feel constrained to work just with methods you are familiar with; research for a thesis will often encourage you to extend your understanding of research methods. But be realistic when developing your research questions; make sure when you are specifying questions that you have or can develop the skills to address them. Over-ambitious research proposals can make for very frustrating research experiences.

The sorts of questions raised above in the example of homelessness among rural youth lend themselves to analysis using qualitative research methods, and students with a good understanding of such methods may be both more interested in and better placed than others to undertake research on this topic. We discuss the development of the methodology section of a research proposal later in this chapter, but it is important to appreciate that the definition of research questions is very much influenced by your preferences for and understanding of different research methods and modes of analysis.

It is useful to cite a further example to illustrate how research questions are formulated and incorporated into a research proposal. Ranjana Chakrabarti (2004), a doctoral student at the University of Illinois at Urbana-Champaign, went to the United States for study after completing a Master's degree in India with a focus on medical/health geography. She is interested in how place shapes women's health care practices. Given her background and current location, she has chosen to study South Asian immigrant women in the United States, specifically in New York where she has personal connections. Her research addresses their low rates of prenatal care utilization, since this has consequences for their health and that of their babies. In the rest of this chapter, we use sections of her dissertation proposal to illustrate some of the key points we seek to make. Box 16.4 summarizes how Ranjana defined the research question and the sub-questions for her dissertation research on the roles of place, context, and culture in shaping South Asian immigrant women's prenatal health care practices in New York.

Once the research questions have been defined in draft (they will be refined as the proposal is developed), the **literature review** and the identification of an

Research Questions for a Health Geography Study Box 16.4

Much of public health research on prenatal care has focused on the role of individual-level maternal risk factors in explaining low levels of utilization of prenatal care. A large number of socio-demographic and structural as well as attitudinal and psychological factors have been identified in previous studies. However, the ways these barriers are experienced and expressed differ among women from different ethnic backgrounds. This difference can be understood through an appreciation of the complex interactions between place, culture, and health (Chakrabarti 2004).

Ranjana's primary research question is: What causes the low rates of utilization of prenatal care facilities by South Asian immigrant women in New York? The sub-questions she identified to assist her frame the study and develop the research proposal are:

- How and why does prenatal care use by South Asian immigrant women vary within New York City?
- How do South Asian women gain access to prenatal services?
- How does the experience of place mediate the prenatal care practices of these women?
- How do these women relate to and gain knowledge from the formal and informal prenatal care resources in their local environment? What geographical, economic, and cultural barriers do they face?
- How does culture affect the prenatal care knowledge and practices of these women? Are such practices place-based and at what scale?
- How do South Asian women create and utilize place-based social networks at the local, national, or transnational scales to gain access to formal and informal prenatal care services and advice? (Chakrabarti 2004)

appropriate theoretical framework and methodology for conducting the research can be completed. We now turn to the critically important component of all research proposals that addresses the theoretical and empirical contexts within which the research questions are situated.

FRAMING YOUR RESEARCH

As you develop your research questions, it is important to think about how you will convey their significance to others and what theoretical and **conceptual frameworks** you will employ. If your audience and study are primarily academic, you will mostly rely on the research literature to justify your project's approach and significance. If your work has a strong applied component, you should also study the perspectives and needs of the group(s) with which you will work. Among the various questions that need to be addressed while framing your research, the following four are especially important:

- Does the project deal with a significant and meaningful problem that lends itself to a substantial research effort?
- Why is the problem of interest to other scholars or practitioners in the field?
- Has a persuasive case been made as to why the problem is worth solving?
- Is it clear who or what will be aided by the research findings?

Source: *Dissertation Proposal Writing Tutorial*. n.d.

We know that identifying the 'right' project is a difficult task and that you may agonize over it. Peggy Hawley (2003, 36) reminds us that there are dozens of 'right' topics and that 'what makes the choice succeed or fail depends mostly on what you do about it after the choice has been made.' As you think about the questions above, it is useful to also follow Hawley's suggestion that you avoid a topic that is broad and unwieldy and ask yourself whether it will be manageable, be in the range of your competence, draw on sound sources of data, and be of interest to you. She also recommends avoiding extremely controversial issues early in your career. This is also the time when the advice of your supervisor(s) can be very helpful. While it is your job to define and justify the topic, it is the responsibility of research mentors (and that is what supervisors are) to assist you in developing confidence in the topic you have chosen. They are likely to have views about the 'significance' of the problem/question you are thinking of addressing, why this problem/issue might be of interest to other scholars in the field, and how to make a case for research on the topic. Do seek advice from your supervisor(s) as you begin to frame your research, but always keep your own interest in the topic to the fore. After all, this is *your* research project.

Writing a review of the literature is essential for preparing the proposal. To situate the research in an appropriate theoretical framework will require careful study of some of the key journals in the discipline (see Box 16.5). This literature review should not be confused with an annotated bibliography or a summary of all you have read. It should be both constructive and critical in tone, identifying the strengths of the pieces you select in order to show how they support your own work and the weaknesses or gaps that demonstrate why your work is fresh and significant. The final report or thesis that you write will always include reference to earlier research on the topic and, if it is a thesis, a substantive assessment of previous findings and how your proposed research furthers understanding of the topic.

Using a literature review to help establish the context for your research is a demanding and time-consuming task. It is not something that is done overnight!

The Research Context	Box 16.5

The key questions that are addressed when describing the context for your research are:

1. What does your work offer to the available field of knowledge that helps us to better understand the issue/topic you have chosen to address in your research?
2. How do existing studies inform your work?

In answering these questions, keep in mind that there are two important dimensions to the context of any research proposal:

1. the theoretical ideas that inform research on the topic (these ideas will tend to come mainly from the existing literature);
2. the empirical or 'real-world' situation that relates to your research topic (there will usually be examples of research on your topic or a closely related one that you can find in the literature).

Make sure that you address both of these dimensions when framing your research. Heath (1997) suggests that when describing the context for your research, you should:

- Use specific language to name and describe the conceptual foundation for your research—that is, the research perspective that informs your approach to the topic. Show clearly why this perspective is appropriate and relevant for your study. For example, you may be using insights from a post-structuralist perspective to provide a theoretical context for your research. Make sure that you explain clearly and simply what the approach means for the way you will do your research and how you will interpret your results. Useful definitions of most of the main theoretical approaches used in human geography can be found in the recent editions of *The Dictionary of Human Geography* (Johnston et al. 2000).
- Cite authors who have already used this approach to address questions/ problems similar to the ones you are researching. An excellent starting point for recent reviews of the theoretical and the empirical literature in most of the main areas of research currently being addressed by human geographers is the journal *Progress in Human Geography*. This journal and its companion, *Progress in Physical Geography*, contain regular updates on new ideas, methods, and findings in contemporary geography.

Box 16.5 contains some suggestions on how you can approach this very important component of a research proposal.

A useful source of examples of research contexts and the literature reviews associated with them is the introductory sections of published articles. A 2004 paper by Marianna Pavlovskaya on her doctoral research on household economies in post-communist Moscow is one such example. She begins by noting that the existing literature on the transition from communism tends to emphasize macroeconomic themes. Next, she draws attention to the theoretical writing by geographers on the importance of scale and connections over space in shaping the social and spatial relations of people and the environments within which they conduct their lives. This geographical literature is used to justify her approach to the topic through an analysis of households and the ways in which they link formal and informal economies in the city. This theoretical perspective leads her to a research design that will enable her to integrate in-depth interviews with households in selected neighbourhoods of inner Moscow (taking into account their gender and class dimensions) with a GIS-based reporting of spatial patterns of neighbourhood characteristics and interactions.

The research context thus establishes the rationale for the particular research methods that will be used to undertake the study. We now turn to this very important component of all research proposals.

DEVELOPING YOUR METHODOLOGY

Identifying how you plan to carry out your research is one of the most demanding aspects of writing a proposal. Reviewers of proposals repeatedly find that the discussion of the methodology is the least satisfactory part of the document. This seems to be a particular problem with research that will involve qualitative methods. The researcher may not go much beyond indicating that data will be collected through focus groups or in-depth interviews, perhaps naming software that will be used for analysis but failing to *show why and how these methods will allow the specific research questions to be answered fully and ethically.*[3]

In Box 16.6, we summarize some of the most important considerations that need to be borne in mind when developing a research methodology. We do not discuss specific methods in this chapter in any detail; these methods are the subjects of other chapters in this volume.

In Boxes 16.7 and 16.8, we draw again on Ranjana Chakrabarti's doctoral research proposal to provide a specific example of a research timetable and a statement of research methodology. These ideas about research methods should be read in the context of her key research question and sub-questions outlined in Box 16.4.

WRITING THE PROPOSAL

Exactly how you organize the final proposal will depend on who will read it. If you are writing it for your supervisor or a class in research methods, you may be

Thinking about a Research Methodology

Box 16.6

- Be realistic. You will have limited time and resources to conduct the project. If your work involves individual interviews, how many, for example, will you be able to complete while still getting sufficient information to explore the range of views people may hold? Consider that you will have to recruit people, schedule the interviews, and deal with some refusals or cancelled meetings. How do you know people will talk to you? What strategies will you use to recruit them? Be careful not to over-commit yourself.

- Will your work be served best by combining methods? Researchers frequently use a combination of quantitative and qualitative approaches (see, for example, Tashakkori and Teddlie 1998 and Chapter 1 of this book). How can you do this and be realistic in relation to your resources and context?

- Assess the strengths of the approach you propose over alternatives. Would focus groups serve your needs better than individual interviews? Why or why not? What are the strengths and limits of the methods you are proposing?

- Show that you know what data you may need to draw from (e.g., public records) and that you will be able to gain access to them and in time to complete the project.

- Consider what you will do if you run into data collection problems. Strong proposals identify potential obstacles and show that you have thought of alternatives. The best-laid plans are not always feasible at the time and place you want to implement them.

- Prepare and include a timeline showing when you will undertake specific tasks. One (but not the only) format for a timeline is a matrix that lists tasks on one axis and time periods (for example, months) on the other. This allows you to show that some components of your work will overlap in time (for example, you may be arranging for interviews at the same time that you are researching background information from other sources on the context). Box 16.7 provides an example of a timeline based on Ranjana Chakrabarti's PhD dissertation.

- If possible, undertake pilot research to demonstrate the feasibility of your study, and refer to it in your proposal.

- Finally, do not forget to allow time to obtain approval from the departmental or university committee responsible for overseeing ethical issues in research.

given guidelines to follow. If you are applying to a funding agency, it will usually have guidelines that should be followed very carefully. A cover sheet or letter and an abstract, for example, may be required. You will probably write these items last to make sure that they are accurate and compelling statements about your project.

Example of a PhD Timetable — Box 16.7

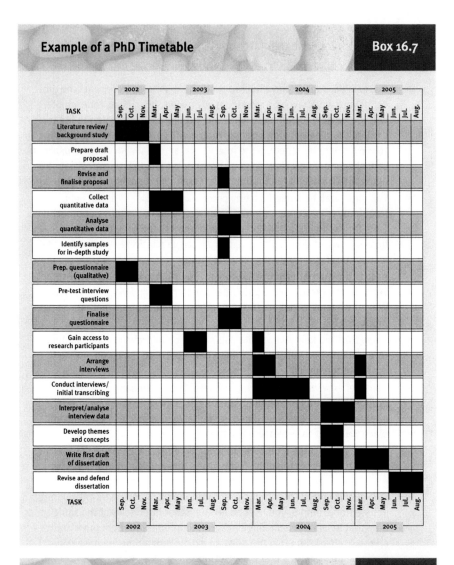

TASK	2002 Sep.	Oct.	Nov.	2003 Mar.	Apr.	May	Jun.	Jul.	Aug.	Sep.	Oct.	Nov.	2004 Mar.	Apr.	May	Jun.	Jul.	Aug.	Sep.	Oct.	Nov.	2005 Mar.	Apr.	May	Jun.	Jul.	Aug.
Literature review/ background study	■	■																									
Prepare draft proposal				■																							
Revise and finalise proposal										■																	
Collect quantitative data				■	■																						
Analyse quantitative data										■																	
Identify samples for in-depth study										■																	
Prep. questionnaire (qualitative)	■																										
Pre-test interview questions				■																							
Finalise questionnaire										■																	
Gain access to research participants								■																			
Arrange interviews													■									■					
Conduct interviews/ initial transcribing													■	■	■	■						■					
Interpret/analyse interview data																			■	■							
Develop themes and concepts																			■								
Write first draft of dissertation																			■	■							
Revise and defend dissertation																									■		
TASK	Sep.	Oct.	Nov.	Mar.	Apr.	May	Jun.	Jul.	Aug.	Sep.	Oct.	Nov.	Mar.	Apr.	May	Jun.	Jul.	Aug.	Sep.	Oct.	Nov.	Mar.	Apr.	May	Jun.	Jul.	Aug.
	2002			2003									2004									2005					

Research Methodology for a Health Geography Study — Box 16.8

In this study, a mix of qualitative and quantitative methods will be used, and these relate to two distinctive, but related, parts of the proposed research into the use of prenatal care services by South Asian mothers. . . . The first phase involves a quantitative analysis of spatial variation in the use of formal prenatal care services by South Asian mothers in New York City (NYC). This work uses data from the births master file of the NYC Department of Health. A geographical information system (GIS) will be used for mapping and analyzing the geographic distribution of South Asian mothers and for exploring geographic variation in prenatal care use for the major South Asian groups. Logistic regression techniques will be used to shed light on the influence of factors such as education and insurance coverage on the use of prenatal care.

The second phase of the research involves an extensive field-based qualitative study to understand how the everyday lives of immigrant South Asian women shape their prenatal care practices. Interviews will be carried out with approximately 60 immigrant South Asian women in New York City. The main goal is to understand the type of barriers women face in using formal prenatal care and how women utilize social networks, both national and transnational, to gain knowledge, social support, and prenatal services during pregnancy. The types of questions that will be used to obtain this information are discussed in the proposal.

The sample women for the in-depth interviews will be chosen to reflect the geographic, economic and cultural diversity amongst South Asian immigrant population in New York City. . . . Based on the quantitative GIS analysis, place-based stratification is proposed to identify neighborhoods for intensive study. The initial research participants in each neighborhood are to be contacted through South Asian organizations, nongovernmental organizations working for immigrant South Asian women, and through informal networks and social gatherings. . . . and through '**snowball-sampling**' procedures.

The qualitative phase of the project began in February, 2004 and is expected to last for 6–8 months. Interviews are being tape-recorded and transcribed, and the interview data will be analyzed using interpretive methods. The qualitative analysis software, NUD*IST, will be used to assist in identifying major themes from the interviews and in aiding the data interpretation. The proposal makes reference to the fact that Ranjana conducted pilot interviews with a small sample of South Asian women in Champaign, Illinois, to pre-test the format and questions for the in-depth interviews. Based on that experience, a detailed set of interview questions were formulated.

The Institutional Review Board [i.e., ethics committee] at the University of Illinois granted final approval for the project in December 2003 (Chakrabarti 2004).

In whatever form the proposal is written, it should address the key questions of *what*, *why*, *how*, and *when*, as this chapter has discussed, and if you are applying for funding, it will need to identify *who* you are, your qualifications, and *how much* (a justified budget) the research will cost. Ancillary information may well include items such as a list of the references cited in your text, the ethics/institutional review clearance, letters of recommendation, and your credentials in the form of a biographical statement or curriculum vitae.[4]

As we noted earlier when outlining the key components of a research proposal, we think that the opening one or two paragraphs of the research proposal should capture the reader's attention by succinctly answering the *what*, *why*, *how* questions. Anthony Heath (1997) provides a useful outline of the content of the introduction to a proposal advocating that you address what it is you want to know or understand, how you became interested in the topic, why the research is needed, referring briefly, for example, to the lack of existing knowledge or limits of other work and to whom the work will be of value. We also suggest that you indicate briefly what conceptual or methodological approach the research will adopt. In Box 16.9, we offer the opening paragraphs from Ranjana Chakrabarti's research proposal for illustrative purposes.

Introducing the Proposal	Box 16.9

Ranjana Chakrabarti's (2004) proposal began with the following two paragraphs:

> The main objective of this PhD dissertation is to understand the roles of place, context and culture in shaping prenatal health care practices of South Asian immigrant women in New York City. These women have very low levels of prenatal care utilization compared to US-born women and women from other immigrant backgrounds (NYC Department of Health and Mental Hygiene, 1999; South Asian Public Health Association, 2002; Coalition for Asian American Children and Families, 2001). This puts at risk the maternal and infant health of one of the most vibrant immigrant populations in the country, because low and inadequate prenatal care use can result in low birth weight babies and other adverse pregnancy outcomes (Kieffer et al., 1992; McDonald et al., 1998).
>
> Little is known about the complexity of circumstances that leads to inadequate utilization of formal prenatal care by South Asian women in New York City. This research is intended to fill this gap by exploring how the geographical contexts of everyday life influence their prenatal care practices. The primary focus is to understand how the experiences of place, culture and context mediate prenatal health care practices. Using quantitative methods and qualitative in-depth interviews I seek to understand how women's situatedness in local and transnational social and geographical networks constrains access to prenatal care and how women draw on such networks in creating new spaces of prenatal care access.

This kind of approach is more effective than beginning with a lengthy 'background' statement and failing to define your project until several pages into the proposal. The latter failing is one of the most common among inexperienced writers. It is especially important to have a strong opening if you are applying for funds in a national competition. Reviewers have many proposals to evaluate in a short time. The easier you make their task, the better chance you have of being successful. In this regard, it is a good idea to specify in the opening paragraphs what is innovative about your research.

We observed at the outset of this chapter that you should allow plenty of time to write your proposal. Just how much time depends on your career stage and the context in which you are writing. Six months preparation is not unusual for a doctoral proposal. Six weeks might be typical for an honours proposal. You should expect to write multiple drafts, to seek critiques from peers, your supervisor(s), and other relevant academic staff members, and then to think carefully about how to address their comments. Solís (2009) provides some useful examples of how geographers have revised their proposals to respond to such critiques. You should seek out models of other successful proposals. Websites that provide helpful examples exist, and we list some of them at the end of this chapter. If program officers in a relevant funding agency are willing to discuss your proposal with you, or if they offer workshops at professional meetings, by all means take advantage of such opportunities.

While your ideas are clearly extremely important, your writing style and the format of the argument are critical features of a successful proposal. Use clear language rather than excessive technical jargon. Using discipline-specific technical jargon is especially problematic if you are applying for funds from agencies that use multi-disciplinary panels to assess the merits of the proposed research. Statements that readers with a good education in any of the sciences can read are much more likely to succeed in competitive funding rounds than ones written in complex prose that is really only understood by researchers working from particular theoretical or ideological perspectives.

Make sure the typeface is easily legible. Make judicious use of underlining, bold, or italic type where appropriate. Proofread carefully, and check/double-check all details. If you have written a proposal for external funding, be especially careful about the funding agency's requirements. Does the proposal have to be mailed by or received by a specific date? How many copies must you send? Is the proposal to be submitted electronically or as paper copy? Do you require approval by your institution before you submit your proposal to an external agency? What does that process involve? How much time does it take? Failure to follow the required guidelines for research proposals can lead to significant delays in getting the project approved or funded or, more likely, to rejection of the proposal.

SEEKING FUNDING FOR RESEARCH

Obtaining funding for research operates very differently in different settings, and for this reason we will deal with the topic only briefly. We suggest that you talk with your supervisor and other students who may have had grants, check the Internet and libraries for information sources, and consult offices in your university that assist in identifying funding sources. You may also wish to check whether funding is available for a **pilot study** that will enable you to demonstrate that your main project is feasible.

If you need funding, start the application process well before you expect to begin the research. You may wait as long as six months to learn whether you have been successful in a national competition. It is sometimes argued that research involving qualitative methods—for instance, collecting detailed information from small numbers of interviewees—has a lower chance of getting financial support from external sources than, say, large quantitative surveys. This was much more of an issue a decade ago than it is today. Well-argued research proposals, even if they rely heavily on qualitative rather than quantitative data, are likely to be funded. A critical issue with regard to funding research is the nature of the research question that is being addressed, the relevance it has for the priorities of the funding agency, and the extent to which the panel assessing your application is excited by and understands your research proposal.

Before applying for funding, check the agency's guidelines *very carefully*: are you eligible for an award, what is required in the proposal (for example, format, page length, type style, budget, bibliography, letters of reference, your credentials), and what is the deadline for receipt of proposals? Writing your research proposal and associated funding applications is time-consuming. Experienced supervisors often suggest beginning your draft up to six months before submission. So start early!

A Concluding Comment

We pointed out in the introduction to this chapter that writing a research proposal is probably the most demanding task any undergraduate or graduate student will experience. We have found that students invariably tell us that the requirement to design and carry out their own research project was the highlight of their geography training at university. One of the hardest things to do is to decide on what you think is a good research topic. As we noted when drawing on our own experiences of graduate and post-graduate research, there is no single source of inspiration for topics. With the wisdom of hindsight, we offer one useful piece of advice: choose a topic that can be specified and explained simply and clearly. A real trap when choosing topics for research is trying to be very clever in order to convince the reader that yours is really an original idea. In reality, very few research ideas are completely novel or 'original'; most research builds on existing ideas and extends them in interesting ways.

The key to success in writing a compelling research proposal is your own excitement about doing the research. If you are interested in the topic, you will think more clearly about all of the issues we have raised in this chapter. Strong commitment to the topic on your part will tend to ensure that the research questions will be well specified, the research will be framed effectively, the methodology will be appropriate, and the proposal will be well written. Undertaking research does not appeal to everyone, and a lack of interest in this aspect of university training usually shows up very quickly in the content of a research proposal. We hope that you will want to carry out research on a topic that is of interest to you; in both our cases, it was a transformative experience that led us on to careers in which research is at the heart of the job. We can think of no better challenge in a university program than the challenge of writing a compelling research proposal.

KEY TERMS

conceptual framework

literature review

pilot study

snowball sampling

REVIEW QUESTIONS

1. Identify two or three possible topics you will consider for your research project, thesis, or dissertation. Describe how these ideas originated. Why do you think they will sustain your interest? What preparation will you need in order to address them?
2. Select a published research article in your area of interest. What does the author identify as the main research questions? Write an opening paragraph that you think would be a strong introduction for a research proposal that will address these research questions.
3. List the information that you will include in a proposal for research you will undertake to demonstrate that you have considered how your methodology will be implemented (for example, how you will select people to interview).
4. Outline the 'big picture' question and up to four related sub-questions for a research topic of your choice. Identify the main elements of the methodology you would need to employ to address these questions.

USEFUL RESOURCES

Bouma, G.D., and R. Ling. 2004. *The Research Process.* 5th edn. New York: Oxford University Press.

Hay, I. 2006. *Communicating in Geography and the Environmental Sciences.* 3rd edn. Melbourne: Oxford University Press.

Heath, A.W. 1997. 'The proposal in qualitative research'. *The Qualitative Report* 3 (1). http://www.nova.edu/ssss/QR/QR3-1/heath.html.

Institute of International Studies, University of California, Berkeley. 2001. 'Dissertation proposal workshop'. http://globetrotter.berkeley.edu/DissPropWorkshop. In addition to sections on conceptualizing, writing, and revising proposals, this site includes sample proposals and comments by the authors reflecting on how they regarded the proposal after they had completed the research.

McIntyre, J. 2007. 'Guide to writing a research proposal'. http://www.education.uts.edu.au/research/degrees/guide.html.

Przeworski, A., and F. Salmon. 1995. 'The art of writing proposals: Some candid suggestions for applicants to Social Science Research Council competitions'. http://fellowships.ssrc.org/art_of_writing_proposals.

Punch, K.F. 2006. *Developing Effective Research Proposals*. 2nd edn. London: Sage.

Sarantakos, S. 2005. *Social Research*. 3rd edn. Melbourne: Palgrave.

Sides, C.H. 1999. *How to Write and Present Technical Information*. 3rd edn. Phoenix: Oryx Press.

Notes

1. Although Fay Gale was shortly to complete her doctoral research in South Australia on regional patterns of Aboriginal communities in South Australia (Gale 1964).

2. Andrew Beer drew our attention to this research idea.

3. Keep in mind that any research involving collection of information from human subjects will necessitate a process of ethical approval for the project. Writing the ethics approval application can be a time-consuming task that should be factored into your timetable. See Chapter 2 for a fuller discussion of ethical issues.

4. For an example of agency guidelines for proposal preparation (and just how specific they can be), see the US National Science Foundation's website section on doctoral dissertation programs, http://www.nsf.gov/pubs/2006/nsf06605/nsf06605.pdf.

Writing Qualitative Geographies, Constructing Meaningful Geographical Knowledges

17

Juliana Mansvelt and Lawrence D. Berg

CHAPTER OVERVIEW

In this chapter, we examine the process of 'writing-up the results' of qualitative research in human geography. Our aim, however, is to contest the simplistic understanding of the relationship between research, writing, and the production of knowledge that arises from describing the process in this manner. Indeed, the very phrase 'writing-up' implies that we are somehow able to unproblematically reproduce the simple truth(s) of our research in our writing. In this framework for understanding research, writing becomes a mirror that serves to innocently reflect the reality of research 'findings'. In contrast, we draw upon post-structuralist approaches to argue that writing is not merely a mechanical process that reflects the 'reality' of qualitative research findings but rather constitutes in part how and what we know about our research. Writing is thus not so much a process of writing-*up* as one of writing-*in*, a perspective that has significant implications for how research is conceptualized.[1]

STYLES OF PRESENTATION

Part of our argument in this chapter revolves around the idea that a number of powerful **dichotomies**—such as subject/object, researcher/researched, data/conclusions, and research/writing—currently exist and that these dichotomies structure our understanding of research. These dichotomies have developed historically as part of a **neo-positivist** framework that, we suggest, limits our understanding of the writing process. We thus feel it is important to present readers with a very brief history of the development of positivist thought in human geography before moving on to discuss an alternative approach to writing human geography.

Positivist and Neo-positivist Approaches: Universal Objectivity

The philosophical approach to scientific knowledge known as **positivism** was founded by Auguste Comte (1798–1857), a French philosopher and sociologist (see Gregory 1978; Kolakowski 1972). Comte argued that scientific knowledge of the world arises from observation only. The notion of a singular truth was associated with this, and positivists of Comte's tradition had significant difficulties with issues and questions arising from the relationship between truth and phenomena such as religion, ethics, morals, and metaphysics.

The logical positivists, whose work developed as a critique of strict positivism during the 1920s, differed from the Comtean positivists in their conception of scientific knowledge. They accepted the validity of more than just empirically verifiable statements. In this sense, Comte allowed only statements arising from empirically verifiable knowledge (that is, information available to the senses) as the basis of factual knowledge (or truth). The logical positivists, on the other hand, acknowledged the existence of two kinds of factual statements: analytical statements, the truth of which is contained in their internal logic (for example, if A = B and B = C, then A = C), and synthetic statements, the truth of which is proven through recourse to empirical observation (for example, if we observe high levels of homelessness in areas with low levels of public housing, our empirical observations could lead us to conclude that the incidence of homelessness is a consequence of particular state housing policies). As you may appreciate from the previous example, numerical correlation is not the same as causation. Assessing the validity of analytical statements derived from empirical observations is thus problematic, and particularly so in social science.

Karl Popper (1959) recognized this and developed an alternative form of neo-positivist thought he termed 'critical rationalism'. Unlike the logical positivists, who focused upon verification as the basis of knowledge, Popper argued that **falsification** should be the basis for making decisions about the validity of factual statements. His argument was based on the notion that we can never know for sure whether a particular hypothesis is true or not (such as whether a lack of state housing does result in more homelessness), but we do have the ability to ascertain whether such statements are false (for example, by finding examples of a low incidence of homelessness in areas with minimal public housing). Under critical rationalism, then, we never prove something, but instead we can either disprove something (or prove it to be false through the process of falsification) or accept it to be contingently 'true' (until it has been proved false). According to Popper, scientific knowledge is much more contingent than conceptualized by the logical positivists, since present 'truths' are always open to future falsification.

Although few geographers have ever fully taken up the ideas of any single school of positivist thought, positivism of one sort or another has played a foundational role in the way that many geographers have come to understand the world. This became most explicit during the so-called **quantitative revolution** of the 1960s (Gregory 1978; Guelke 1978), which saw geography develop as what we might broadly term a neo-positivist 'science'. Geographers maintained a strict distinction between facts and values (Gregory 1978), giving emphasis to observational statements over theoretical ones (Berg 1994a) and often universalizing their findings across all contexts (Barnes 1989). They also made a strict distinction between objective and subjective knowledge.

Objective knowledge is seen as 'scientific', rigorous, and detached and consequently valid. It is constituted in opposition to **subjective** knowledge, which is personal, value-based, non-scientific, and non-academic (and therefore unacceptable as a basis for establishing 'the' truth). Objective knowledge is founded on interrelated and highly gendered notions of rationality, disembodied reason, and universality (Berg 1994a, 1997; Bondi 1997). Trevor Barnes and Derek Gregory (1997, 15) suggest that 'scientific geographers' thus imagined themselves:

> as a person—significantly, almost always a man—who had been elevated above the rest of the population, and who occupied a position from which he could survey the world with a detachment and clarity that was denied to those closer to the ground (whose vision was supposed to be necessarily limited by their involvement in the mundane tasks of ordinary life).

Knowledge from the disembodied vantage point of the objective researcher looks the same from any perspective—it is monolithic, universal, and totalizing. This concept of un-located and disembodied rational knowledge draws on powerful metaphors of mobility (the researcher can move to any and all perspectives) and transcendence (the researcher is not part of the social relations she or he is examining but instead can rise above them to see everything) for its rhetorical power to convince readers of its claims to the truth (see Barnes and Gregory 1997; Haraway 1991).

In adopting a broadly neo-positivist model for their work, geographers also developed a specific approach to writing their research. They developed commonly used approaches or forms of writing (or what we call **tropes**) in their academic studies that attempted to erase the authorial self from their written work. Similarly, they tried to create, through their writing, distance between themselves as researcher/author and their research objects. These tropes are most evident in the practice of writing in the third person—a practice that is still prevalent in much academic writing today. Many geography undergraduate students, for example, are still required to use the formal **third-person narrative** form in their essay assignments. The notion of universal knowledge underpins writing in the third person, and it is also intended to erase the imprint of the author on the text and to ensure the distancing of the author from the research. In other words, third-person narratives are needed to maintain impartiality and objectivity, two cornerstones of the neo-positivist model. The third-person narrative constructs an **objective modality** (Fairclough 1992). This effectively removes the author from his or her writing while at the same time implying the author's full agreement with the statements being made. In so doing, it does the work of transforming interpretative statements into factual statements (see Box 17.1).

Writing in the objective mode is often accompanied by **nominalization**, a process that further removes the writer from the text (see Box 17.1). Nominalization involves the transformation of adjectives and verbs into nouns. It occurs, for example, when in the process of researching, researchers and research subjects are rhetorically transformed into things such as 'the research' or when actions and processes are given particular kinds of argumentative power as nouns, such as can be found in statements such as 'the analysis suggests' (Fowler 1991). Nominalization deletes a great deal of helpful and important information from sentences. For example, it deletes information about the participants—normally the agent, or the person 'doing' something, and the affected participant, or the person having something 'done to' them. It also changes the **modality**—that is, the implicit indication of the writer's degree of agreement with the statement—and it tends to hide details about the circumstances relating to a set of actions. Nominalization creates **mystification** because it permits concealment, hiding the participants, indicators of time or place, and stance. It also results in **reification**, whereby complex, uncertain, and often contradictory processes, such as are involved in human geographic research, assume the much more certain status of 'things'. Thus, the complex processes of research are smoothed over into a one-dimensional and simple thing: 'the research' (see, for example, Fowler 1991, 80). Ironically, while positivist epistemology is founded on the idea that any statement can be contested through recourse to empirical evidence and logic, the language used to communicate research findings within this framework implies an unquestionable accuracy that is a very poor approximation of the messiness of social life.

Interestingly, writers drawing on neo-positivist approaches have shown ambivalence with regard to their own writing practices. On the one hand, they have explicitly acted as if language has no impact on meaning, yet ironically, they have implicitly acknowledged—through their insistence on writing in the third person—the significant role that language plays in constructing knowledge and meaning.

Post-positivist Approaches: Situated Knowledges

Although the positivist-oriented science model dominated during the quantitative theoretical period of the 1960s and 1970s, its **hegemony** (or conceptual dominance) has been contested in geography by post-positivist approaches such as humanism, Marxism, political economy, and post-structuralism (for example, Barnes 1993; Berg 1994b, 2001; Dixon and Jones 1996). In particular, feminist **post-structuralist**

Removing the Writer—Third Person and Nominalization   Box 17.1

Writing in third person and a process called nominalization help to create a more formal written text. This has the effect of distancing the author from the narrative being constructed and creating a sense of neutrality and objectivity.

As you read these sentences, consider how the use of third person, by removing terms that describe oneself (such as 'I') and substituting descriptors of others (for example, the investigator) creates a sense of distance, from the research and the research subjects, and makes the statement appear more factual.

'The investigator examined five types of place-based knowledge.'
'The researcher demonstrated that understandings of home are complex.'
'One tries not to influence the responses gained from interview questions.'

Nominalization is the process of turning verbs into nouns. It involves taking actions and events and changing them into objects, concepts, or things—making sentences seem more formal and abstract. Like writing in third person, nominalization contributes to a sense of detachment from the text (and the research), removing both personal reflections and the context of the subject being discussed.

For example, the verb 'developed' is nominalized to the noun 'development':

'I *developed* a framework for analyzing interviews' (**first-person narrative**) might become 'the *development* of a framework for analyzing interviews'.

Below are two more examples of text. In the informal text, few verbs have been nominalized. The second paragraph is an example of more formal research writing in which verbs and verb phrases have been nominalized. As you read these two texts, think how the agency of the researcher appears to be removed in the second, concealing the researcher's emotions and experiences. Note, for example, how the first sentence in the formal text suggests that there is much to be learned from the research (now assumed to have the status of a thing) rather than from the researcher-led processes that constituted it, as implied by the more informal excerpt.

Informal text:

Designing and conducting the research was a learning process for me. Potential participants often refused to be involved in the study, and I found this very frustrating. It was also troubling to know that other research subjects felt as though they were being coerced into participating. Analyzing the research findings produced limited insights into the research question. I concluded that my research project was ill-conceived!

Formal text:

Much was learned from the design and conduct of the research. Potential participants' refusal to be involved in the study caused frustration. It was a matter of concern that some individuals felt their participation was subject to researcher coercion. Research analysis provided limited insights into the research question. The research was ill-conceived.

The formality achieved through the use of third person and processes of nominalization may have the effect of producing an author as an all-seeing and all-knowing (seemingly objective) surveyor of the research rather than an active participant, complicit in the production of both text and research. Such distancing is an illusion, negating the ways in which power and subjectivity are constituted through the research process.

writers (Anzaldúa 1987; Frankenberg and Mani 1993; Massey 1993; Mohanty 1991) have been keen to confront the universalism, mastery, and disembodiment inherent in positivist notions of objectivity, criticizing masculinist and Eurocentric concepts of universal knowledge (see also Berg 2004; Berg and Kearns 1998; Bondi 1997; England 1994; Henry and Berg 2006). Such ideas, they suggest, play a significant role in marginalizing those who do not fit into dominant conceptions of social life.

Donna Haraway's (1991) evocative metaphor of **situated knowledges** provides perhaps the most useful approach to contest universalist forms of knowledge. She argues that within dominant ideologies of scientific knowledge, objectivity must be seen as a 'God Trick' of seeing everything from nowhere. She proposes a different concept of 'objectivity', one that attempts to situate knowledge by making the knower accountable to their *position* (see also Chapters 2 and 11). All knowledge is the product of specific embodied knowers, located in particular places and spaces: 'there is no independent position from which one can freely and fully observe the world in all its complex particulars' (Barnes and Gregory 1997, 20). Research that draws upon situated knowledges is thus based on a notion of objectivity much different from that posed by positivists, and this conception of objectivity also requires a different form of writing practice.

Work by Isabel Dyck (1997) provides an excellent example of the importance of 'position' and the kind of 'truths' situated knowledge might produce (also see England 1994). Dyck reflects on the importance of her own position as a white middle-class Canadian academic and the impact it had on two different research projects she undertook to examine the time–space strategies adopted by Indo-Canadian immigrant women in Vancouver. She found that immigrant women were more willing to speak with her about certain aspects of their lives than about others because she was not seen as a threat to their own social networks and relationships within the Indo-Canadian community in Vancouver. At the same time, as an outsider she was occasionally excluded from aspects of Indo-Canadian women's lives that were defined as culturally sensitive. Her research thus points to the specificity of position and the importance of recognizing the politics of position in research processes.

Post-structuralists and feminists also contest approaches to inquiry that conceptualize writing and language as simple reflections of 'reality'. They argue that 'language lies at the heart of all knowledge' (Dear 1988, 266). It should be made clear, however, that such arguments do not assume that language and ideas are the same as 'real' phenomena, objects, and material things. Instead, arguments about the centrality of language express the fact that all processes, objects, and things are understood by humans through the medium of language. Thus, while we might experience the very material process of hitting our 'funny bone' on a table in ways that do not necessarily entail language (for example, as a very visceral sensation we know as 'pain'), we come to understand the process and the objects involved through language (with categories such as table, funny bone, pain, and so on). Accordingly, language must be seen as not merely reflective but instead as *constitutive* of social life (for example, Barnes and Duncan 1992; Bondi 1997; Dear 1988; see also Chapter 11 of this book).

The post-structuralist critiques of both the **mimetic** concept of language and of disembodied concepts of universal objectivity have significant implications for writing practices. If language is constitutive of knowledge and meaning, then it would seem to matter *how* we write our knowledges of the world. Likewise, if we are to *locate* our knowledge, then we must locate ourselves as researchers and writers within our own writing. Accordingly, post-structuralist writers reject the ostensibly 'objective' modality of writing their work in the third person. Instead, they opt for locating their knowledge-defining objectivity as something to be found not through distance, impartiality, and universality but through contextuality, partiality, and **positionality**. However, as Gillian Rose (1997) has argued, given the difficulty in completely understanding the 'self', it may be virtually impossible for authors to fully situate themselves in their research. Notwithstanding such difficulties, it is possible for authors to go some way towards locating themselves within their work. Certainly, the first step is to reject the third-person narrative, replacing it with a first-person narration of our essays.[2] At the same time, it is not enough to merely adopt the first-person narrative form. Instead, it is important to both reflect upon and analyze how one's position in relation to the processes, people, and phenomena we are researching actually affects both those phenomena and our understanding of them. Again, Isabel Dyck's (1997) study of 'two research projects' discussed above provides an excellent example of how we might undertake such analyses. Thus, rather than writing ourselves *out* of our research, we write ourselves back *in*. This is, perhaps, one of the most important distinctions to be made between what we will refer to as the writing-*up* (distanced, universal, and impartial) and the **writing-in** (located, partial, and situated knowledge) models. Another significant difference arises from the ways that post-positivists conceptualize the relationship between observation and theory.

BALANCING DESCRIPTION AND INTERPRETATION—OBSERVATION AND THEORY

The Role of 'Theory' and the Constitution of 'Truth'

We argue in this section that there currently exist a number of powerful dichotomies—observation/theory, subject/object, researcher/researched, data/conclusions, and research/writing—that structure our understanding of research. It is important to remember that these dichotomies are not recent developments in Western thought. Instead, they arose within the long history of dualistic thinking in Western philosophy (Berg 1994a; Bordo 1986; Derrida 1981; Foucault 1977b; Jay 1981; Le Doeff 1987; Lloyd 1984; Nietzsche 1969). These dichotomies became racialized and gendered through a long historical process of developing a singular Eurocentric and masculine concept of rational thought. For example, through a process that Susan Bordo (1986) terms the 'Cartesian masculinisation of thought', Descartes's mind–body distinction came to define appropriate forms of knowledge. The mind was conceptualized as rational, and it came to be a property of European men. The body was seen as irrational (have you ever heard the phrase 'mind over matter'?) and was associated with everything that was not European or masculine: women, racial minorities, and sexual dissidents, for example. In other words, the mind was unmarked, but the body was a mark of difference. Further, Descartes's dualistic philosophy of knowledge formed the basis for the dominant present-day conception of objectivity as impartial, distanced, and disembodied knowledge. Accordingly, Cartesian dualistic thinking forms the foundation of positivist thought (Karl Popper, of whom we spoke earlier, for example, was a well-known adherent of the mind–body dualism).

As we have already discussed, positivist-inspired geographers make a rigid distinction between objective and subjective knowledge and between theory and observation. As with the observation/theory binary, the so-called 'objective' is valued at the expense of the subjective. Such hierarchically valued dichotomies form parts of a whole series of other binary concepts—including (but not limited to) mind/body, masculine/feminine, rationality/irrationality, and research/writing—that are interlinked through complex processes of signification (Derrida 1981; Jay 1981; Le Doeff 1987; Lloyd 1984). In this way, for example, observation is equated with objectivity, mind, masculinity, rationality, and research. Theory is constituted as lacking, and it is associated with all those other negatively valued concepts: the subjective, the body, the feminine, and the irrational.

Conceptualizing one side of the binary as a *lack* of the other leads to a devaluation of the subordinate term. Thus, in the case of the positivists who conceive of theory as a lack of empirical observation, the ways in which theory constitutes our understanding of empirical 'reality' (in addition to explaining it) are underestimated.

Indeed, positivist constructions of factual knowledge as phenomena that are available to the senses (empirically observable) tend to efface the very theoretical nature of positivist thinking itself. As we have already suggested, positivism has both a history and a geography associated with Europe. It is an epistemology—a theory of knowledge—that has developed relatively recently and has come to dominate contemporary intellectual life in the West. Nonetheless, it is not the only theory of knowledge; rather, it is one of many competing theories of knowledge. However, because it is dominant, or hegemonic, it rarely has to account for its own epistemological frameworks. With this in mind, we argue for 'recognition that we cannot insert ourselves into the world free of theory, and neither can such theory ever be unaffected by our experiences in the world' (Berg 1994a, 256). Observations are thus *always already* theoretical, just as theory is always touched by our empirical experience. Recognition of this relationship has important consequences for the way we write-*in* our work (and *work*-in our writing).

Writing and Researching as Mutually Constitutive Practices

Metaphors that allude to research as 'exploration' or 'discovery' are hard to avoid, so taken for granted are their meanings (how often have you heard lecturers speak of their research in terms of 'exploring', 'examining', 'discovering', and 'uncovering'?). This is particularly the case with the so-called 'writing-up' of the research. The term writing-up powerfully articulates the written aspect of the research process in a way that engenders it as somehow less significant and/or less problematic than other aspects of the research process. Writing-up is usually seen as a phase that occurs at the end of a research program; indeed, many textbooks about conducting research (this one included) include the section on 'writing-up' at the end of the book (for example, Flowerdew and Martin 2005; Kidder, Judd, and Smith 1986; Kitchin and Tate 2000). Writing is also often seen as a neutral activity, although ironically, scientific and positivist modes of writing and thinking are subject to rhetoric, creativity, and intuition (Bailey, White, and Pain 1999a).

The 'writing-up phase' may be discussed in such a way that it appears to be merely a matter of presenting the results and conclusions in an appropriate format at the *end* of a research program. We argue here that in writing research, the researcher is not so much presenting her or his findings as *re-presenting* the research through a particular medium. Re-presentation speaks of the mediated character of the process of writing research.

Rather than reflecting the outcome of a particular research endeavour, we believe the act of writing is a means by which the research is constituted—or given form—and that this process occurs throughout the research process. For example, keeping in mind the constitutive character of language, we can see that any attempt to write research involves a process of selecting categories and language to describe complex phenomena and relationships. But we are getting at much more than that here. Research and writing are iterative processes, and writing helps to shape the research as much as it reflects it. At the larger scale of disciplinary practices and epistemological conventions, as we have suggested above, the way we conceptualize the author (as distant and impartial or as involved and partial, for example) has significant implications for the ways that the very processes of research itself can be understood. Perhaps just as important, writing involves very clear decisions to include some narratives and to exclude others. It thus brings some ideas into existence while (implicitly) denying the existence of others (Ely 2007). For us, then, writing is not so much a matter of writing-*up* as of writing-*in*, a perspective that has considerable implications for how qualitative research is conceptualized and undertaken.

Writing is not devoid of the political, personal, and moral issues that are a feature of undertaking research, nor is writing devoid of our embodied emotions as we sense and feel the narratives we construct. As a creative process, writing can be enjoyable for authors and readers (Bradford 2003), opening up new possibilities for representation, narrative, and engagement with those who may participate in and reflect on the research. Further, the separation between 'fieldwork' and writing is artificial (Denzin 2008). Whatever the qualitative research technique utilized, some means of recording the researcher's interpretations, impressions, and analysis must be used, and although such accounts may be recorded on tape, memory stick, or video, the words with which they are constructed are an integral part of the research, not simply a result, recollection, or recording of it. The research cannot be separated from the labels, terms, or categories used to describe it and interpret it, because it is through them that the research is made meaningful.

For example, after I (Juliana) had conducted several qualitative interviews with local authority economic development officers for my PhD research, I realized that using the word 'traditional' as a label for a certain form of local economic initiative was problematic. This was because definitions of 'traditional economic initiatives' were contested by officers and because the term appeared to position local authorities who undertook this type of initiative as 'old-fashioned' or 'not progressive'. I learned a valuable lesson about the power embodied in words I had simply drawn from the literature on local economic development and about the kind of assumptions embedded in my own research agenda. This understanding enabled me to construct my questions (and consequently my entire research project) in a different way.

I (Lawrence) had a similar experience of reorienting research agendas in a participatory action research project I undertook with a number of colleagues (Berg et al. 2007). This project involved the fusion of indigenous methodologies, participatory action research, and white studies (Evans et al. 2009) to understand the exclusion of urban Aboriginal people from the so-called 'universal health care system' in Canada. However, before we even started our formal research, we had to come to grips with the hegemonic ideology among white Canadians that Aboriginal people do not live in cities in Canada but instead are to be found out in the countryside, on rural reserves, disengaged from modernity. Such categories are especially problematic given that as of the 2006 census, more than 51 per cent of all Aboriginal people now live in urban areas. What it means to be 'Aboriginal' in Canada is thus contested (Berg et al. 2007) and so was our own understanding of our research project as a 'participatory' partnership with urban Aboriginal people (Berg et al. 2007; Evans et al. 2009). It should be clear that any attempts to write about such work involve significant politics in the construction of our categories and in the decisions about what to say (and not to say) about any specific events. Accordingly, the 'writing-up' of the research is only the initial phase of an iterative process of negotiating the production of knowledge between the researchers and the participants. I am not suggesting here, however, that negotiated knowledge is intrinsically better than other forms of knowledge. Instead, what I want to point out is that the process of making explicit the act of negotiation helps to make the research accountable in ways that are appropriate given the specificity of positions and power relations. All research is caught up with power relations, and to deny this is to deny an important aspect of knowledge production (see Chapters 2, 3, and 13). Taking this process seriously has enabled me to rethink the role of writing in my intellectual endeavours.

Thoughts, observations, emotions, and interpretations that occur during the research thus become important components of any research endeavour, not because they record events or ideas but because they are **signifiers** of them (in this sense, they act to 'define' complex constellations of ideas and thoughts about the research in more simplified categories of knowledge). Whether interpretations are noted by way of a personal diary, log, video, or audio recording, they can provide insight into the researcher's own speaking position and how it is articulated, challenged, and modified through the research journey.

Writing-*in* is not a matter of 'telling'—it is about knowing the world in a certain way. The process of writing constructs what we know about our research, but it also speaks powerfully about who we are and where we speak from. As we suggested in a previous section, the detached third-person writing style so common in academic journals and reports implies that the researcher is omnipotent—that he or she has a perspective that is all-seeing and all-knowing. However, what may appear to be the truth spoken from 'everywhere' is actually a partial perspective spoken from some*where* and by some*one*. Knowledge does not, according to a post-structuralist perspective, exist independently of the people who created it—knowledges are partial and geographically and temporally located. As the researcher writes and inscribes meaning in the qualitative text, she or he is actually constructing a particular and partial story. Richardson and St Pierre

(2008) suggest that writing creates a particular view not only of what we are talking about (and what we do not say) but also of ourselves. Power is connected with speaking position through text, so a qualitative researcher should consider not only the standpoint spoken from in constructing a research account but also the implications of their interpretations for those who may have been involved in the research and on the structuring and power relations in everyday life.

Because the practice of writing is not neutral, the voices of qualitative researchers do not need to hide behind the detached 'scientific' modes of writing. Such modes of writing position the researcher as a disembodied observer of the truth rather than as a (re)presenter and creator of a particular and partial truth. The researcher is an instrument of the research, and accordingly, we suggest that researchers should acknowledge their position in ways that demonstrate the connection between the processes of research and writing. **Reflexivity** is the term often used for writing self into the text. Kim England (1994, 82) defines this as 'self-critical sympathetic introspection and the self conscious analytical scrutiny of self as researcher'. A reflexive approach can make researchers more aware of their necessary connection to the research and their effects on it and of asymmetrical (i.e., where a researcher has more social power and influence than their participants) or exploitative relationships, but it cannot remove them (England 1994, 86; also see Rose 1997 and Chapters 2 and 3 of this volume). One way in which reflexivity can be encouraged in the writing-*in* of qualitative research is by the use of personal pronouns (for example, I, we, my, our).

However, it is important that the use of personal pronouns does not become merely an emotive tool. Alison Jones (1992) suggests that in academic spheres, the use of 'I' may result in the insertion of 'emotion' as a replacement for 'reason', thereby creating a work of 'fiction' hiding power relations as much as it might make them explicit. Employing reflexivity through use of first person should instead make explicit the politics associated with the personal voice and draw attention to assumptions embedded in research texts. Reflexivity is also concerned with constructing research texts in a way that gives consideration to the voices of those who may have participated in the research. Reflexivity is about writing critically, in a way that reflects the researcher's understanding of their position in time and place, their particular standpoint, and the consequent partiality of their perspective. Writing situated accounts may involve acknowledging the role of embodied emotions in research (Bondi, Davidson, and Smith 2005; Widdowfield 2000), thinking about how one's positionality is 'mutually constituted through the relational context of the research process' (Valentine 2003, 377) and thinking about the ways in which our texts might constitute performances, enacting and valorizing certain metaphors and meanings and silencing others in the spaces where they might circulate (Denzin 2003). This understanding of the **dialogic nature of research and writing** (in the sense of a 'dialogue' between various aspects of the research process) enables qualitative researchers to acknowledge in a meaningful way how their assumptions, values, and identities constitute the geographies they create. It also provides an opportunity to play and experiment with writing as a way of knowing and representing.

Just as there are many ways of knowing, there are also numerous ways in which qualitative researchers may construct their research narratives as 're-presentations'—constructions that evoke what we as researchers have lived and learned (Ely 2007). Richardson and St Pierre (2008) believe that to write 'mechanically' shuts down the creativity and sensibilities of the researcher. They encourage researchers to explore text and genre in the (re)presentation of qualitative research through a variety of media, including oral and visual (see, for example, Latham and McCormack 2007 and Sanders 2007 on how students can use digital photography in presenting their research) and to experiment with diverse forms of the written word (prose, poetry, play, autobiography). We support Richardson and St Pierre's (2008) metaphorical construction of writing as **'staging a text'** and encourage geographical researchers to consider how they are (re)presenting the research 'actors', creating the plot, action, and dialogue of a research 'tale'; how they are constructing the stage, the setting of the 'research' play; and at whom (i.e., the audience) the 'production' is aimed.

Of course, most undergraduate geography students will be required to write their research within a given format—the essay or report (for some advice on the conventions associated with these forms of writing, see Bradford 2003; Hay 2006; Hay, Bochner, and Dungey 2006; Kitchin and Tate 2000; and Chapter 18 of this book). Nevertheless, it may be possible to persuade your lecturer to let you produce another form of geographic representation: a play, a video, a poem, a short story, a poster-board, to name a few options. For example, I (Lawrence) recently had third-year students in my culture, space, and politics course create two- to three-minute videos involving critical analyses of 'landscapes' as part of their major assignment for the class and then upload these videos to YouTube for public viewing (some of these videos are still available on the GEOG359 channel on YouTube). Despite writing constraints and 'staging' conventions imposed by self and audience (such as for an academic publication or a thesis), we argue that there is no single correct way to 'stage' a text (see Box 17.2). By exploring the varied ways in which the text can be staged and how in such staging different stories may be emphasized and other voices may come to the fore, the researcher has the potential to create dynamic and interesting research pieces that engage and challenge both writer and reader.

ISSUES OF VALIDITY AND AUTHENTICITY

The interpretative nature of qualitative research has given rise to a considerable amount of debate concerning how the **validity** and authenticity of qualitative research accounts might be assessed (see Bailey, White, and Pain 1999a; Baxter

Writing-in—Alternative Representations   Box 17.2

The unconventionality of Marcus Doel and David Clarke's (1999) staging of an 'academic' article is conveyed by its unusual title, 'Dark Panopticon. Or, attack of the killer tomatoes'. In this paper, which appeared in *Environment and Planning D: Society and Space*, the authors write as if in a dream world. Their piece is simultaneously fragmentary and powerful. The form of writing is cleverly indicative of post-structuralist perspectives through which various arguments and thoughts are constructed. In it, narratives of dream, text from other academics, and authorial thoughts are presented as a complex reflexive tapestry. The narratives contained within the article are disjointed and struggled over, and the readers are invited on hallucinatory journeys through real and imagined spaces. These are spaces that reflect the ways in which depersonalization, de-individuation, and fatigue characterize both ways of thinking about the world and commodity relations. As you read the following excerpt, consider how theoretical stance and textual staging are integrated through this 'alternative' approach:

> Flash-darks touch the flesh of the world, giving rise to all manner of ontic unease. To be touched by the Dark Panopticon is exemplified for us in the display of greengrocery that greets countless millions as they flow into the world's markets, supermarkets, and hypermarkets. And what we feel here—is it only us?—is an unprecedented and almost unbearable brutality against the flesh of the world: not simply the flesh of humans or animals, but above all else the flesh of vegetables and soft fruit. In particular we are horrified by violence against tomatoes. Yet as we shall tease out in due course, tomatoes, like all objects in the consumer society and data in the surveillance society, take their revenge. The tomatoes strike back . . . (Doel and Clarke 1999, 429)

and Eyles 1997). The reflexive writing-*in* of research experiences and assumptions is not a licence for sloppy research or monographs based solely on personal opinion. Validity, integrity, and honesty in writing-*in* are no less important in qualitative research than they are in quantitative research. Works by writers such as Mike Davis (1990), Gillian Rose (1993), and Cole Harris (1997) provide telling examples of critical qualitative research and carefully considered analysis of socio-spatial relations.

The truth and validity of knowledge arising from geographical and social science research has been the subject of discussion for a considerable period of time. Three decades ago, for example, Harrison and Livingston (1980) suggested that the conduct of research is necessarily linked to what it is possible to know and how we can know it. Post-structuralist thinking has also challenged the assumption of a singular truth and the privileging of certain claims to knowledge. Associated with this is what has been termed the 'crisis of representation' (Marcus and Fisher 1986). That is, doubts have arisen over researchers' authority to speak for others in the conduct and communication of research (Alvermann, O'Brien, and Dillon 1996). In recent years, an increased sensitivity to power and control on the part of some qualitative researchers has encouraged a rethinking of research design and implementation (Glesne and Peshkin 1992, 10; Evans et al. 2009). It has also meant a growing concern over the textual appropriation (how researchers appropriate participants' voices) of data in the writing of research accounts (Opie 1992). Moreover, it has become important to acknowledge that one's research writing might have an active role in constructing and reconstructing research relations in unintended ways and that research accounts are in themselves a kind of creative invention (Freeman 2007)

Baxter and Eyles (1997) have argued that geographers need to be more explicit about how rigour has been achieved throughout the research process. The process of writing-*in* qualitative research requires, therefore, that writers explicitly state the criteria by which a reader may assess the 'trustworthiness' of a given piece of research (see Chapters 4 and 5 for a discussion). Addressing this issue is difficult, because it involves qualitative writers grappling with the tensions between the complexity and richness of information that emerges from qualitative research and the need to produce some sort of 'standardized' evaluation criteria (Baxter and Eyles 1997).

The post-structuralist challenge to a singular notion of truth and a growing awareness of issues of representation do present difficulties for assessing the validity of qualitative research. Much of the debate surrounding the validity of qualitative writing rests upon how terms such as 'rigour', 'validity', 'reliability', and 'truthfulness' are defined. A related issue is whether these definitions, which have often been used as 'objective' measures of the quality of quantitative research, are applicable to qualitative endeavours (see Chapters 4 and 5). Post-structuralist thinking casts doubt on foundational arguments that seek to anchor a text's authority in terms such as reliability, validity, and generalizability.

A debate between two geographers in 1991 and 1992 issues of *Professional Geographer* highlights some of the issues surrounding reading and evaluating texts and the validity of qualitative approaches. The exchange between Erica Schoenberger (1991; 1992), and Linda McDowell (1992b) is significant because it was written at a time when in-depth interviewing was not used extensively in industrial and economic geography. The debate is interesting not only because it centred upon contested definitions of validity but also because it raised issues of meaning and interpretation, audience, and representation through the language that constituted these articles. It demonstrates powerfully the multiple reading of texts and the care needed in constructing qualitative interpretation. Schoenberger sought to argue for the legitimacy of qualitative forms of interviewing but in doing so suggested how quantitative definitions of reliability (defined by Schoenberger as the stability of methods and findings) and validity (accuracy and truthfulness of findings) might be applied to qualitative research. McDowell disagreed with Schoenberger's concept of validity as interpretation that is verifiable and that corresponds to some external truth. She suggested that Schoenberger's definition was based on a continued adherence to a positivist-inspired quantitative research model, as demonstrated by the language through which her arguments were constructed. Schoenberger (1992) addressed McDowell's criticisms by referring to issues of intended audience (positivist geographers), power, and interpretation, suggesting that McDowell had misinterpreted her claims.

If concepts such as validity are contested, how then is it possible to construct truthful research texts? Clifford Geertz (1973) has argued that good qualitative research comprises **thick description**. Such descriptions take the reader to the centre of an experience, event, or action, providing an in-depth study of the context and the reasons, intentions, understandings, and motivations that surround that experience or occurrence. While it may not be possible to assess the authenticity of such partial descriptions, the interpretations upon which they are constructed can be articulated. Mike Davis (1990, 253–7), for example, provides a particularly compelling description of the makeshift prisons and holding centres for the urban underclass who make up the more than 25,000 prisoners in 'the carceral city' found in a three-mile radius of Los Angeles city hall (see Box 17.3). His descriptions, meticulously researched and referenced, are as much novel-like evocations of city life as they are academic descriptors. In the end, we are trying to understand the *meaning* of particular aspects of social and spatial life, and meaning is always a matter of interpretation.

Communicating qualitative research is thus about choices—for example, about how we show the workings of our research, how we present and convey others' voices, how we locate our own subject position, and what effects our writing may have for those who may read it and who may want to engage with it (Denzin 2008; Holliday 2007). Although such choices are not always conscious and not necessarily made in circumstances of our own choosing, researchers can attempt to take responsibility for their perspective and position in research (sometimes referred to as 'transparency') through reflexively acknowledging and making explicit those choices that have influenced the creation, conduct, interpretation, and writing-*in* of the research. Such choices are likely to be guided by principles of ethics and truthfulness (see Chapters 2 and 3). Transparency may make researchers, and the audiences for whom they write, more aware of the constraints on interpretation, of the limitations imposed by the 'textual staging', and of the implications of the former for research participants. For example, the choices surrounding the use of research participants' quotes in a written text comprise far more than a simple matter of how,

Writing Fortress LA

Box 17.3

The demand for law enforcement *lebensraum* in the central city, however, will inevitably bring the police agencies into conflict with more than mere community groups. Already the plan to add two highrise towers, with 200–400 new beds, to County Jail on Bauchet Street downtown has raised the ire of planners and developers hoping to make nearby Union Station the center of a giant complex of skyscraper hotels and offices. If the jail expansion goes ahead, tourists and developers could end up ogling one another from opposed highrises. One solution to the conflict between carceral and commercial redevelopment is to use architectural camouflage to finesse jail space into the skyscape. If buildings and homes are becoming more prison- or fortress-like in exterior appearance, then prisons ironically are becoming architecturally naturalized as aesthetic objects. Moreover, with the post-liberal shift of government expenditure from welfare to repression, carceral structures have become the new frontier of public architecture. As an office glut in most parts of the country reduces commissions for corporate highrises, celebrity architects are rushing to design jails, prisons, and police stations.

An extraordinary example, the flagship of an emerging genre, is Welton Becket Associates' new Metropolitan Detention Center in Downtown Los Angeles, on the edge of the Civic Center and the Hollywood Freeway. Although this ten-story Federal Bureau of Prisons facility is one of the most visible new structures in the city, few of the hundreds of thousands of commuters who pass it by every day have any inkling of its function as a holding and transfer center for what has been officially described as the 'managerial elite of narco-terrorism'. Here, 70% of federal incarcerations are related to the War on Drugs. This postmodern Bastille—the largest prison built in a major US urban center in generations—looks instead like a futuristic hotel or office block, with artistic charms (like the high-tech trellises on its bridge-balconies) comparable to any of Downtown's recent architecture. But its upscale ambience is more than mere facade. The interior of the prison is designed to implement a sophisticated program of psychological manipulation and control—barless windows, a pastel color plan, prison staff in preppy blazers, well-tended patio shrubbery, a hotel-type reception area, nine recreation areas with Nautilus workout equipment, and so on. In contrast to the human inferno of the desperately overcrowded County Jail a few blocks away, the Becket structure superficially appears less a detention than a convention center for federal felons—a 'distinguished' addition to Downtown's continuum of security and design. But the psychic cost of so much attention to prison aesthetics is insidious. As one inmate whispered to me in the course of a tour, 'Can you imagine the mindfuck of being locked up in a Holiday Inn?' (Mike Davis 1990, 256–7).

where, how many, and in what form participants' voices are to be included. The inclusion of quotations raises issues of representation, authority, appropriation, power, and participation. Kay Anderson (1999, 83–4), for example, provides a particularly compelling description of the juxtaposition of identities and spaces in Redfern, an inner-city suburb of Sydney closely identified with spaces of 'Aboriginality' (see Box 17.4). Anderson weaves together quotes from research participants, evidence gathered from archival research, and rich theoretical writing to argue for understanding Redfern as a hybrid space of porous and fluid identities and spaces. Her work manages quite subtly to acknowledge issues of power in research and representation while simultaneously providing us with a vivid description of life on the Block.

Rethinking Redfern—Writing Qualitative Research   Box 17.4

On the Block [a small area of Redfern, an inner-city suburb of Sydney] itself, in 1994 the dominant language group was Banjalang, but a wide range of other place-based dialect groups were present, including Eora (Sydney region), Wiradjuri (Nowra), Kamlaroi (Dubbo and Moree), and many other groups from throughout New South Wales and Queensland. In addition, approximately two-thirds of the total number of rent-payers on the block were women, a significant minority of whom (among those interviewed) were married or partnered to Tongans, Fijians, Torres Strait Islanders, and members of other ethnicities. In most cases the men did not live with the women, who supported their children and funded the periodic visits of husbands or partners, friends and relatives. Sociability has always been both fractious and friendly. Some tenants saw the Block as 'home'; others perceived their place of birth as home; most considered they had multiple homes, including Redfern. In the words of a tenant known on the Block as a community elder and who has since refused to leave it: 'Redfern is an Aboriginal meeting place. People come from all over the country to get news of friends and family'. Now a member of the housing coalition which has been formed to fight the AHC [Aboriginal Housing Company], the same woman recently stated that Redfern has long been a place where children taken from their relatives encountered relatives or information that would lead to those relatives (*Melbourne Age*, 1 February 1997). Another woman had this to say in an interview for a recent documentary: 'People who come from interstate or wherever make to Sydney. Their first aim is the Block because they gotta know who's who and where's where and where to go from here' (ABC TV 1997).

The 'traffic of relating' on the Block opens up fresh ways of thinking about home and community in a context of more widely invoked images of 'flight', 'flow', 'crossings', 'travel' and transnational exchange in contemporary cultural geography. Models of mixing that work with the idea that cultures are porous and fluid are particularly apt in relation to this case study. There is, as I have suggested, no 'pure' culture at Redfern, no crisp boundaries of inside and outside, even for so stigmatised an area. This is not only a methodological issue. It is also an epistemological problem in that the boundaries of researchable communities are not secure and areas never exist as discrete entities (Anderson 1999, 83–4).

Jamie Baxter and John Eyles (1997) suggest that the criteria of credibility, transferability, dependability, and confirmability are useful general principles for guiding an evaluation of the rigour (trustworthiness) of a piece of qualitative research. They see these categories as broadly equivalent to the concepts of validity, generalizability, reliability, and objectivity that have been used to evaluate the quality of quantitative research endeavours. We believe it is important to keep in mind the constructedness of these concepts, to avoid using them as universal assessment criteria, and to avoid engaging in comparative analysis of qualitative studies. Notwithstanding this, Baxter and Eyles's principles have much to commend them in that they may assist in evaluating the internal consistency and trustworthiness of a piece of research. It is important to note, however, that we avoid using the term 'rigour' in our approach to constructing pieces of qualitative research, since we believe the term is too closely tied to quantitative approaches that are predicated on systematic exposure of pre-existing truth. Instead, we prefer to use the term 'trustworthiness', which speaks much more directly to qualitative geographical research as a reflexive practice that constitutes and understands meanings in place: one that actively recognizes that knowledge is constructed, open-ended, and fluid. As reflexive processes, ways of seeing, doing, and writing qualitative research are powerful in that they leave room for negotiation and debate around the meanings that researchers give to the world and the practices by which they do so (see Chapter 4, for example, for a different interpretation of rigour). Writing qualitatively should encourage researchers to explore and make explicit their own research agendas and assumptions and to elaborate on how they believe their research text constitutes the 'truth' about a particular subject (see Box 17.5). While transparency is not in itself ultimately achievable, if conscious reflexive writing produces qualitative texts that are open to scrutiny by research participants and audience, and if they present challenges to taken-for-granted ways of seeing and knowing and provoke and promote questions about 'place' in the world, then perhaps this goes some way towards establishing the 'validity', credibility, and trustworthiness of a qualitative research text. Communicating qualitative research is as much about how we know as it is about what we know, as examples of **auto-ethnographic** writing demonstrate (see Box 17.5).

Auto-ethnography: A Method for Reflexive Writing   **Box 17.5**

Ellis (2004, xix) defines auto-ethnography as

> research, writing, story, and method that connect the autobiographical and personal to the cultural, social, and political. Auto-ethnographic forms feature concrete action, emotion, embodiment, self-consciousness, and

introspection portrayed in dialogue, scenes, characterization, and plot. Thus auto-ethnography claims the conventions of literary writing.

Ian Cook has encouraged his students to use auto-ethnography as a means of writing reflexively and critically, bringing the insights from their reading, visual, and aural material research and their felt experiences, emotions, and knowledges into their assessment for a course on geographies of material culture (Cook et al. 2007). Similarly, David Butz and Katharine Besio (2004) have used auto-ethnography to better understand the implications (for research and knowledge production) of their own positionality as white, Western researchers working in the Karakoram. Auto-ethnographies provide one way by which the emotions, experiences, contradictions, and inconsistencies in our research journeys (and those whose lives and experiences intersect with them) can be re-presented. The use of literary rather than objective modalities and the active inclusion of self means that one's writing,

> far from being a dispassionate process of producing what was, is instead a product of the present, and the interests, needs, and wishes that attend it. This present, however—along with the self whose present it is—is itself transformed in and through the process at hand. [Freeman 2007, 137–8]

Writing thus becomes a method of knowing. Examine this piece of auto-ethnographic writing by Sarah, one of Ian Cook's students:

> It's evening. To go out-on-the town I need an image refinement. Understatement! My cybernetic mask thickens with the make-up and lip-gloss, hair spray. . . . A reach for the wardrobe and a selection of T-shirts. Removing my glasses I read the small print—a third 'gaze': no glasses—close-up-focus but not *understanding*. Black tops from Hong Kong and EEC. Or should I go to Greece? Or with the red tops— more classy—to Thailand, England (no?!), Indonesia? I must emphasize never having looked at the origin of my clothes before. Think I will wear the silver one which has no label. Safer to not know where it comes from. Would 'made in Syria' really make a difference? I still wouldn't know how, where, when it was made and by whom.[7] And it goes with my chain— connecting my outfit as well as me with the world.[8]

---

[7] Again, the concept of 'commodity fetishism' is important. As a cybernetic being I am embedded in global networks and yet I am unaware of those that I connect.

[8] Through the clothes that I wear I am connected not only to the people who made them but to the machines themselves that manufactured my outfit. As a cyborg, are those machines a part of me? The amalgam of man [*sic*] and machine, to which Haraway's (1991) 'cybernetic organism' alludes, where does it end? What is the extent of my hybridity?

What effects might such writing have? Sarah's writing, like Doel and Clarke's earlier 'staging' (see Box 17.2), might be seen as opaque, diffuse, and uncertain. While auto-ethnographic pieces might seem disconcerting, even confusing, relative to conventional objective modalities of writing research, such narratives can provoke an awareness of the situated problematics and politics of knowing, representing, and transforming knowledge. Ian Cook believes this approach can have radical effect—an effect that arises out of a less straightforward or didactic connection between what is known and how it is interpreted. What do you think?

CONCLUSION

We have discussed the importance of language in the social construction of knowledge, how power is articulated through dichotomies, and how meaning is inscribed in language. We believe models of writing that construct the writer as a disembodied narrator are inappropriate for communicating qualitative research. Our focus has been on written rather than visual texts, since writing remains the predominant means of communicating qualitative research. Breaking down dichotomies—through an interpretative understanding of writing and researching as mutually constitutive processes—and an understanding of which principles might guide valid qualitative research are critical to writing 'good' qualitative research. It is also crucial to understand how power and meaning are inscribed in the words that we use (and those we choose not to use) to constitute the research process, to recognize our subjectivities, standpoint, and locatedness (shifting and partial though they might be), and to acknowledge the voices of those with whom we undertake research. We believe that doing this enables qualitative researchers to have confidence in the 'validity' and truthfulness of their interpretations. Consequently, this chapter has not been a 'how to' guide but a means of raising important issues that are inherent in the writing-*in* process.

KEY TERMS

auto-ethnography

dialogic nature of research and writing

dichotomy

falsification

first-person narrative

hegemony

mimetic

modality

mystification

neo-positivism

nominalization

objective

objective modality	situated knowledge
positionality	staging a text
positivism	subjective
post-structuralism	thick description
quantitative revolution	third-person narrative
reflexivity	trope
reification	validity
signified/signifier	writing-*in*

REVIEW QUESTIONS

1. What implications do post-structuralist perspectives have for writing qualitative research?
2. In what ways is the term 'writing-up' misleading?
3. Why should writing be seen as an integral part of the entire research process?
4. How can a researcher endeavour to produce 'trustworthy' research?
5. Why can writing be seen as a method of knowing?

USEFUL RESOURCES

Baxter, J., and J. Eyles. 1997. 'Evaluating qualitative research in social geography: Establishing "rigour" in interview analysis'. *Transactions of the Institute of British Geographers* 22 (4): 505–25.

Bondi, L. 1997. 'In whose words? On gender identities, knowledge and writing practices'. *Transactions of the Institute of British Geographers* 22: 245–58.

Chandler, D. n.d. 'Semiotics for beginners'. http://www.aber.ac.uk/media/Documents/S4B/semiotic.html. This site is a comprehensive introduction to some post-structuralist understandings of the power of language.

Cook, I. 2009. 'Geographies of food—afters'. http://food-afters.blogspot.com. This is Ian Cook's blog for co-authoring an article for the journal *Progress in Human Geography*. Ian is an associate professor of geography at the University of Exeter and has been active in various forms of participatory qualitative research. The blog gives a wonderful insight into many of the issues discussed in this chapter.

Cook, I., et al. 2007. '"It's more than just what it is": Defetishising commodities, expanding fields, mobilising change . . .'. *Geoforum* 38: 1113–26.

Jones, A. 1992. 'Writing feminist educational research: Am "I" in the text?'. In S. Middleton and A. Jones, eds, *Women and Education in Aotearoa*. Wellington: Bridget Williams Books.

McNeill, D. 1998. 'Writing the new Barcelona'. In T. Hall and P. Hubbard, eds, *The Entrepreneurial City: Geographies of Politics, Regime and Representation*. London: John Wiley.

Richardson, L. 2003–7. 'Laurel Richardson'. http://www.sociology.ohio-state.edu/lwr. A key advocate for creative writing in research, Richardson refers to a substantial list of publications.

Richardson, L., and E.A. St Pierre. 2008. 'Writing: A method of inquiry'. In N.K. Denzin and Y.S. Lincoln, eds, *Collecting and Interpreting Qualitative Materials Research*, 3rd edn. Los Angeles, London, New Delhi, and Singapore: Sage.

Rose, G. 1997. 'Situating knowledges: Positionality, reflexivities and other tactics'. *Progress in Human Geography* 21: 305–20.

Notes

1. Readers seeking specific 'how to' advice on stylistic conventions associated with the presentation of research are advised to consult Chapter 18 of this volume, Hay (2006), Hay, Bochner, and Dungey (2006), Kneale (1999), or Stanton (1996) in conjunction with the conceptual material of this chapter.

2. It is appropriate to note, however, that it can be difficult institutionally to present some forms of research this way. For instance, social or environmental impact statements prepared by government departments or consulting firms will frequently not list authors, in which case the 'we' would be poorly defined. Moreover, neither group would want individuals to be sued for their opinions.

From Personal to Public: Communicating Qualitative Research for Public Consumption

18

Dydia DeLyser and Eric Pawson

CHAPTER OVERVIEW

This chapter aims at helping beginning qualitative researchers effectively communicate their research to wider audiences. To do so, it explores different modes of communication, including written work, public talks, and web pages, and considers audiences, including supervisors and examiners, respondents, the public at large, and ourselves. It examines 'writing' as part of the process of doing research as well as one of the ways of communicating that research and shows how communicating findings is part of the social networks of academia. It describes some of the structural elements of works (written, spoken, or on the web) that help to focus the author's creativity and details some techniques that help him or her to express ideas imaginatively. It offers suggestions on presenting qualitative research in ways that will convince an audience and concludes with an outline of the interactive nature of writing and research.

INTRODUCTION

How do qualitative researchers set about communicating their findings? In our case, it begins on a cold, wet Monday morning. Sitting in a quiet room with a large whiteboard, we talk about books and articles relevant to writing this chapter that we've read in the weeks before. But what exactly is the focus of what we are to write about, and what will our argument be?

We use the whiteboard for brainstorming and sorting out our ideas as we strive to identify the focusing themes (they now appear in the 'chapter overview' guide above) that will structure our argument. We talk about the things we would really like to say and where, if anywhere, they will 'fit'. We revisit what we have read and talk through how it can be used to substantiate, elaborate, or illustrate the points on the whiteboard. Our arguments become clearer as we modify them, revise them, and flesh them out.

What is the purpose of these opening paragraphs? They tell a story, or **anec-dote**, which is designed to get your attention and address the chapter's major themes. Since you're still reading, perhaps the strategy is working. And if this strategy worked for us, then it will probably work for you when you're communicating the results of your own research. Telling a story, one that encapsulates the main points you wish to make, is a good way to start. It need not be personal: in fact, successful anecdotes often come directly from the writer's own qualitative data.

Our opening anecdote (from data on our experience with writing) describes a writing and communication *process*, one that has parallels with what you write for your instructors or supervisors, your web readers, or the public to whom you give a talk. Our story emphasizes interaction, demonstrating that writing is a *social process*, intimately linked with thinking, reading, and discussing. It implies that we cannot know what we want to say or how we will say it until we have written and talked through a plan, elaborating it with notes and reading. Nor is this a new idea: 'How can I know what I think till I see what I say?' wrote the novelist E.M. Forster back in the 1920s (in Forster and Stallybrass 2005, 99). What this reveals is that writing to communicate is not just the end-point of research; rather, it is central to the thinking process that *is* research. Writing can and should be part of the different phases of the research process—from conceptualization to final 'product' (Richardson and St Pierre 2005; see also Chapter 17 of this volume for additional discussion of the dialogic nature of research and writing).

This is significant, because if we can communicate our ideas to others and learn more about those ideas by talking them through, then there is a greater chance of success of communicating them in formal situations. Good preparation leads to greater confidence about what we're saying and why we are taking the risk of saying it. In other words, we will have something we want to communicate.

UNDERSTANDING AUDIENCES

Before beginning to write, it is necessary to understand the audience by asking, 'With whom do we wish to communicate the results of our research?' The initial answer may be surprising: we first communicate our research with ourselves. As the above paragraph suggests, we write first for ourselves and to ourselves, because the very act of writing reveals what it is that we think and know and what we do not (Richardson and St Pierre 2005). In other words, writing is an *iterative process* (Kitchin and Tate 2000), with reading, writing, and research being interlinked phases and each activity clarifying the direction that needs to be taken in the others. Reading makes clearer what to write and vice versa. Writing helps to elucidate what we think, and being clearer on what we think enables more effective communication of those thoughts to others.

Understood in this way, it becomes evident that the writing process itself is *formative*; writing is an essential part of clarifying thinking (Becker 2007). 'Writing *is* thinking,' says Harry Wolcott (2001, 22). And since a good grade, positive feedback on a website, or a good public reception for a talk is dependent on how clearly we can communicate our thoughts, this means that these thoughts themselves have to be clear. To achieve clarity in thinking, it helps to practise writing by writing to oneself. Write (whether a rough plan, a first draft, or something more complete), and then put that writing aside. You will be a different person when you return to it an hour, a day, or a week later. Your mind will have reflected on the ideas; you will have a fresher perspective.

Although we are our own **audience** and critic for all that we write, it is necessary to have an external audience as well. So we have to ask, 'Who is the audience for our research?' As scholars, we have a wide public: we are accountable within the academy (to meet course or degree requirements or to our peers), and we have obligations beyond it (for example, to those who worked with us as respondents or gave us access to networks). It is best not to jump straight in front of these wider audiences without a trial, or pilot. A pilot study 'in the field' is widely accepted as a good research method; piloting one's writing or talk is less frequently discussed but equally helpful.

Further, even though many people save 'writing up' for the last minute, if writing is understood as part of the thinking and research process, then there will already be *something* written on paper, or outlined in our heads, that can be piloted with an external audience. Try the grandmother test. Can your grandmother (or another family member or a flatmate) understand your thoughts, your draft paper, your thesis outline? She may be a Nobel laureate or a lifelong homemaker, but she will likely be a sympathetic audience with whom you should be able to communicate your ideas. If she is not clear on what you are saying and why, restate things so she will be. This is also the role of a supervisor, dissertation adviser, or instructor, but explaining your work to somebody outside the field is a sure way to clarify what it is that has to be said and how best to say it.

Because academic life takes place within social networks and involves social responsibilities, each of us must be able to explain our ideas to ourselves as the foundation for explaining them to wider audiences. Grandmothers and supervisors are not the final arbiters of our work. Sharing research—with examiners, with those who helped in its compilation, and with wider networks of people who have a stake in similar areas of investigation—is an essential part of the role of researcher. In fact, in order to fully participate in the academic world, it is necessary to contribute back to both the academic life and the communities that have provided the research opportunities (Becker 2007).

Depending on which audience or which part of the audience we are communicating with, the ground rules vary. Universities, departments, or instructors will have publicized their expectations for term papers, dissertations, or theses, and such rules are best adhered to. If you are in doubt, academic advisers are there to help. More challenging, however, is the issue of obligations to wider audiences. How do we share our research, or how do we 'share the field', as Julie Cupples and Sara Kindon (2003, 224) put it? Their view is unequivocal: '**sharing the field** is essential both in terms of the training of future researchers and as a means of giving back something to those who have helped us' (see also DeLyser 2001; Routledge 2001).

How to share the field is an important issue. Some populations feel heavily 'over-researched' by people from outside. This can arise in small towns that are popular for university field trips or with particular suburbs that have attributes that attract frequent attention, such as locality-based environmental schemes. It has long been an issue in indigenous communities. Inuit residents of Arctic Canada, for example, have been known to compare researchers from the south with the annual visit of the snow geese: they arrive en masse, carry on, then depart leaving behind a mess (Stewart 2009). Linda Tuhiwai Smith sees this sort of behaviour as a reproduction of colonizing attitudes. So don't be a snow goose: 'sharing is a responsibility of research' (Smith 1999, 161).

In other words, we have obligations to those who have helped us get to where we are. Different strategies will be necessary depending on the context. Making a digital copy of a thesis available can be sufficient in some cases, but often something quite different will be appreciated.

COMMUNICATING WITH A WIDER AUDIENCE

We can communicate with audiences in a number of ways: the written word, the spoken word, with visual representations, and on-line. Although discipline-specific written presentations of our research may fulfill course or degree requirements, are they the best way to give something back to the participants in one's research network? There may be more accessible ways of passing on the insights. Vehicles such as short articles for local newsletters or reports abstracted from a dissertation for, say, local government enable you to tailor the language, length, and tone of the feedback to the audience concerned. Julie Cupples, for example, found that the women who had worked with her in Nicaragua particularly appreciated receiving copies of interview transcripts (Cupples and Kindon 2003).

Releasing written material through the media can be effective too. But a word of caution: reporters may have a different interest in and reading of the research that you have done and may represent your work in ways that have unforeseen consequences. Iain Hay's class on fear of crime in a suburb in the Australian city of Adelaide found results consistent with other studies—namely, low levels of fear. But the local press, after receiving copies of the class report, saw things differently, with one paper headlining 'Grim nights in a suburb of fear' and another 'Fear of crime shackles Unley', thus undoing a lot of good work with public agencies that had helped with the research. Although over time such agencies began to pay more attention to how fear of crime constrained people's behaviours, when it happened, a 'student project intended to demonstrate [this] became an instrument of oppression' by raising fear levels as a result of the sensational reporting (Hay 1995, 269).

If a topic is potentially sensitive or if it could be misconstrued by different parties, then it may be better to give targeted feedback, using talks. Talks are an excellent means of involving your audience and giving them the opportunity to participate with you by listening and asking questions. Still, whether given off campus or on, they do need careful preparation (see Hay 2006), even if only to help you to overcome the natural fear that many of us hold about public speaking. Practice helps, as does the grandmother test. Both underline the things that can draw in an audience as well as those that will put them to sleep (Box 18.1).

For Successful Public Presentations   Box 18.1

Consider:

- *Preparation.* If you know your stuff, you will be less nervous about presenting it, and that will give your audience more confidence in what you are saying.
- *Content.* Your audience has come primarily to hear what you have to say and only secondarily to hear how you will say it.
- *Visuals.* Make sure they *add to* what you have to say. Use them when and where you need them, as illustrations of your point(s).
- *Practice.* Even in an empty room, practice will build confidence and help to ensure that your presentation is the appropriate length.
- *Interaction.* Engage with an audience by including them in the picture. Start with an anecdote, and relate what you are saying to the people in the audience and the place where you are talking. Leave time for questions.

Avoid:

- *Rambling.* Both you and your audience will understand what is being said if you have a clear structure and keep to it.
- *Poor time-keeping.* An audience likes to know how long a talk will be so that they can anticipate where it is going. If you say 'I will talk for 20 minutes and take questions', don't then talk for 30.
- *Mumbling.* Ask your audience whether they can hear you. They will not thank you if they cannot. Speak as clearly as you can (see Strunk and White 1959, xi, who advise, 'If you don't know how to pronounce a word, say it loud!'), pace yourself, and look at your audience as much as you can.

A common fall-back for public presentations is **PowerPoint**. But many Power-Point presentations are lengthy and largely redundant, which is why the phrase 'death by PowerPoint' entered the language. Too many slides, badly designed, distract from a presentation and remove the audience's need to pay attention to what you have to say. Indeed, they may well undermine what you are trying to say. So the best advice is to keep PowerPoint slides to a minimum, with little text and without silly illustrations. In a well-known piece, Tufte (2003) argues that overloaded PowerPoint presentations actually contributed to the series of misunderstandings that led to the loss of the Space Shuttle Columbia. But PowerPoint also has advantages if used well. It can be an excellent way of providing a transparent structure for a talk and 'signposting' it to the audience. Keeping it simple is the key, as Garr Reynolds shows in his evocatively titled book *presentationzen* (2008) (Box 18.2).

Another means of communicating research results is via a website, a tool effective in a range of circumstances and with varying audiences. Setting up web pages while doing research is another aspect of the formative process of writing. Weblogs are useful as a means of enabling research respondents to give feedback on the initial

---

For Success with PowerPoint	Box 18.2

Use it creatively:

- Minimize the number of slides: simplicity is the ally of the audience.
- What is the absolutely central point? If the audience will remember just one thing, what do you want it to be?
- Use the slides to reinforce your words, not to repeat or anticipate them. Slides should convey your message in fewer words than you speak.
- Don't use cheesy images. You might be mesmerized by your baby, cat, or dog, but don't inflict them on others.
- Avoid dissolves, spins, or other transitions.

Death by PowerPoint:

Think back to all the bad presentations that you have witnessed in lecture halls and at conferences:

- with far too many slides, at odds with the message, or merely repeating each slide word for word;
- with complex background designs and silly colour choices that say nothing about the subject at hand;
- that confuse quantity of information with clarity of message.

and do things differently, avoiding those pitfalls.

results, assisting in their verification. They have proved valuable in northern Canada (e.g., northern Manitoba and Nunavut) where community-based research, consultation, and feedback are now expected (Stewart and Draper 2009). They provide an opportunity to let you help respondents feel that they are valued for their roles and are part of the research process rather than merely being 'the researched'. In other words, such outreach can demystify the research process and contribute to the maintenance of meaningful research relationships in particular places.

Some choose to describe their research not through individually designed websites but through social networking sites like **Facebook**. Others elect to make information about their research topics available through publicly editable on-line encyclopedia sites like **Wikipedia**. Carefully and attractively done, such sites and postings may draw an audience—they can give you contacts working in similar fields or potential respondents; they may also help you to make your research available to those outside your department and to give something back to those you worked with in the field. A web presence is also a way of giving something back your department: it may attract others to think about studying there in the future.

Writing for the web requires design as careful as any other means of communication. Balance words with images, but be aware of the politics of visual representation: it is only possible to use images for which you have permission to do so, and research subjects should not be misrepresented by picturing them insensitively—i.e., picturing them in ways devoid of context or placing them in inappropriate juxtapositions (Rose 2008).

STRUCTURE AND CREATIVITY

Today, few people write in longhand, directly on paper. The personal computer has revolutionized how we write and make revisions to our writing. But the notion of 'word processing' can be a dangerous misnomer, implying that the computer does all the work, taking agency away from the writer and stifling creativity. However we choose to write, remember that *we* control the process—not the machines.

One way to facilitate that control is to keep the structural basics for any piece of communication firmly in mind. Box 18.3 illustrates these structural basics for written work, but **structure** is just as important for a successful poster, talk, or web page.

Writers facing projects like term papers or theses may feel initially that any externally imposed form or structure hinders creativity, forcing the work into 'boring' moulds and 'stifling' formats. Pausing to consider highly restrictive forms like the *haiku* or the sonnet, however, reveals rather that such structures are in fact stimulants to creativity—after all, no writer today would deny the creativity in Shakespeare's work. More productive, then, is an understanding that *structure harnesses creativity*. Indeed, writers use structure to *focus* their creative energies.

Using Structure to Communicate	**Box 18.3**

- Begin with clear research questions: they narrow the world of possibilities and allow us as writers to focus attention, thus sharpening our creative abilities.
- Develop an outline. This allows us to communicate with ourselves, reminding us of points we want to make and quotes we wish to engage. Allow the outline to change and grow as the paper changes and as the writing project progresses.
- Use subheadings. When ideas need to shift, call attention to such important transitions with descriptive subheadings. Use them in first drafts, or later, when there is a shift in our argument or the need to call more attention to a specific point.
- Develop clear paragraphs, advancing usually just one idea each. Locate the thesis statement (the 'point' of the paragraph) at the beginning or end of the paragraph and use the middle to build that idea.

As an example, consider film. Because most Hollywood films (and many others besides) must run between 90 and 110 minutes in length and because in standard screenplay format one page becomes one minute in the final film, nearly every script must be between 90 and 110 pages long—the novice screenwriter who writes drastically more or less will likely not be taken seriously. Further, because many film audiences grant a new movie just 10 minutes of their time before deciding whether to stay or to walk out, screenwriters face a 10-minute (and therefore 10-page) limit in which they *must* hook their audiences, demonstrating to them what and who the film is about, showing them why they would want to watch it to the end. Finally, screenwriters must accomplish everything primarily with dialogue and visuals—descriptive or explanatory narration is seldom successful. Since nearly every film is likely to fit these rigid restrictions, we can begin to understand the creativity that such limitations can produce.

Successful academic writers and speakers also use structure in carefully considered ways. Box 18.4 highlights some of the key techniques that enable the presentation of the richness of qualitative research and the expression of originality within clear structures. Using these techniques helps us to clarify communication, giving us the best chance of convincing our audience of the value of what we wish to say.

CONVINCING THE AUDIENCE

How do we convince an audience—and ourselves—of the value of the research that we have done? Everything said about communication in this chapter so far

---

Presenting Qualitative Research　　　　　　　　**Box 18.4**

- 'Show, don't tell.' Use quotes from your empirical material (participant observation, archives, and so on) to tell your story (there is an example at the start of this chapter). Whenever making an assertion, *show* the audience how you know this is so.
- Outline how you have collected the data (for example, using focus groups or interviews) and how you have interpreted it (e.g., discourse analysis, grounded theory).
- Rather than speaking of 'generalizability', engage social theory to speak beyond the nuances of your empirical results: this may give you and your audience insight into places, people, and situations very different from those originally studied (see Chapters 4, 5, and 17 in this volume and DeLyser 2009).
- Use anecdotes that are directly on point to draw your reader in and powerfully present your argument.

---

will help: clear thinking, good preparation, transparent structure. Most audiences today are well past the stage where they need to be convinced of the worth of qualitative research per se, but it is still necessary to describe the methods that have been used, demonstrating why these methods were the most appropriate for the questions we have set out to answer.

Qualitative research, unlike some quantitatively based reports (often designed to be interpreted through their tables and summaries), cannot be 'skimmed', because it is in the very details, richness, and diversity that the author's points become clear (Richardson and St Pierre 2005). But how much data is 'enough', and to what extent should respondents be quoted: briefly or at length? Audiences will only follow longer quotes when the words relate directly to the point the writer seeks to make. Limit the length of quotes by omitting unneeded passages, and consider paraphrasing certain sections, quoting only the relevant words. You may find it useful to consult journals that commonly report the results of qualitative research, like *Cultural Geographies* or *Social and Cultural Geography*, to see examples of how best to make a point with emphatic, colourful, and telling quotations. In spoken presentations, it may be helpful to put an abridged version of the text of a long quote on a slide so that the audience can follow along without getting lost and so that it is clear that a respondent is 'talking' and not the presenter.

Whenever a point is being made, whether in a written or spoken presentation, there must be data to back it up. Whether that point is nuanced or boldly central to the argument, it can be made evident with empirical material reflecting the issues. Using several short quotes from interviews, for example, can show a variety of respondents' views. Again, there is no need to show them all—a sampling of a few will indicate that breadth to the reader or listener. Indeed, we convey our points best when staying close to our empirical materials, avoiding making claims the data will not support.

In spoken presentations, convincing the audience continues into the question and answer period through the ongoing use of powerful and accurate evidence, properly acknowledged. When answering questions in any format, respond honestly, drawing on the research and on background knowledge, without bluffing an answer. This is the best way to retain credibility. Moreover, it is an essential part of ethical research conduct—a theme that has been taken up repeatedly in this book.

Finally, on the subject of honesty and ethical practice, it bears noting that in all forms of communication with audiences, using the words of others without proper acknowledgement, whether those words are from interviews, focus groups, books, articles, newspapers, or websites, has the potential to undermine both our own credibility as well as that of other researchers. It is not worth the risk.

CONCLUSION

It is now two weeks since we started to write this chapter, although in reality the writing began months ago with reading and exchanges of e-mails about developing ideas. On that wet Monday when we began, we tried to focus the task, then allowed ourselves time for more reading; we accepted the messiness of our first drafts and let our thoughts gel over days spent doing other things. Finally, we scheduled enough time for revisions. So where have we got to?

In general, for us as qualitative researchers, the best way to produce strong and polished work is to use a transparent structure (one that is clear to the audience) to harness creativity and to stay close to the data as we build the arguments. Still, when reading the published works of others, it is easy to lose sight of all the iterations their efforts involved. Instead, the finished product of someone else's work often appears as a streamlined whole, and we may miss the complex and not-always-forward-moving steps in the process. In fact, this is a common mistake found far beyond the bounds of qualitative research.

So, for example, when people today quote American astronaut Neil Armstrong's famous words upon setting the first foot on the moon in 1969, 'one small step for man; one giant leap for mankind', they are noting *his* accomplishment, *his* moment of glory, perhaps forgetting that preceding his Apollo 11 mission came years of design and training, as well as many other missions—none without risk—each leading in different increments to that first footfall. Like the process leading up to Armstrong's footfall, our research and communication proceed incrementally, in an iterative process of reading, writing, thinking, and talking. These seemingly small steps add up to greater accomplishments, and it is these steps that can be so satisfying and engaging to qualitative researchers and their audiences.

KEY TERMS

anecdote

audience

Facebook

PowerPoint

sharing the field

structure

Wikipedia

REVIEW QUESTIONS

1. Identify four ways to communicate research. Which audience might each best serve?
2. How is writing also thinking?
3. How is writing a social process?
4. How can structure serve to harness creativity?
5. How can we ensure that our work is geared properly for our audience?
6. How are the processes of writing, reading, and doing research all related?

USEFUL RESOURCES

A wide and constantly changing array of academic and commercial websites is available to help with writing and presentation skills. An Internet search will yield many. Some sites that we found helpful are listed below, together with useful published resources.

Becker, H.S. 2007. *Writing for Social Scientists: How to Start and Finish Your Thesis, Book, or Article.* 2nd edn. Chicago: University of Chicago Press.

Budge, G. 2008. 'Writing for the Reader'. http://www.lhds.bcu.ac.uk/english/writing4reader/index.html.

Cupples, J., and S. Kindon. 2003. 'Returning to university and writing the field'. In R. Scheyvens and D. Storey, eds, *Development Fieldwork: A Practical Guide*, 217–31. London: Sage.

DeLyser, D. 2009. 'Writing qualitative geography'. In D. DeLyser et al., eds, *Handbook of Qualitative Geography*. London: Sage.

Giesbrecht, W., and W. Denton. 2008. 'Academic writing guide'. http://www.library.yorku.ca/ccm/rg/preview/academic-writing-guide.en. This is York University's helpful guide to writing for a variety of academic purposes and includes advice on subjects such as preparing literature reviews, research reports, and essays.

Hay, I. 2006. *Communicating in Geography and the Environmental Sciences*. 3rd edn. South Melbourne: Oxford University Press.

Manalo, E., and J. Trafford. 2004. *Thinking to Thesis: A Guide to Graduate Success at All Levels*. Auckland: Pearson Education.

Presentationhelper.co.uk. 2009. 'Presentation helper'. http://www.presentationhelper. co.uk. Offers guidance and resources for effective speeches and PowerPoint presentations.

Proctor, M. 2009. 'Advice on academic writing'. http://www.utoronto.ca/writing/advise. html. A very comprehensive review of academic writing.

Reynolds, G. 2008. *presentationzen: Simple Ideas on Presentation, Design and Delivery*. Berkeley, CA: New Riders.

Smith, L.T. 1999. *Decolonizing Methodologies: Research and Indigenous Peoples*. London: Zed Books.

Stewart, E.J., and D. Draper. 2009. 'Reporting back research findings: A case study of community-based tourism research in northern Canada'. *Journal of Ecotourism* 8 (2): 128–43.

Tufte, E. 2003. *The Cognitive Style of PowerPoint*. Cheshire: Graphics Press LLC.

Weebly. 2009. 'Weebly—Web creation made easy'. http://www.weebly.com.

Wolcott, H. 2001. *Writing up Qualitative Research*. 2nd edn. Newbury Park: Sage.

Glossary

abstracting Reducing the complexity of the 'real world' by generating summary statements on the basis of common processes, experiences, or characteristics in the data. (See also *data reduction*.)

accidental sampling See *convenience sampling*.

action–reflection Periods of action followed by times when participants reflect on what they have done and what can be learned. The learning informs the next phase of action, creating an iterative cycle of action and reflection. This process enables change to occur throughout the research process. (See also *reflexivity*.)

action research A term coined by sociologist Kurt Lewin in the 1940s to talk about the idea of an iterative (or repeated) cycle of action and reflection in research, oriented towards solving a problem or improving a situation.

activism Political and practical action usually intended to bring about social, economic, or other change. (See also *applied people's geography* and *critical geography*.)

aide-mémoire A list of topics to be discussed in an interview. May contain some clearly worded questions or key concepts intended to guide the interviewer. Alternative term for *interview guide*.

analytical generalization A strategy for creating in-depth, rich, and credible concepts/theory. Rather than achieving generalization through large probability samples, analytical generalization is focused on the qualitative notion of transferability, specifically: (a) the careful selection of informative cases and (b) the creation of theory that is neither too abstract nor too case-specific. Readers of research narratives must be able to see how the concept might apply to other phenomena or in other contexts.

analytical log Critical reflection on substantive issues arising in an interview. Links are made between emergent themes and the established literature or theory. (See also *personal log*.)

analytic code A code that is developed through analysis and is theoretically informed; a code based on themes that emerge from relevant literature and/or the data. (See also *interpretive code*; compare with *descriptive code*.)

anecdote A story, often personalized to the author or presenter and directly related to the point of the paper or presentation, that captures the attention of an audience and persuades them of the importance, relevance, and/or interest of what they are reading or hearing.

applied people's geography Term coined by David Harvey to refer to geographical research that is consciously 'part of that complex of conflictual social processes which give birth to new geographical landscapes' (Harvey 1984, 7). (See also *activism* and *critical geography*.)

archival research Research based on documentary sources (for example, public archives, photographs, newspapers).

archives Narrowly defined as the non-current records of government agencies but also includes company and private papers. Typically managed by a specialist in a government agency dedicated to the records' long-term use and preservation. A distinction can be made between archives as surviving records and archives as the institution dedicated to their preservation.

archivist Professional curator of non-current records who has expertise in the accession, arrangement, and preservation of such records (in contrast to the current files of a central or local government agency that are controlled by records managers).

asymmetrical power relation A research situation characterized by an imbalance in power or influence between researcher and participant. Sometimes used to refer specifically to relationships in which informants are in positions of influence relative to the researcher. (See also *potentially exploitative power relation*, *reciprocal power relation*, and *studying up*.)

asynchronous groups On-line discussions in which participants contribute at different times. Used in focus group research with groups lasting from several days to several months and with up to 40 participants. (See also *on-line focus group* and *synchronous groups*.)

asynchronous interviewing When the answers in an interview do not occur immediately after the questions—that is, there is some extended delay between the putting of a question and the receipt of answer. It is most common in interviews undertaken using e-mail in which there is a delay between the sending of a question and the making of a reply, since the researcher and informant are unlikely to be undertaking the interview at the same time. (Compare with *synchronous interviewing*.)

attribute database A set of information compiled from measurable characteristics, such as the census.

audience The people with whom you wish to communicate the results of your research. Identifying who constitutes the audience is fundamental to the success of communicating with them. In discourse analysis, the term audience refers to the social process through which different collective social categories are forged, say, 'American' or 'tourist'. Audiences are conceived to emerge through how texts are produced, circulated, intersected, and interpreted. Hence, discourse analysis sees an audience 'taking shape' rather than being a pre-given category. For example, a presidential inauguration speech may help to create an audience fashioned by shared values of the collective 'we' of a nation. Alternatively, infomercials selling beauty products may rely on cultural norms of attractiveness and femininity to generate an audience (market) interested in purchasing their products.

auto-ethnography A qualitative method involving the explicit writing of an embodied and situated self into the research. It often entails writing in literary fashion, involving autobiographical narratives in which the author/researcher actively reflects on her or his choices, emotions, and knowledges as a vital part of the construction of the research.

bias Systematic error or distortion in a data set that might emerge as a result of researcher prejudices or methodological characteristics (for example, case selection, non-response, question wording, interviewer attitude).

border pedagogy Henry Giroux's concept of teaching that challenges, crosses, and reconfigures taken-for-granted boundaries and creates borderlands where students—and intellectuals—create new identities, histories, learning, and intellectual spaces.

bulletin board An electronic medium devoted to sending and receiving messages for a particular interest group (for example, the about.com Geography Bulletin Board at http://geography.about.com/mpboards.htm?once=true&).

canon A body of work, such as texts, held by some critics to be the most important of their kind and therefore worthy of serious study by all interested in the field.

CAQDAS See *computer-assisted qualitative data analysis software.*

CAQ-GIS (computer-aided qualitative geographical information systems) The integration of computer-assisted qualitative data analysis software (CAQDAS) with geographic information systems (GIS) using techniques that bridge the two software systems and move towards the goal of creating complementary analysis of qualitative and spatially referenced data.

case Example of a more general process or structure that can be theorized. (See also *case study.*)

case study Intensive study of an individual, group, or place over a period of time. Research is typically done in situ.

CATI See *computer-assisted telephone interviewing.*

chain sampling See *snowball sampling.*

Chicago School of Sociology Both a body of work and a group of University of Chicago researchers from the 1920s and 1930s involved in pioneering work in urban ethnography. The Chicago School helped to establish the in-depth case study as a legitimate and powerful means for conducting relevant social science. Notable researchers from this era are William Thomas (1863–1947), Robert Park (1864–1944), Ernest Burgess (1886–1966), and Louis Wirth (1897–1952).

closed questions Questions for which respondents are offered a limited series of alternative answers from which to select. Respondents may be asked, for example, to select one or more categories, to rank items in order of importance, or to select a point on a scale measuring the intensity of an opinion. (Compare with *open questions* and *combination questions.*)

CMC interviews See *computer-mediated communications* (CMC) *interviewing.*

co-constitution of knowledge The idea, informed by post-structural theory, that knowledge does not exist 'out there' ready to be discovered by objective researchers using discerning research methods but instead is built intersubjectively through interaction between the researcher and the research subjects. Qualitative research encounters are thus seen as social relationships (however fleeting) in which the researcher and the researched are closely involved in the process of construction of knowledge. (See also *co-learning.*)

code and retrieve software Specialist software packages that allow text to be segmented or grouped for coding and display. They facilitate electronic marking up, cutting, sorting, reorganizing, and collecting traditionally done with scissors, paper, and sticky tape. Examples include Kwalitan and The Ethnograph.

codebook An organizational tool for keeping track of the codes in a project, including their meanings and applications, as well as notes regarding the coding process.

coding The processes of assigning qualitative or quantitative 'values' to chunks of data or categorizing data into groups based on commonality or along thematic lines for the purposes of describing, analyzing, and organizing data.

coding structure The organization of codes into meaningful clusters, hierarchies, or categories.

co-learning A philosophy of teaching and learning that positions researchers and participants as equals in a process of mutual inquiry and education. Co-learning challenges traditional assumptions that the researcher has more knowledge and is dominant in the research relationship. It requires considerable self-reflexivity on the part of the researcher and a genuine desire to facilitate a process in which researcher and participants collectively create a learning community. (See also *co-constitution of knowledge.*)

collaborative research Research designed, conducted, interpreted, and disseminated by a team of local and non-local members, with local members directing the process or sharing equally in decision-making (see also *participatory action research* and *participatory research*).

collective action A process in which a group of individuals take action together to affect social or political change. This action may be the outcome of research and analysis in which individuals have come to understand the ways in which they are each implicated or affected by wider inequalities. The desire to take collective action reflects an interpersonal commitment and what some authors have called a 'we intention'.

colonial research Imposed, often exploitative research in both imperial and non-imperial contexts that maintains distance from, and domination of, the marginalized 'others' that it seeks to study and which denies the validity of their knowledge, ways of knowing, experience, and concerns. (Compare with *decolonizing research*, *inclusionary research*, and *post-colonial research*.)

combination questions Questions made up of both *closed* and *open* components. Their closed component offers respondents a series of alternative answers to choose between, while their open component allows respondents to suggest an additional answer not listed in the closed component or to elaborate on the reason why a particular option was selected in the closed component.

common questions Asked of each participant in oral history interviews. They build up varying views and information about certain themes. (Compare with *orientation questions*, *specific questions*, and *follow-up questions*.)

community supervision Procedures through which indigenous peoples and other communities authorize research and oversee researchers and research programs.

comparative analysis A form of analysis used in case study research that compares similarities and differences across multiple instances of a phenomenon to enhance theoretical/conceptual depth. It is also known as *comparative case study* or *parallel case study*.

comparative case study See *comparative analysis*.

complete observation A situation in which observation is overwhelmingly one-way and the researcher's presence is masked such that she or he is shielded from participation.

complete participation A situation in which the researcher's immersion in a social context is such that she or he is first and foremost a participant. As a result of this level of immersion, the researcher may need to adopt critical distance to achieve an observational stance. That critical distance might be gained by reflection out-of-hours in the field or through short-term exits from the field.

computer-assisted cartography Any hardware or software that is used to facilitate map-making. *Geographic information systems* (GIS) belong under this heading.

computer-assisted qualitative data analysis software (CAQDAS) Both a general acronym and the specific acronym for the CAQDAS network based in Surrey, United Kingdom.

computer-assisted telephone interviewing (CATI) Questionnaire/interview conducted by telephone with questions read directly from a computer file and responses recorded directly onto a computer file.

computer-mediated communications (CMC) **interviewing** Interviews that use the medium of the Internet. They can be asynchronous, taking the form of either e-mail exchanges or postings to a web-based platform. Less commonly, they can be synchronous through the use of chat-rooms or instant messaging boards. They do not have the direct access that a face-to-face interview has or the direct voice contact of a telephone interview. (See also *asynchronous interviewing* and *synchronous interviewing*.)

concept building Refers to the process of entering and *coding* data in a systematic way that relates to the research question being asked. The ability of a software package to support the systematic organization of concepts leads to that software being categorized as theory-building software. (See also *conceptual mapping*.)

conceptual framework Intellectual structure underlying a research project that emerges from an integration of previous literature, theories, and other relevant information. The conceptual framework provides the basis for framing, situating, and operationalizing research questions.

conceptual mapping As for *concept building* but refers to the specific ability to visually represent data in some form. For example, QSR NUD*IST™ software uses *hierarchical tree* structures, and ATLAS/ti uses network diagrams. Inspiration and Decision Explorer are purpose-built conceptual mapping programs for qualitative research.

conditions of use form Sometimes known as an informed consent form, a form outlining what will happen to the material research participants share with you—what their rights are, who will own copyright, where recordings will be stored and for how long, what they will be used for, and so on (See also *memorandum of understanding.*)

confederate Someone thought by other research study participants to be another participant but who is in fact part of the research team.

confirmability Extent to which results are shaped by respondents and not by researcher's biases.

constructionist approach An approach for challenging assumptions of coherence and truth within positivist knowledge (sometimes referred to as either rationalist, objectivist, or Cartesian knowledge (see *positivism*). Draws attention to social practices in the production of all knowledge, including scientific knowledge.

content analysis See *latent content analysis* and *manifest content analysis.*

controlled observation Purposeful watching of worldly phenomena that is strictly limited by prior decisions in terms of scope, style, and timing. (Compare with *uncontrolled observation.*)

convenience sampling Involves selecting cases or participants on the basis of expedience. While the approach may appear to save time, money, and effort, it is unlikely to yield useful information. Not recommended as a *purposive sampling* strategy.

copperplate script A writing style used in the nineteenth century (and earlier) produced using a sharp metal-nibbed pen. It is characterized by an elegant looping script in which the lettering is thicker on the heavy downward stroke and thinner on the upward loops. Typewriters typically replaced copperplate handwritten records in most government records from the 1880s and 1890s. (See also *modern hand* and *secretary hand.*)

co-researcher In cross-cultural research, an individual of different, 'other' social/cultural identity and positionality with whom one carries out research collaboratively with equal participation in decision-making. (See also *Other*, *positionality*, and *collaborative research.*)

corroboration A strategy for guarding against threats to the credibility of a theory or concept. Corroboration often involves the process of checking that a concept/theory makes sense to the participants in a case study (i.e., participant checking). Longitudinal studies provide a useful context for corroboration whereby concepts developed through an intensive case study are checked for enduring relevance in later time periods.

creative interviewing Defined by Douglas (1985) as informant-focused and -driven. This approach to interviewing has been positioned as the opposite of professional interviewing in which the researcher uses the pre-interview to establish status and respect. Douglas's creative interviewing style involves the humbling of the researcher and the ceding of power and status to the informant. (Compare with *professional interviewing.*)

credibility The plausibility of an interpretation or account of experience.

criterion sampling Choosing all cases that satisfy some predetermined standard.

critical geography Various ideas and practices that are committed to challenging unequal power relationships, developing and applying critical theories to geographical problems, and working for political change and social justice. (See also *activism* and *applied people's geography.*)

critical inner dialogue Constant attention to what an informant is saying, including in situ analysis of the themes being raised and a continual assessment of whether the researcher fully understands what is being said.

critical reflexivity See *reflexivity*.

cross-case comparison A strategy for comparative analysis that compares different case studies. Such comparisons help to develop richly detailed conceptual explanations of phenomena.

cross-sectional case study A case study conducted at one point in time. Contrast with *longitudinal case study*.

culturally safe Having knowledge of the history, beliefs, and practices of minority groups and maintaining awareness of these factors.

cultural protocols Local-, community-, or group-defined codes of appropriate behaviour, interaction, and communication to which outsider researchers are expected to adhere.

data cleaning Identifying and correcting errors of *coding* in a data set.

data management software A generic term for any software that facilitates the entry, organization, retrieval, and/or coding and mapping of input data.

data reduction Using categorization and qualification to lump data together into larger packages, thereby reducing the complexity and number of data points but increasing the level of understanding of trends, processes, or other insights. (See also *abstracting*.)

data retrieval Refers to the process of getting access to data that has already been entered into a computer system.

data storage Refers to the process of introducing data into a computer system so that it may be archived in some way (for example, as a document, spreadsheet, or graphics image).

debriefing Procedure by which information about a research project (some of which may have been withheld or misrepresented) is made known to participants once the research is complete.

decode To analyze in order to understand the hidden meanings in a text.

decolonizing research Research whose goals, methodology, and use of research findings contest imperialism and other oppression of peoples, groups, and classes by challenging the cross-cultural discourses, *asymmetrical power relationships*, and institutions on which they are based. (See also *applied people's geography*, *inclusionary research*, and *post-colonial research*.)

deconstruction A method for challenging assumptions of coherence and truth within a *text* by revealing inconsistencies, contradictions, and inadequacies (for example, when matters that are problematical have been naturalized).

deduction Reasoning from principles to facts. (Compare with *induction*).

deep colonizing Colonialism perpetuated through relationships and practices that are embedded in supposedly decolonizing institutions and practices.

dependability Minimization of variability in interpretations of information gathered through research. Focuses attention on the researcher-as-instrument and the extent to which interpretations are made consistently.

dependent variable A study item with characteristics that are considered to be influenced by an *independent variable*. For example, flooding is heavily dependent on rainfall.

descriptive code A *code* describing some aspect of the social data, typically aspects that are fairly obvious. (See also *manifest message* and *initial code*. Compare with *analytic code*.)

deviant case sampling Selection of extraordinary cases (for example, outstanding successes, notable failures) to illuminate an issue or process of interest.

dialogic nature of research and writing Research and writing are dialogic in that they are relational. In this sense, research is always informed and constructed by writing, just as writing is always already informed and constructed by research. The dialogic nature of research and writing extends beyond their relationality to each other and also alludes to their relationality to the wider set of social and spatial relations within which academic researcher-writers work.

diary interviews Diary interviews and diary photographs are interrelated techniques encouraging respondents to take photos of their daily routines and record their experiences, encounters, feelings, and other thoughts, the specific and the general, in written or montage form. These texts constitute the basis for one or more in-depth interviews. Commenting on this approach, Latham (2003; 2002) asserts that the diaries and interviews themselves become 'a kind of performance' enabling deeper insights into the lived reality of specific spatial engagements.

diary photographs See *diary interviews.*

dichotomy A division or binary classification in which one part of the dichotomy exists in opposition to the other (for example, light/dark, rich/poor). In most dichotomous thinking, one part of the binary is also more positively valued than the other.

disclosure When a researcher reveals information about herself or himself or the research project or when research participants reveal information about themselves.

disconfirming case Example that contradicts or calls into question researchers' interpretations and portrayals of an issue or process.

discontinuous writing A form of longitudinal journal kept by a research respondent but which does not require entries to be made on a strict daily basis. Rather, the respondent records certain events and their reactions to and/or feelings about them as they occur or in reflection. The events/feelings to be recorded are determined by the researcher, typically in negotiation with the respondent. Such diaries are also referred to as *solicited diaries,* denoting the relationship between researcher and respondent. Meth (2003) provides a thorough discussion of the uses and limitations, especially matters of ethics, associated with discontinuous writing. See also Bell (1998) on solicited diaries.

discourse There is no fixed meaning for discourse. The term has accrued a number of meanings that are in circulation in both academic and popular cultures. Even among cultural theorists, whose ideas human geographers draw upon, discourse is used in differing ways (for example, Mikhail Bakhtin's double-voiced discourse). In this book, discourse is generally understood as it was used by Michel Foucault. To make matters more complex, Foucault employed the term in at least three different ways: (1) as written/visual texts or statements that have meaning and effect, (2) an individual system or group of texts or statements that have meaning, and (3) a regulated practice of rules and structures that govern particular texts or statements. Particular attention in this book is given to his third definition because it is this rule-governed quality of discourse that is of primary importance to geographers. This definition of discourse evokes how it shapes social practices, influencing our actions, attitudes, and perceptions. (See also *discourse analysis.*)

discourse analysis Method of investigating rules and structures that govern and maintain the production of particular written, oral, or visual *texts.* (See also *discourse.*)

discursive structures or formations A key concept of Foucauldian *discourse analysis.* The rules and structures governing the production of *discourse* that affect the way individuals think, act, and express themselves—for example, through travel, comportment, clothes, make-up.

document In archival research, an individual archived item such as a memorandum or letter, handwritten or typed, that constitutes a single item or part of a larger file.

eclecticism An approach characterized by extensive borrowing of ideas from different *discourses* and their incorporation into a single argument.

emancipatory A term used to describe a process by which someone is freed from political or other restrictions.

embodied knowledge Refers to how bodies are experienced. The ways people make sense of their experiences, and themselves, cannot be separated from competing and contradictory discourses through which bodies are given meaning. For example, in the context of the uneven gendered social relationships of a public bar, there are normative assumptions about what women should do and wear to become 'attractive' to men. Hence, because of the possibilities of feeling self-confident, attractive, and feminine, some women may make a deliberate decision to make their bodies more visible by wearing a low-cut top and push-up bra, for example. Others may make deliberate decisions to make their bodies less visible to avoid being marked as sexual objects, experiencing inappropriate sexist behaviour, and being disrespected.

emoticon Used commonly in Internet communications as well as texting or SMS (short messaging service). The word is an amalgam of the concepts of emotion and icon. Emoticons began as text symbols that approximate crude graphics, such as a colon and a right-side parenthesis, to approximate a smiley face, if looked at from side-on :). Most word processing systems now convert those two characters into the smiley graphic ☺. These crude graphics can efficiently convey emotions that would otherwise take some time to convey in written word. Popular abbreviations and acronyms have similarly become popular in Internet and SMS communication (lol for laugh out loud).

episteme In the writings of Michel Foucault, episteme refers to the whole sets of discursive structures/ formations within which a culture thinks. An episteme refers to the social processes by which certain statements about the world are considered as knowledge and others are dismissed. Episteme therefore requires critically addressing the range of methodologies that a culture employs at a particular time as 'common sense' to allow certain statements to become knowledge about particular people, events, and places. (See also *discursive structures/formations*.)

epistemology Ways of knowing the world and justifying belief. (See also *ontology*.)

essentialism The idea that words (language) have some clear, apparent, and fundamental meaning rather than being given a meaning by users (readers, writers).

ethics Refers to the moral conduct of researchers and their responsibilities and obligations to those involved in the research.

ethnography A research method dependent on direct field observation in which the researcher is involved closely with a social group or neighbourhood. Also an account of events that occur within the life of a group, paying special attention to social structures, behaviour, and the meaning(s) of them for the group.

extensive research Research typically involving large scale questionnaires or other standardized or semi-standardized methods to identify regularities, patterns, and distinguishing features of a population and to yield descriptive generalization. (Compare with *intensive research*.)

external validity See *generalizability*.

extreme case sampling See *deviant case sampling*.

Facebook A web-based social networking service that connects people with one another and allows people to post messages about themselves and their activities, including their research.

facilitator The person who encourages or moderates the discussion in a *focus group*. In *participatory action research*, a person who helps others to learn by guiding an appropriate process rather than imposing their own agenda, using techniques that enable people's self-reflection and analysis.

faction A blend of 'fact' and 'fiction'. Usually refers to the tendency to weave interesting imaginary elements (for example, dialogue) into an otherwise factual account.

factoid A 'fact' that is considered to be of dubious origins or accuracy.

falsification Derived from the writings of Karl Popper (1902–94), the concept of falsification suggests that it is possible to demarcate scientific theory from non-scientific theory on the basis that scientific theories could be falsified or proven untrue by empirical observation and testing. Popper noted that it was not possible to prove a scientific theory through recourse to empirical evidence, since scientific knowledge is always contingent (that is, since scientific observations cannot include all aspects of a phenomenon, we can never be sure that a scientific theory covers the total population of a phenomenon). Accordingly, Popper developed his concept of falsification, which allowed for the contingency of scientific knowledge and allowed scientists to state that a particular theory had yet to be proven false and thus that theory could still be considered valid. Popper's famous example is that it only takes one black swan to falsify a theory that 'all swans are white'. Although this term is used largely by quantitative researchers, in the context of qualitative studies falsification may help to develop more robust concepts and/or open up new areas of inquiry by exploring negative cases.

FAQ Acronym for 'frequently asked question'.

field notes An accumulated written record of the fieldwork experience. May comprise observations and personal reflections. (Compare with *research diary*.)

field relations Refers to the ways in which we relate to those individuals and groups we encounter while undertaking fieldwork. One challenge in maintaining integrity in field relations can be achieving a balance between ethical conduct and rigorous procedures in the collection of qualitative data.

fieldwork diary See *field notes*.

files A set of archived papers—typically held together by a paperclip—created by an official agency and relating to a common topic or theme. Usually organized in reverse chronological order with the newer material overlaying older documents.

finding aids *Archives* equivalent to a library catalogue, typically taking the form of a list of accession of *files* by name as originally organized by creating agencies. Some archives now have computer-based systems that allow material to be located by use of keywords.

first-person narrative In the context of a research narrative (e.g., research report, journal publication, public talk), the presentation of the researcher(s) as 'I' or 'we'. In first-person narrative, researchers or 'narrators' are able to insert their stance, beliefs, emotions, and assumptions explicitly into the text. In so doing, they become more accountable, and their knowledge can be better 'situated'. (Compare with *third-person narrative*.)

focused interview Technique in which an interviewer poses a few predetermined questions but has flexibility in asking follow-up questions.

focus group A research method involving a small group of between 6 and 10 people discussing a topic or issue defined by a researcher, with the researcher facilitating the discussion.

follow-up questions Sometimes known as *prompts*, these are questions that permit the interviewer to ask the participant to elaborate on certain elements of an earlier response.

funnelling Interview question ordering such that the topics covered move from general issues to specific or personal matters. (Compare with *pyramid structure*.)

genealogy Refers to the ideas of Michel Foucault that question histories of human societies, especially how certain knowledge excludes certain individuals from civil society through, for example, particular portrayals of sexuality, insanity, and illness. To conduct this critique, Foucault (1980, 50) refers to the term genealogy as an interpretation that can account for the knowledges, discourses, and objects of human societies without reference to the notion of the Cartesian subject, an individual, unified self with agency and control over itself through thought and reason. Instead of a unified self, Foucault's writing focuses on the processes that he thought were crucial in the constitution of a person's subjectivity.

generalizability The degree to which research results can be extrapolated to a wider population group than that studied. (See also *transferability*.)

genre In discourse analysis, a subcategory of source forms. For example, oral sources can be subdivided into genres including oral histories, semi-structured interviews, focus groups, conversations, and life narratives. Remaining alert to the genre of the source form is crucial because particular accounts of the world are articulated through differences between the intended audience, social relationships, and technologies of production.

geographic information system (GIS) A generic title for a number of integrated computer tools for the processing and analyzing of geographical data, including specialized software for input (digitizing) and output (printing or plotting) of mapable data. GIS is not the name of a specific software package.

GIS See *geographic information system.*

go along interviews As the name suggests, the go along interview involves the researcher accompanying the respondent within the 'field' and engaging in a direct discussion of spatial engagement. This technique combines aspects of well-established techniques such as *interviewing, oral history,* and *participant observation.*

graphic mapping In CAQDA, the graphical representation of concepts, often using nodes and links; conceptual network–builders have graphic mapping capacity.

graphics software Any computer program for the display and manipulation of pictures.

grounded theory A systematic inductive (data-led) approach to building theory from empirical work in a recursive and reflexive fashion. That is, using a method of identifying themes or trends from the data, then checking through the data (or collecting more), then refining the themes using repeated checks with the data to build theory that is thoroughly 'grounded' in the real world. Initiated by Glaser and Strauss (1967) and reinterpreted and refined greatly since then by them and other authors.

haptical quality The immediate impact on the senses of a *text,* such as a film.

hegemony A social condition in which people from all sorts of social backgrounds and classes come to interpret their own interests and consciousness in terms of the *discourse* of the dominant or ruling group. The hegemony of the dominant group is thus based, in part at least, on the (unwitting) consent of the subordinate groups. Such consent is created and reconstituted through the web of social relations, institutions, and public ideas in a society.

hermeneutic circle The circle (or more broadly, process) of interpretation of qualitative information, which accounts for the point that no such interpretation is free from the values, experiences, attitudes, and ideas of the observer or researcher. Implicit in this realization is a need for the researcher to be clear about his or her position and to ensure that interpretation is participatory and iterative—that is, involves participants and is done in one or more collaborative rounds.

hermeneutics The study of the interpretation of meaning in *texts,* whether there is assumed to be a single dominant meaning or a multiplicity of meanings.

hierarchical tree-building The system for graphically representing coded and categorized data in a software program such as NUD*IST™.

iconography The identification and description of symbols and images.

iconology A search for deep symbolical images that provide a representation of the values of an individual or a group.

idiographic An approach to knowledge that highlights the particular, subjective, and contingent aspects of the social world. Cases are understood holistically. Credible and authentic descriptions are emphasized instead of statistical *generalizability.* Contrast with *nomothetic.*

image processing Refers to the digital manipulations to which images are subjected in electronic systems.

inclusionary research *Decolonizing research* projects that empower marginalized and oppressed peoples, groups, and classes with training and tools that they can use to transform their situations and conditions. (See also *post-colonial research.*)

independent variable A study item with characteristics that are considered to cause change in a dependent variable. For example, the independent variable rainfall may promote flooding (the *dependent variable*).

indigenous methodologies In many indigenous settings, the term research is deeply enmeshed in colonizing processes and experiences. In developing approaches to contemporary research, many indigenous groups are exploring ways of conceptualizing, executing, disseminating, and evaluating research against their own explicit criteria. These methodologies affirm indigenous peoples' ways of knowing, research purposes, and protocols as well as critiquing and adapting 'Western' research paradigms. Adoption of indigenous methodologies shifts the balance of power, responsibility for ethical oversight, and judgment about the value, meaning, and utility of research away from traditional research institutions such as universities and funding agencies and towards contemporary structures of indigenous governance, decision-making, and accountability—including new structures and processes within some universities and agencies.

induction Process of generalization involving the application of specific information to a general situation or to future events. (Compare with *deduction*.)

informant Person interviewed by a researcher. Some refer to those who are interviewed as 'subjects' or 'respondents'. Others argue that someone who is interviewed, as opposed to simply observed or surveyed, is more appropriately referred to as an informant. That is because an interview informant is likely to have a more active and informed role in the research encounter.

informed consent Informant/subject agreement to participate in a study having been fully apprised of the conditions associated with that study (for example, time involved, methods of investigation, likely inconveniences, and possible consequences).

informed consent form See *conditions of use form*.

initial codes Codes that are pre-determined in some way, usually because they are a prominent theme in the research questions or are inherent in the topic of the research. (See also *descriptive codes*.)

insider A research position in which the researcher is socially accepted as being 'inside' or a part of the social groups or places involved in the study. (Compare with *outsider*.)

intensive research Research typically of individual agents or small groups, involving semi-standardized or unstructured methods (e.g., interviews, oral histories, ethnography). Focuses on causal processes and mechanisms underpinning events and particular cases. (Compare with *extensive research*.)

interpretive codes Codes that emerge from analysis and interpretation of the data along themes and toward theory generation. (See also *analytic codes*.)

interpretive community Involves established disciplines with relatively defined and stable areas of interest, theory, and research methods and techniques. Influences researchers' choice of topics and approaches to and conduct of study.

intersubjectivity Meanings and interpretations of the world created, confirmed, or disconfirmed as a result of interactions (language and action) with other people within specific contexts. (See also *subjectivity* and *objectivity*.)

intertextuality The necessary interdependence of a *text* with those that have preceded it. Any text is built upon and made meaningful by its associations with others.

interview A means of data collection involving an oral exchange of information between the researcher and one or more other people.

interviewer effects When a person is interviewed, they are not doing their normal everyday activities. An interview is a formal data-gathering process. This formality, and the unusual discursive style of an interview, can have an influence on what an informant says and how they say it. This is one example of an interviewer effect. There could also be effects that flow from the demeanour, dress, accent, and physiology of an interviewer. More broadly, in ethnographic work, they are referred to as researcher effects.

interview guide A list of topics to be covered in an interview. May contain some clearly worded questions or key concepts intended to guide the interviewer. (Compare with *interview schedule*.)

interview schedule Ordered list of questions that the researcher intends to ask informants. Questions are worded similarly and are asked in the same order for each informant. In its most rigid form, an interview schedule is a questionnaire delivered in face-to-face format. (Compare with *interview guide*.)

in vivo **codes** Codes that emerge from the body of the work being examined; phrases and terms used by respondents in the course of ethnographic research or already appearing in examined texts that suggest a theme worthy of analysis. (See also *coding*.)

landscape Landscape is used broadly to mean a built, cultural, or physical environment (and even the human body), which can be 'read' and interpreted.

latent content analysis Assessment of implicit themes within a text. Latent content may include ideologies, beliefs, or stereotypes. (Compare with *manifest content analysis*.)

latent message The underlying or implied meanings of data; compare with *manifest message*.

legitimacy Approval and respect accorded researchers, research projects, and research methodologies and methods that are considered appropriate and are valued and welcomed by the people(s) and communities with whom researchers work.

life history An interview in which data on the experiences and events of a person's life are collected. The aim is to gain insights into how a person's life may have been affected by institutions, social structures, relations, rites of passage, or other significant events. (Compare with *oral history*.)

listening posts Small columns, poles, or other parts of a designed display containing audio devices or speakers—such as one might find in a museum—that allow playback of recordings (e.g., from interviews) and other sounds.

literature review Comprehensive critical summary and interpretation of resources (for example, publications, reports) and their relationship to a specific area of research.

local research authorization Review of researchers, research projects, and research methodologies and methods by indigenous peoples, communities, or local groups whose permission is sought or required prior to research with them or in their territories.

longitudinal case study A case study that involves a revisit whereby the researcher returns to the case after an intervening time period during which no appreciable research is done.

lurking Observing—typically anonymously and perhaps with the purpose of gathering data—and not contributing to chat-rooms, listserves, or other on-line platforms. Tends to be regarded as anti-social behaviour.

manifest content analysis Assessment of the surface or visible content of text. Visible content may include specific words, phrases, or the physical space dedicated to a theme (for example, column centimetres in a newspaper or time in a video). (Compare with *latent content analysis*.)

manifest message The plainly visible content of the data; compare with *latent message*.

margin coding A simple system of categorizing material in transcripts. Typically involves marking the transcript margin with a colour, number, letter, or symbol code to represent key themes or categories.

masculine gaze This term speaks to the ways in which a viewer looks upon the people either present or represented (e.g., via photography, painting). Feminist theory has added to understanding by speaking of the masculine gaze to express an asymmetric (or unequal) power relationship between viewer and the person or population viewed.

maximum variation sampling Form of *sampling* based on high diversity aiming to uncover systematic variations and common patterns within those variations.

member checking See *participant checking.*

memoing In qualitative software systems, a process whereby the researcher may write memos or reflections on the research process as she or he works and then incorporate these memos as electronic data for further investigation.

memorandum of understanding (MoU) A document specifying the aims, process, roles, responsibilities, and rights of parties involved in a research project. (See also *conditions of use form.*)

metaphor An expression applied to something to which it is not literally applicable in order to highlight an essential characteristic.

method The means by which data are collected and analyzed (e.g., in-depth interviewing).

methodology The philosophical and theoretical basis for conducting research that is much broader and sometimes more politically charged than method alone (e.g., feminist methodology).

mimetic Miming or imitating.

mixed methods A combination of techniques for tackling a research problem; the term is often used specifically to mean a combination of quantitative and qualitative methods.

modality See *objective modality* and *subjective modality.*

moderator See *facilitator.*

modern hand A way of referring to the array of more simplified ways of forming linked letters constituting writing, as opposed to printing, from the later nineteenth century onwards. Initially produced by metal-nibbed pens dipped in ink and later with fountain pens. From an archival researcher's perspective, this is often a sort of 'dark age' when handwriting standards slip and file material can be very difficult to read and may be badly smudged. It is, however, important not to ignore handwritten notes and comments in favour of typewritten records because the former can often provide vital clues as to the concerns underpinning decisions. (See also *copperplate script* and *secretary hand.*)

multiple interviews A series of interviews recorded with the same interviewee in separate sessions over a period of time.

multiple voices A reference to the need to listen to alternative literatures, texts, expressions, or opinions and therefore avoid the assumption that there is only one view of merit.

mystification In a research narrative (e.g., research report, journal publication, public talk), the concealment of particular details in ways that give other details legitimacy and/or coherence. For example, scientific third-person writing and grammatical structure may sometimes involve obscuring details of thought, feeling, emotion, time, place, embodiment, and stance in ways that mystify the social agents and participants being described. Third-person writing is particularly useful for mystifying the social location, partiality, and embeddedness of authors (researchers). Haraway refers to this as the 'god trick' of being everywhere and nowhere simultaneously.

neo-positivism Variants of positivism as they have evolved from the original axiom of 'logical positivism' developed by the Vienna Circle in the 1920s and 1930s. Neo-positivists have responded to various critiques (especially Popper's critique that logical positivism's reliance on 'verifiability' was too strong a criterion for science and instead we should rely on falsifiability as a primary criterion for scientific knowledge) and incorporated these critiques into their work. Neo-positivism was most strongly represented in geography during the quantitative revolution of the 1960s and 1970s.

nominalization The transformation of verbs and adjectives into nouns. Nominalization reduces information available to readers, and it mystifies social processes by hiding actions and the identity of actors.

nomothetic An approach to knowledge that emphasizes generalizability for understanding the social world. Social phenomena are reduced to variables for the purposes of generating statistically generalizable findings. The credibility, authenticity, and holistic understanding of each subunit studied is of lesser importance. (Contrast with *idiographic.*)

NUD*IST™ A software system for qualitative data analysis developed by Richards and Richards at Qualitative Research Solutions in Australia. The acronym stands for Nonnumerical, Unstructured Data: Indexing, Searching, Theorising. (See also *NVivo.*)

NVivo A software package to help organize and analyze qualitative data. This is a specific form of computer-assisted qualitative data analysis software (**CAQDAS**). Allows importing and coding of textual data, text editing, coded data retrieval and review, word and coding pattern searches, and data import/export to quantitative analysis software. NVivo™ and N6™ are contemporary versions of QSR's (Qualitative Solutions and Research Pty Ltd) NUD*IST™ software. (See also *computer-assisted qualitative data analysis software* and NUD*IST™.)

objective/objectivity Unaffected by feelings, opinions, or personal characteristics. Often contrasted with *subjectivity*. (See also *intersubjectivity.*)

objective modality A form of writing that implicitly hides the writer's presence in the text (for example, third-person narrative form) but which clearly signals agreement with the statement being made. (Compare with *subjective modality.*)

observation Most literally, purposefully watching worldly phenomena. Increasingly broadened beyond seeing to include apprehending the environment through all our senses (for example, sound, smell) for research purposes.

observer-as-participant A research situation in which the researcher is primarily able to observe but in so doing is also participating in a social situation. (Compare with *participant-as-observer.*)

on-line focus group Focus groups that are conducted on-line using real-time technology such as chat-rooms or asynchronous technology such as bulletin boards. (See also *synchronous groups* and *asynchronous groups.*)

ontology Beliefs about the world. Understanding about the kinds of things that exist in the universe and the relations between them. (See also *epistemology.*)

open questions Questions in which respondents are able to formulate their own answers, unrestricted by having to choose between pre-determined categories. (Compare with *closed questions* and *combination questions*).

opportunistic sampling Impromptu decision to involve cases or participants in a study on the basis of leads uncovered during fieldwork.

oral history A prepared *interview* conducted in question-and-answer format with a person who has firsthand knowledge of a subject of interest. (Compare with *life history.*)

oral methods Verbal techniques, such as *interviews* or *focus groups*, as opposed to written methods for seeking information.

Orientalism As used by Edward Said (1978), a key post-colonial term (see *post-colonialism*) referring to Western (mis)representations and construction of an imagined Orient in discourses that serve to produce and legitimize imperialism.

orientation questions Used in *oral history* interviews to establish the participant's background. (Compare with *common questions, specific questions,* and *follow-up questions*).

Other The non-Self, groups and peoples perceived as fundamentally different from one's self and against which a person might compare themselves and establish their own social position, meaning, and identity. Also taken to mean that which is oppositional to the mainstream—marginal or outside the dominant ideology. Initially developed by Simone de Beauvoir in her 1949 book *The Second Sex* to characterize patriarchal representations and subjugation of women, the term was extended by Franz Fanon (1967) and Edward Said (1978) to the cross-cultural representations and relationships that underlie colonialism.

outsider A research position in which the researcher is rendered 'outside' a social circle or feels 'out of place' on account of differences such as visible appearance, unfamiliarity, or inability to speak the language or vernacular used. (Compare with *insider*.)

over-disclosure Can occur in an interview or focus group when research participants reveal personal, sensitive, or confidential information that goes beyond the scope of the research or that they may regret having mentioned after the interview or focus group. (Contrast with *under-disclosure*.)

panopticon A circular prison with cells surrounding a central guards' station. In the panopticon, inmates may be observed at any time, but they are not aware of the observation.

paradigm Set of values, beliefs, and practices shared by a community (e.g., members of an academic discipline) that provides a way of understanding the world.

paralinguistic clues Tacit signs perceptible in face-to-face interviewing. The tone of speech used by an informant is an important indicator of their emotional disposition when answering a question. It can also indicate an informant's level of comfort and the degree of rapport between the researcher and the informant. Other non-spoken clues include eye contact, fidgeting, furtive glances, and aggressive or defensive postures.

parallel case study The study of multiple cases at the same time for the purposes of *comparative analysis*.

participant Person taking part in a research project. Usually the *informant* rather than a member of the research team.

participant-as-observer A research situation in which the researcher is primarily a participant in a social situation or gathering place but in so doing can maintain sufficient critical distance to observe social dynamics and interactions. (Compare with *observer-as-participant*.)

participant checking Informant's review of the transcript of their contribution to an *interview* or *focus group* for accuracy and meaning. May also involve the informant reviewing the overall research output (for example, thesis, report). Also serves as a means of continuing the involvement of informants in the research process.

participant community One's research participants; the community may be known as such to one another—members of a formal or informal grouping—or may be seen as a community by the researcher.

participant observation A fieldwork method in which the researcher studies a social group while being a part of that group.

participation A process in which people play active roles in decision-making and other activities affecting their lives.

participatory action research (PAR) An umbrella term covering a range of participatory approaches to action-oriented research involving researchers and participants working together to examine a situation and change it for the better. (See also *action research*, *participatory research*, *collaborative research*, and *decolonizing research*.)

participatory diagramming A technique whereby a group of people, with support from a *facilitator*, collectively produces a visual representation (for example, drawing, diagram, chart, mind-map, sketch) for subsequent analysis using locally appropriate materials (for example, stones, leaves, chalk, ground, pens, paper, whiteboards), criteria, and symbols. Diagrams usually convey relationships between key stakeholders, institutions, or resources, sometimes over different time periods. (See also *participatory mapping.*)

participatory geographic information systems (PGIS) A process that adapts GIS software to incorporate local expertise and knowledge, usually within a participatory action research framework to enable mapping and data analysis by non-professionals. PGIS emerged in response to criticisms that the high financial, time, and training requirements of GIS can discourage grassroots groups from using it to inform their own research and development. (See also *participatory mapping.*)

participatory mapping A technique whereby a group of people, with support from a *facilitator*, collectively produces a 'map' for subsequent analysis using locally appropriate materials, criteria, and symbols. Maps usually focus on material aspects of life such as a watershed, a village, a body, or the distribution of particular resources within a particular area. (See also *participatory diagramming* and *participatory geographic information systems.*)

participatory research A community-based approach to research involving local people and their knowledges as a foundation for social change. Participatory research was developed by educators like Paulo Freire in Brazil and others in the global South to support consciousness-raising and political action. It now informs participatory action research and processes of participatory development within the global South.

pastiche A *text* that is a medley drawn from various sources.

'pencil only' rule A typical archival convention to ensure the protection of file materials, it requires that researchers make notes in pencil only. The rule is rendered somewhat redundant by the increasing use of personal computers.

personal log Recorded reflections on the practice of an interview. Includes discussions of the appropriateness of the order and phrasing of questions and of the informant selection. Also contains assessments of matters such as research design and ethical issues. (See also *analytical log.*)

PGIS See *participatory geographic information systems.*

photo-elicitation A technique related to diary photographs and *diary interviews.* However, according to Harper (2002), the principal objective of this technique is to elicit alternate ways of seeing and understanding the same image among respondents. The outcome of this is to invoke a deep discussion of values and meaning.

photovoice A methodology that was developed within health education and has now been adopted and adapted within the social sciences, including geography. Participants are asked to represent their perspective by taking photographs, which subsequently offer a window into how they conceptualize their circumstances. Photography 'gives voice' to otherwise marginalized groups and has been associated with social action.

pilot study Abbreviated version of a research project in which the researcher practises or tests procedures to be used in a subsequent full-scale project.

plagiarism Failure to acknowledge the source of ideas, illustrations, or text. In the worst cases, sections of text are lifted verbatim and reproduced, without quotation marks, as if the author's own. In many universities, plagiarized work—which is often easy to spot—automatically fails.

population The larger group from which a *sample* has been selected for inclusion in a study. In quantitative research, based on *probability* (random) *sampling*, it is assumed that the sample has been selected such that the mathematical probability of sample characteristics being reproduced in this broader population can be calculated. In qualitative research where *purposive sampling* is used, no such assumption is made.

positionality A researcher's social, locational, and ideological placement relative to the research project or to other participants in it. May be influenced by biographical characteristics, such as class, race, and gender, as well as various formative experiences.

positionality statement A record of how a researcher is situated in the production of knowledge over time. This requires careful reflection on the researcher's points of connection and disconnection with a project. Positionality statements are not stand-alone testimonials. Instead, they document the co-constitution of researcher and project. (See also *co-constitution of knowledge*.)

positivism An approach to scientific knowledge based around foundational statements about what constitutes truth and legitimate ways of knowing. There are a number of variants of positivist thought, but central to all is the construction of a singular universal and value-free knowledge based on empirical observation and the scientific method.

post-colonialism An approach to knowledge that seeks to represent voices of the *Other*, especially colonized peoples and women, and to recognize knowledge that has been ignored through processes of colonization and patriarchy.

post-colonial research Research that rejects imperialism and the goals, attitudes, representations, and methods of imposed, 'colonial' research and instead seeks to conduct research that is welcomed and that fosters egalitarian relationships and openness, values local knowledge and ways of knowing, and contributes to self-determination and locally defined welfare. (See also *decolonizing research* and *inclusionary research*. Contrast with *colonial research*.)

postmodernism A movement in the humanities and social sciences that includes *post-colonialism* and embraces the pluralism of multiple perspectives, knowledges, and voices rather than the grand theories of modernism. Individual interpretation is considered partial because it is to some degree socially contingent and constituted.

post-structuralism A school of thought that endeavours to link language, subjectivity, social organization, and power.

potentially exploitative power relation A research situation in which the researcher is in a position of power relative to the research participant. (See also *asymmetrical power relation*, *reciprocal power relation*, and *studying up*.)

power In Foucauldian discourse analysis, power is central to thinking about *discourse* as something that has an impact. It is through power that the elements of discourse have effects on what people do and think and how they express themselves. Yet power is not conceptualized in terms of acting upon people in an oppressive way. Rather, the individual is seen as an effect of power. That is, power makes things possible as well as restricting possible actions and attitudes. These possibilities are instances of power/knowledge relationships. (See also *power/knowledge*.)

power/knowledge A key concept of Foucauldian discourse analysis. Foucault argues that the relationship between power and knowledge is essential to thinking about the effects of *discourse*. He argues that statements that are accepted as knowledge are themselves the outcome of power struggles. For example, what has constituted geographical knowledge in universities has been a constant struggle over different versions of what constituted space/place.

PowerPoint Proprietary presentation graphics software, originally designed for business presentations, that has been widely adopted in classrooms and the academy. Be aware that its use risks promoting style over substance and that audiences can generally spot this very easily.

praxis The use of research findings by researchers to make constructive social and environmental change.

preliminary meeting In *oral history*, a first meeting between interviewer and *respondent* designed to establish rapport, clarify ethical responsibilities and rights, explore the parameters of the pending interview(s), and work through other matters of mutual interest or concern; an important step for both parties after which additional preparations may be made for the interview.

pre-testing See *pilot study*.

primary observation Research in which the investigator is a participant in and interpreter of human activity involving her or his own experience. (Compare with *secondary observation*.)

primary question Interview question used to initiate discussion of a new topic or theme. (Compare with *probe question*.)

primary sources From a historical geography perspective, primary sources in the narrower sense are public and private records created in an earlier period of time that is of interest to the researcher. They include letters, diaries, and journals, original census returns, minutes of organizations, and files of government departments. In addition, original maps, survey plans, and photographs would be included. In some circumstances, period newspapers and published official documents might also be regarded as primary sources. (Compare with *secondary literature*.)

privileged discourse While there are always competing and sometimes contradictory cultural discourses to make sense of the world, privileged discourse is one that takes priority over others in shaping social, cultural, and political meanings. For example, there exists a range of conflicting and competing discourses about Hawai'i. Yet in Western society, that which is privileged portrays Hawai'i as an earthly paradise.

probability sampling Sampling technique intended to ensure a random and statistically representative *sample* that will allow confident generalization to the larger *population* from which the sample was drawn. (Compare with *purposive sampling*.)

probe question A gesture or follow-up question used in an interview to explore further a theme or topic already being discussed. (See also *prompt*. Compare with *primary question*.)

professional interviewing A style of interviewing relationship in which the researcher seeks to establish and maintain the respect of the informant. The interview has a formal feel to it, and the relationship between the researcher and informant is intended to be professional. This style of interviewing relationship has been considered important when researchers are interviewing powerful people and the researcher believes that in order for him or her to be taken seriously, it is important to seek and maintain the respect of the informant (Schoenberger 1992) (Compare with *creative interviewing*.)

prompt A follow-up question in an *interview* designed to deepen a response (for example, 'why do you say that?', 'what do you mean?'). (See also *probe question*. Compare with *primary question*.)

provenance Organizational principle for *archives* that stresses the importance of the original internal arrangement of a collection of *files* and the order of information in files as devised by their creating agencies as a means of understanding past events.

purposeful sampling See *purposive sampling*.

purposive sampling *Sampling* procedure intended to obtain a particular group for study on the basis of the specific characteristics they possess. (Compare with *probability sampling*.) Aims to uncover information-rich phenomena/participants that can shed light on issues of central importance to the study.

pyramid structure Order of interview questions in which easy-to-answer questions are posed at the beginning of the interview while deeper or more philosophical questions/issues are raised at the end. (Compare with *funnelling*.)

quantitative methods Statistical and mathematical modelling approaches used to understand social and physical relationships.

quantitative revolution A period in the mid-twentieth century (particularly from the late 1950s) during which transformations in information and computer technologies, developments in mathematical modelling, and sophisticated quantitative techniques influenced the form and nature of research being conducted. The quantitative revolution led to an increased use of statistical techniques for collating and analyzing large amounts of data, with these techniques being linked to empirical testing of models, theory, and hypotheses.

quota sampling Selecting sampling elements on the basis of categories known or assumed to exist in the universal population.

random sampling See *probability sampling.*

rapport A productive interpersonal climate between informant and researcher. A relationship that allows the informant to feel comfortable or confident enough to offer comprehensive answers to questions.

reactive effects The influence a research method or researcher has on the individuals or phenomena under observation.

reciprocal power relation A research situation in which researcher and *informant* are in comparable social positions and experience relatively equal costs and benefits of participating in the research. (See also *asymmetrical power relation, potentially exploitative power relation,* and *studying up.*)

records A generic term for files, maps, plans, and other documents held in an *archive.*

recruitment The process of finding people willing to participate in a research project. Recruitment strategies can range from asking people 'on the street' (perhaps to fill in a questionnaire) to inviting key individuals to participate (in a focus group, for example).

reflexivity Self-critical introspection and a self-conscious scrutiny of oneself as a researcher. (See also *action–reflection.*)

regime of truth A key concept of Foucauldian discourse analysis. Foucault argues that within a particular time and society, particular statements become accepted as truths or common sense. Foucault stressed that these statements must be examined with these social circumstances. It is this social-cultural context that both gives texts their sense and makes them compelling. (See also *discourse* and *discourse analysis.*)

reification When the complexity of social life is reduced to concrete and simplified 'things' in the construction of texts. For example, one might talk about 'the research' or 'the participant', thereby obscuring the way in which these 'nouns' come to be constituted and expressed through a variety of social and spatial relations.

relativism An approach to knowledge in which it is held that there is no means of significantly differentiating between the merits of arguments. Suggests that there are no absolute, unequivocal standards of true/false or right/wrong.

reliability Extent to which a method of data collection yields consistent and reproducible results when used in similar circumstances by different researchers or at different times. (See also *validity.*)

replicability Ability to be repeated or tested to see how general the particular findings of a study are in the wider *population.*

representation The way in which something (the world, human behaviour, a city, the landscape) is depicted, recognizing that it cannot be an exact depiction. An important insight from post-structuralist thinkers is that representations not only describe the social world but also help to shape or constitute it.

research diary A place for recording observations in the process of being reflexive. Contains thoughts and ideas about the research process, its social context, and the researcher's role in it. The contents of a research diary are different from those of the field notes, which more typically contain qualitative data, such as records of observations, conversations, and sketch maps.

respondent See *informant.*

rigour Accuracy, exactitude, and trustworthiness.

sample Phenomena or participants selected from a larger set of phenomena or a larger *population* for inclusion in a study.

sampling Means of selecting phenomena or participants for inclusion in a study. A key difference between qualitative and quantitative inquiry is in the logic underpinning their use of *purposive* and *probability* (random) *sampling* respectively.

sampling frame A list or register (for example, electoral roll, phone directory) from which respondents for a questionnaire are drawn.

satisficing behaviour Conduct in which the decision-maker or agent acts in ways that yield satisfactory outcomes rather than optimal or 'maximizing' outcomes.

saturation The point in the data-gathering process when no new information or insights are being generated. This is one method used by researchers to determine when to stop gathering data.

secondary data Information collected by people/agencies and stored for purposes other than for the research project for which they are being used (for example, census data being used in an analysis of socio-economic status and water consumption).

secondary literature From an historical geography perspective, an existing set of published material in the form of books, essays, and articles produced after the events that they discuss and based on primary sources and/or oral testimony and personal observations. In terms of a research exercise, it also refers to the existing academic writing on a specific topic. This literature may be obviously divided into groups that offer contrasting interpretations of the same events. (Compare with *primary sources*.)

secondary observation Research in which the data are the observations of others (for example, photographs).

secondary question Interview prompts that encourage the informant to follow up or expand on an issue already discussed. (See also *follow-up questions*.)

secretary hand A style of writing used by professional clerks that became increasingly widespread in England in the sixteenth and seventeenth centuries. Twenty-first-century researchers require some special skills to be able to decipher the calligraphy and the now archaic English prose. (See also *copperplate script* and *modern hand*.)

semiology See *semiotics*.

semiotics The system or language of signs (sometimes referred to as semiology). (See also *signifier* and *signified*.)

semi-structured interview Interview with some predetermined order but which nonetheless has flexibility with regard to the position/timing of questions. Some questions, particularly sensitive or complex ones, may have a standard wording for each *informant*. (Compare with *structured interview* and *unstructured interview*.)

series lists A list of individual files from a particular organization held in an archive. Typically, they replicate the referencing system used by the creating organization; they are usually numerical or alphanumeric and can include sub-series as well as large files that are broken into several individually recorded parts.

sharing the field Ways of sharing our work with participants and respondents and the decisions that have to be made concerning reciprocity and confidentiality (i.e., what do we give back to them in return for their involvement, and how do we protect confidences?)

sign Written or other image/mark that represents something else. Comprises *signifier* and *signified*.

signified The meaning derived from a *signifier* (or from a set of signifiers, such as a text).

signifier Images such as written marks or features of the landscape with which meaning is associated.

silence In discourse analysis, silence refers to how frameworks of understanding always conceal as much as they reveal about the world. Concealment inevitably occurs when all frameworks of understanding are conceived of as the outcome of a highly social process. For example, life narratives often frame events surrounding discriminatory legislation around personal accounts, often silencing the role of political organizations. In contrast, a spokesperson for a political organization may choose to frame the same events through ideas about the collective nation. Such official accounts may erase differences in lived experiences based on various intersections of gender, sexuality, class, race, and so on.

situated knowledge A metaphor that evokes recognition of the positionality (or contextual nature) of knowledges. The inscription and creation of knowledge is always partial and 'located' somewhere.

situated learning Learning that occurs in the same context as that in which it is applied.

snowball sampling A sampling technique that involves finding participants for a research project by asking existing informants to recommend others who might be interested. From one or two participants, the number of people involved in the project 'snowballs'. Also known as *chain sampling*.

social justice A situation in which there is an equitable and respectful negotiation of social and spatial difference and a fair distribution of resources.

social structure See *structure*.

solicited diary See *discontinuous writing*.

sound document The outcome and output from an *oral history* interview.

soundscape A compound word joining sound and landscape to create a term that captures the way a sound or combination of sounds can contribute to the sensory environment. Sounds can be 'natural' (e.g., water) or contrived (e.g., music), and their apprehension can be understood as a non-visual form of observation.

specific questions Relate to interviewees' individual experiences and are developed through follow-up work. (Compare with *orientation questions*, *common questions*, and *follow-up questions*.)

staging a text A theatrical metaphor for *writing-in* research that encourages geographical researchers to consider how the construction of a research text is actually a form of cultural production. The author of the text is a creator, director, and performer in the particular narrative that he or she is constructing.

stakeholder Any individual or group that has an interest in a project because of the way it may benefit, harm, or exclude them.

standardized questions A uniform set of questions that are repeated for all *interviews* or *focus groups* in a research project, in contrast to spontaneous questions that develop out of the conversational flow of an interview or focus group.

strategy of conviction Refers to the processes by which particular social realities or ways of making order of the world become accepted within a particular spatial and historical context as common sense or as truth. The term is derived from the work of Michel Foucault. In particular, the term strategy of conviction is derived from his argument that knowledge and power are inseparable. According to Foucault, the process by which particular social realities are produced, circulated, and maintained requires thoughtful consideration of the intersection between authorship, technologies, and the type of text.

structure A functioning system (for example, social, economic, or political) within which individuals are located, within which all events are enacted, and which is reproduced and transformed by those events.

structured interview Interview that follows a strict order of topics. Usually the order is set out in an *interview schedule*. The wording of questions for each interview may also be predetermined. (Compare with *unstructured interview* and *semi-structured interview*.)

studying up An asymmetrical power relationship in which the research participant is in a position of power relative to the researcher. (See also *asymmetrical power relation*, *potentially exploitative power relation*, and *reciprocal power relation*.)

subaltern Oppressed, exploited, marginalized minority peoples and groups. The term derives from the work of Antonio Gramsci and from the 'subaltern studies' project undertaken by Indian historians since the early 1980s that endeavoured to write history from 'below' but is often used in a wider sense.

subject See *informant.*

subjective/subjectivity Refers to the insertion of the personal resources, opinions, and characteristics of a person into a research project. Often contrasted with objectivity. (See also *intersubjectivity.*)

subjective modality A form of writing that explicitly acknowledges the writer's presence in the text (e.g., *first-person narrative* form) and clearly signals their agreement or disagreement with the statement being made. (Compare with *objective modality.*)

surveillance Involves monitoring the behaviour of people or objects so as to observe (and sometimes record) deviation from conformity. In French, surveillance means 'watching over', and in one sense this covers all types of all observation. Contemporary technology such as closed-circuit television epitomizes surveillance as social control.

synchronous groups Real-time on-line discussions used in focus group research. Groups last for several hours and have 6 to 10 participants. (See also *asynchronous groups* and *on-line focus group.*)

synchronous interviewing Interview in which the informant's answers immediately follow the questions. Face-to-face and telephone interviews are common examples. Some forms of Internet-based interviewing can also be synchronous, such as conversations that occur within chat-rooms and other on-line platforms. (Compare with *asynchronous interviewing.*)

synergistic effect A key feature of focus groups that occurs when participants are prompted and provoked by the things others say. Generally results in a lively discussion as participants respond to each other and add new thoughts and ideas. Best summed up by the phrase 'the whole is greater than the sum of its parts'.

text Traditionally synonymous with the written page but now used more broadly to refer to a range of source forms such as oral texts (including *semi-structured interviews* and *oral histories*) and images (including painting, photographs, and maps) as well as written and printed texts (including newspapers, letters, and brochures).

text-based managers Software with capacity for managing and organizing data, creating subsets of data for further analysis, and searching and retrieving combinations of words, phrases, coded segments, memos, or other material. Examples include askSam, FolioVIEWS, and MaxQDA.

text retriever software Computer program for recovering data by category on the basis of keywords that appear in the data, for finding words, phrases, or other character strings, and for finding things that are misspelled, sound alike, mean the same thing, or have certain patterns. Examples include Metamorph, The Text Collector, WordCruncher, ZyINDEX, and Sonar Professional.

textual analysis Reading and constant reinterpretation of texts as a set of *signs* or signifying practices.

textual community Group of individuals who share certain understandings of the meaning of texts.

theme In *coding*, an important process, commonality, characteristic, or theory that emerges from the data and can be used to analyze and abstract the data.

theory building software Computer programs that deal with relationships between data categories to develop higher-order classifications and categories and to formulate and test propositions or assertions. Examples include AQUAD, ATLAS/ti, HyperRESEARCH, and N6.

theory generation The inductive process of identifying concepts to explain a phenomenon. One of the major advantages of qualitative research is that it facilitates the generation of new or revised theory. The theory may be as simple as a few loosely connected concepts or as complex as a large number of tightly integrated concepts (e.g., Marx's theory of capitalism).

theory testing A process whereby concepts or theories are compared to empirical data to test for congruence between the concepts and the empirical world. It is often conducted formally as quantitative statistical hypothesis testing but may also be done qualitatively using non-quantified textual data. (See also *deduction* and *induction*.)

thick description A term made common by anthropologist Clifford Geertz, involves not just describing an event, occurrence, or practice but the detailed context in which it occurs. This allows researchers to interpret the situated nature of the 'event', reflecting on the ways in which the subject or object of study is constructed symbolically in relation to broader cultural and social relations and discourses.

third-person narrative In the context of a research narrative (e.g., research report, journal publication, public talk), involves constructing the narrative without reference to the researcher's thoughts, opinions, or feelings. This consequently conveys a distanced and seemingly neutral, omniscient, and 'objective' point of view. Third-person narratives have dominated in the presentation of 'scientific' research. (Compare with *first-person narrative*.)

timed tape logs A summary of the contents of an interview that notes the topics discussed and key words and the time at which they occur within the recording. These logs allow users to locate desired audio or video segments easily.

transcript Written record of speech (for example, interview, focus group proceedings, film dialogue). May also include textual description of informant gestures and tone.

transferability Extent to which the results of a study might apply to contexts other than that of the research study. (See also *generalizability*.)

transformative reflexivity A process through which a researcher and researched group reflect on their (mis)understandings and negotiate the meanings of the information generated together. The shared process has the potential to transform each person's own understandings.

triangulation Use of multiple or mixed methods, researchers, and information sources to confirm or corroborate results.

trolling A form of anti-social behaviour within on-line environments. There are two aspects to trolling. First, there is the non-participatory collection of data from on-line communities (see also *lurking*). Second, and more commonly, trolling refers to the baiting of participants by posting controversial or purposefully incendiary statements that are intended to incite heated exchanges (e.g., racist or sexist comments). Internet trolls appear to delight in generating discord, and their comments distract user groups and on-line communities from their common topics of discussion. Researchers using the Internet need to monitor for the presence of trolls within their own discussions and must ensure that their own postings and prompts are not seen as trolling.

trope Figure of speech that allows writers or producers of other forms of *text* to say one thing but mean something else. May involve use of *metaphor* or metonymy.

trustworthiness A principle in ethics and description applying to those who are given trust by others and who do not violate that faith; qualitative data and insights gained from them need to be trustworthy in order to be reliable and generalizable.

typical case sampling Selection of samples that illustrate or highlight what is considered typical or normal.

uncontrolled observation Purposeful watching of worldly phenomena that is relatively unconstrained by restrictions of scope, style, and time. (Compare with *controlled observation*.)

under-disclosure When participants in an interview or in focus groups provide very little information. (Contrast with *over-disclosure*.)

unstructured interview Interview in which there is no predetermined order to the issues addressed. The researcher phrases and raises questions in a manner appropriate to the informants' previous comment. The direction and vernacular of the interview is informant-driven. (Compare with *structured interview* and *semi-structured interview*.)

validity The truthfulness or accuracy of data compared with acceptable criteria. (See also *reliability*.)

vernacular Occurring in the location where it originated. Vernacular language is the language of a place.

voice capture Voice capture software can be used in questionnaire survey work conducted by telephone or face-to-face mode and is particularly useful for recording responses to open-ended questions. Software packages are available to digitally record responses in the respondents' actual voice, capturing information about, for instance, the intensity, emotion, or intonation of the response. The responses can then be played back for electronic coding to produce numeric tables and charts for further analysis.

voice recognition computer software Computer software that converts spoken words into text. It has three principal uses: (1) to support dictaphone use; (2) for accessibility purposes (e.g., so that people who are vision-impaired or who cannot use keyboards can use Internet communications); and (3) for the 'transcription' of research interviews into text. Examples include Dragon Naturally Speaking™, Windows Speech Recognition™, and MacSpeech Dictate™. (See also *transcript*.)

warm-up A set of pre-interview techniques intended to enhance rapport between interviewer and *informant*. May include small talk, sharing food, or relaxed discussion of the research.

Wikipedia A web-based encyclopedia that is written collaboratively by volunteers and may be edited by users.

word processing A generic concept that includes the use of computer capacity to create, edit, and print documents.

word searching The process of looking for individual words in an electronic text.

writing-*in* The active and situated process of writing in which the author engages with the ways in which meanings are constructed through the creation of his or her text.

References

Abraham, F. 1982. *Modern Sociological Theory: An Introduction*. New Delhi: Oxford University Press.

Adelman, C., ed. 1981. *Uttering, Muttering: Collecting, Using and Reporting Talk for Social and Educational Research*. London: Grant McIntyre.

Agar, M. 1986. *Speaking of Ethnography*. Beverly Hills: Sage.

Agar, M., and J. MacDonald. 1995. 'Focus groups and ethnography'. *Human Organization* 54 (1): 78–86.

Agius, P., et al. 2004. 'Comprehensive native title negotiations in South Australia'. In M. Langton et al., eds, *Honour among Nations? Treaties and Agreements with Indigenous People*. Melbourne: Melbourne University Press.

Agius, P., R. Howitt, and S. Jarvis. 2003. *Different visions, different ways: Lessons and challenges from the native title negotiations in South Australia*. Paper presented to the Native Title Conference, Alice Springs, June.

Aitken, S.C., and L.E. Zonn. 1993. 'Weir(d) sex: Representation of gender–environment relations in Peter Weir's *Picnic at Hanging Rock* and *Gallipoli*'. *Environment and Planning D: Society and Space* 11 (2): 191–212.

Alder, P.A., and P. Alder. 1994. 'Observational techniques'. In N.K. Denzin and Y.S. Lincoln, eds, *Handbook of Qualitative Research*. Thousand Oaks, CA: Sage.

Alexander, C., et al. 2007. 'Participatory diagramming: A critical view from north east England'. In S. Kindon, R. Pain, and M. Kesby, eds, *Participatory Action Research Approaches and Methods: Connecting People, Participation and Place*, 112–21. London: Routledge.

Aluka. http://www.aluka.org.

Alvermann, D.E., D.G. O'Brien, and D.R. Dillon. 1996. 'On writing qualitative research'. *Reading Research Quarterly* 31: 114–20.

American Historical Association. 2003. *Statement on Standards of Professional Conduct*. http://www.historians.org/PUBS/Free/ProfessionalStandards.htm.

Anderson, K.J. 1993. 'Place narratives and the origins of inner Sydney's Aboriginal settlement, 1972–73'. *Journal of Historical Geography* 19 (3): 314–35.

———. 1995. 'Culture and nature at the Adelaide Zoo: At the frontiers of "human" geography'. *Transactions of the Institute of British Geographers* 20 (3): 275–94.

———. 1999. 'Reflections on Redfern'. In E. Stratford, ed., *Australian Cultural Geographies*. Melbourne: Oxford University Press.

Anderson, K.J., and F. Gale, eds. 1992. *Inventing Places: Studies in Cultural Geography*. Melbourne: Longman Cheshire.

Andrews, G.J., et al. 2006. 'Their finest hour: Older people, oral histories, and the historical geography of social life'. *Social and Cultural Geography* 7 (2): 153–77.

Anfara, V.A., K.M. Brown, and T.L. Mangione. 2002. 'Qualitative analysis on stage: Making the research process more public'. *Educational Researcher* 31 (7): 28–38.

Anzaldúa, G. 1987. *Borderlands/La Frontera: The New Mestiza*. San Francisco: Spinsters/Aunt Lute Press.

Askew, L., and P.M. McGuirk. 2004. 'Watering the suburbs: Distinction, conformity and the suburban garden'. *Australian Geographer* 35: 17–37.

Association of American Geographers. 2005. *Statement on Professional Ethics*. http://www.aag.org/Publications/Other%20Pubs/Statement%20on%20Professional%20Ethics.pdf.

Atkinson, P., and M. Hammersley. 1984. 'Ethnography and participant observation'. In N.K. Denzin and Y.S. Lincoln, eds, *Handbook of Qualitative Research*. Thousand Oaks, CA: Sage.

Ayres, L. 1997. 'Defining and managing family caregiving in chronic illness: Expectations, explanations and strategies'. (University of Illinois at Chicago, PhD thesis).

Babbie, E. 1992. *The Practice of Social Research*. 6th edn. Belmont, CA: Wadsworth.

———. 1998. *The Practice of Social Research*. 8th edn. Belmont, CA: Wadsworth.

———. 2001. *The Practice of Social Research*. 9th edn. Belmont, CA: Wadsworth.

Bailey, C. 2001. 'Geographers doing household research: Intrusive research and moral accountability'. *Area* 33 (1) 107–10.

Bailey, C., C. White, and R. Pain. 1999a. 'Evaluating qualitative research: Dealing with the tension between "science" and "creativity"'. *Area* 31 (2): 169–83.

———. 1999b. 'Response'. *Area* 31 (2): 183–4.

Baker, A.R.H. 1997. 'The dead don't answer questionnaires: Researching and writing historical geography'. *Journal of Geography in Higher Education* 21 (2): 231–43.

Bampton, R., and C.J. Cowton. 2002. 'The e-interview'. *Forum: Qualitative Social Research Sozialforschung* 3 (2): article 9.

Banks, M. 2001. *Visual Methods in Social Research*. London: Sage.

Barbour, R. 2007. *Doing Focus Groups*. Los Angeles: Sage.

Barnes, T.J. 1989. 'Place, space and theories of economic value: Contextualism and essentialism'. *Transactions of the Institute of British Geographers* NS 14, 299–316.

———. 1993. 'Whatever happened to the philosophy of science?' *Environment and Planning A* 25: 301–4.

Barnes, T.J., and J. Duncan. 1992. 'Introduction: Writing worlds'. In T.J. Barnes and J. Duncan, eds, *Writing worlds: Discourse, Text and Metaphor in the Representation of Landscape*. London: Routledge.

Barnes, T.J., and D. Gregory. 1997. 'Worlding geography: Geography as situated knowledge'. In T.J. Barnes and D. Gregory, eds, *Reading Human Geography: The Poetics and Politics of Inquiry*. London: Arnold.

Barrett, M. 1991. *The Politics of Truth: From Marx to Foucault*. Cambridge: Polity Press.

Barry, C. 1998. 'Choosing qualitative data analysis software: ATLAS/ti and NUDIST compared'. *Sociological Research Online* 3 (3). http://www.socresonline.org.uk/3/3/4.html.

Barthes, R. 1973. *Mythologies* tr. A. Lavers. London: Paladin.

Baxter, J. 1998. 'Exploring the meaning of risk and uncertainty in an environmentally sensitized community' (Department of Geography, McMaster University, Ontario, PhD thesis).

Baxter, J., and J. Eyles. 1997. 'Evaluating qualitative research in social geography: Establishing "rigour" in interview analysis'. *Transactions of the Institute of British Geographers* 22 (4): 505–25.

———. 1999a. 'The utility of in-depth interviews for studying the meaning of environmental risk'. *Professional Geographer* 51 (2): 307–20.

———. 1999b. 'Prescription for research practice? Grounded theory in qualitative evaluation'. *Area* 31 (2): 179–81.

Baxter, J., and D. Lee. 2004. 'Explaining the maintenance of low concern near a hazardous waste treatment facility'. *Journal of Risk Research* 6 (1) 705–29.

Bazeley, P. 1997. 'NUD*IST 4: Survey research: How do I set up data input for a survey to link structured responses (e.g. in SPSS) with qualitative data in NUD*IST 4?' *User Support Notes* http://kerlins.net/bobbi/research/nudist/resources/bazeley1.html.

———. 2002. 'Computerized data analysis for mixed methods research'. In A. Tashakkori and C. Teddlie, eds, *Handbook of Mixed Methods for the Social and Behavioral Sciences*. London: Sage.

———. 2007. *Qualitative Data Analysis with NVivo*. London: Sage.

Becker, H.S. 2007. *Writing for Social Scientists: How to Start and Finish Your Thesis, Book, or Article*. 2nd edn. Chicago: University of Chicago Press.

Beckett, C., and S. Clegg. 2007. 'Qualitative data from a postal questionnaire: Questioning the presumption of the value of presence'. *International Journal of Social Research Methodology* 10: 307–17.

Bedford, T., and J. Burgess. 2002. 'The focus-group experience'. In M. Limb and C. Dwyer, eds, *Qualitative Methodologies for Geographers: Issues and Debates*. London: Edward Arnold.

Bell, D. 1991. 'Art and land in New Zealand'. *New Zealand Journal of Geography* 92: 15–17.

Bell, D., and J. Binnie. 2000. *The Sexual Citizen: Queer Politics and Beyond*. Cambridge: Polity Press.

Bell, D., et al. 1994. 'All hyped up and no place to go'. *Gender, Place and Culture* 1 (1): 31–48.

Bell, L. 1998. 'Public and private meanings in diaries: Researching family and childcare'. In J. Ribbens and R. Edwards, eds, *Feminist Dilemmas in Qualitative Research: Public Knowledge and Private Lives*, 72–86. London: Sage.

Bennett, K. 2002. 'Interviews and focus groups'. In P. Shurmer-Smith, ed., *Doing Cultural Geography*. London: Sage.

Berg, B.L. 1989. *Qualitative Research Methods for the Social Sciences*. Boston: Allyn and Bacon.

Berg, L.D. 1993. 'Between modernism and postmodernism'. *Progress in Human Geography* 17: 490–507.

———. 1994a. 'Masculinity, place, and a binary discourse of theory and empirical investigation in the human geography of Aotearoa/New Zealand'. *Gender, Place and Culture* 1 (2): 245–60.

———. 1994b, 'Masculinism, power and discourses of exclusion in Brian Berry's "scientific" geography'. *Urban Geography* 15: 279–87.

———. 1997. *Banal geographies*. Paper presented to the Inaugural International Conference of Critical Geographers, Vancouver, 10–13 August.

———. 2001. 'Masculinism, emplacement and positionality in peer review'. *Professional Geographer* 53 (4): 511–21.

———. 2004. 'Scaling knowledge: Towards a critical geography of critical geography'. *Geoforum* 35 (5): 553–8.

Berg, L.D., M. Evans, D. Fuller, and the Okanagan Urban Aboriginal Health Research Collective. 2007. 'Ethics, hegemonic whiteness, and the contested imagination of "Aboriginal community" in social science research in Canada'. *ACME: An International E-Journal for Critical Geographies* (special issue on participatory research ethics) 6 (3): 395–410.

Berg, L.D., and R.A. Kearns. 1998. 'America Unlimited'. *Environment and Planning D: Society and Space* 16: 128–32.

Berg, L., and J. Mansvelt. 2000. 'Writing in, speaking out: Communicating qualitative research findings'. In I. Hay, ed., *Qualitative Research Methods in Human Geography*. Melbourne: Oxford University Press.

Berger, T.R. 1991. *A Long and Terrible Shadow: White Values, Native Rights in the Americas*. Vancouver and Seattle: Douglas and McIntyre and University of Washington Press.

Bernard, H.R. 1988. *Research Methods in Cultural Anthropology*. Newbury Park: Sage.

Berreman, G.D. 1972. *Hindus of the Himalayas: Ethnography and Change*. 2nd edn. Berkeley: University of California Press.

Bertrand, J.T., J.E. Brown, and V.M. Ward. 1992. 'Techniques for analyzing focus group data'. *Evaluation Review* 16 (2): 198–209.

Bhowmick, T. 2006. 'Building an exploratory visual analysis tool for qualitative researchers'. *Proceedings, AutoCarto*, Vancouver.

Bijoux, D., and J. Myers. 2006. 'Interviews, solicited diaries and photography: "New" ways of accessing everyday experiences of place'. *Graduate Journal of Asia-Pacific Studies* 4 (1): 44–64.

Billinge, M., D. Gregory, and R.L. Martin. 1984. 'Reconstructions'. In M. Billinge, D. Gregory, and R.L. Martin, eds, *Recollections of a Revolution: Geography as Spatial Science*. London: MacMillan.

Bingley, A. 2002. 'Research ethics in practice'. In L. Bondi et al., eds, *Subjectivities, Knowledges and Feminist Geographies: The Subjects and Ethics of Research*. London: Rowman and Littlefield.

Binnie, J., and G. Valentine. 1999. 'Geographies of sexuality—A review of progress'. *Progress in Human Geography* 23: 176–87.

Bishop, P. 2002. 'Gathering the land: The Alice Springs to Darwin rail corridor'. *Environment and Planning D: Society and Space* 20: 295–317.

Bishop, R. 1994. 'Initiating empowering research?' *New Zealand Journal of Educational Studies* 29 (1): 175–88.

Black, I. 2006. 'Analysing historical and archival sources'. In N. Clifford and G. Valentine, eds, *Key Methods in Geography*, 477–500. London: Sage.

Blaikie, P., and H. Brookfield. 1987. *Land Degradation and Society*. London: Methuen.

Blake, K. 2001. 'In search of Navajo sacred geography'. *The Geographical Review* 91 (4): 715–24.

Blumen, O. 2002. 'Criss-crossing boundaries: Ultraorthodox Jewish women go to work'. *Gender, Place and Culture* 9 (2): 133–51.

Blunt, A. 2003. 'Home and identity'. In A. Blunt et al., eds, *Cultural Geography in Practice*. Euston: Arnold.

Bogdan, R. 1974. *Being Different: The Autobiography of Jane Frey*. London: Wiley.

Bogdan, R.C., and S.K. Biklen. 1992. *Qualitative Research for Education: An Introduction to Theory and Methods*. 2nd edn. Boston: Allyn and Bacon.

Bondi, L. 1997. 'In whose words? On gender identities, knowledge and writing practices'. *Transactions of the Institute of British Geographers* 22: 245–58.

Bondi, L., J. Davidson, and M. Smith. 2005. 'Introduction: Geography's "emotional turn"'. In J. Davidson, L. Bondi, and M. Smith, eds, *Emotional Geographies*, 1–16. Aldershot: Ashgate.

Bonnet, A. 1996. 'Constructions of "race", place and discipline: Geographies of "racial" identity and racism'. *Ethnic and Racial Studies* 19 (4): 864–83.

Bordo, S. 1986. 'The Cartesian masculinization of thought'. *Signs* 11: 439–56.

Botes, L., and D. van Rensburg. 2000. 'Community participation in development: Nine plagues and twelve commandments'. *Community Development Journal* 35 (1): 53–4.

Bouma, G.D. 1993. *The Research Process*. rev. edn. New York: Oxford University Press.

———. 1996. *The Research Process*. 3rd edn. New York: Oxford University Press.

Bouma, G.D., and R. Ling. 2004. *The Research Process*. 5th edn. New York: Oxford University Press.

Bourdieu, P. 2003. 'Participant observation'. *Journal of the Royal Anthropological Institute* 9: 281–94.

Bowes, A. 1996. 'Evaluating and empowering research strategy: Reflections on action-research with South Asian women'. *Sociological Research Online* 1 (1). http://www.socresonline.org.uk/1/1/1.html#top.

Boyle, M., and R. Rogerson. 2001. 'Power, discourse and city trajectories'. In R. Paddison, ed., *Handbook of Urban Studies*. London: Sage.

Boyle, P. 1997. 'Writing up—Some suggestions'. In R. Flowerdew and D. Martin, eds, *Methods in Human Geography. A Guide to Students Doing a Research Project*. Harlow: Addison Wesley Longman.

Bradford, M. 2003. 'Writing essays, reports and dissertations'. In N.J. Clifford and G. Valentine, eds, *Key Methods in Geography*, 515–32. London: Sage.

Bradshaw, M. 2001. 'Contracts and member checks in qualitative research in human geography: Reason for caution?' *Area* 33 (2): 202–11.

Brannen, J., ed. 1992a. *Mixing Methods: Qualitative and Quantitative Research*. Aldershot, US: Avebury.

———. 1992b. 'Combining qualitative and quantitative approaches: An overview'. In J. Brannen, ed., *Mixing Methods: Qualitative and Quantitative Research*. Aldershot, US: Avebury.

Bridge, G., and R. Dowling. 2001. 'Microgeographies of retailing and gentrification'. *Australian Geographer* 32 (1): 93–107.

Brockington, D., and S. Sullivan. 2003. 'Qualitative research'. In R. Scheyvens and D. Storey, eds, *Development Fieldwork: A Practical Guide*. London: Sage.

Bryant, R., and S. Bailey. 1997. *Third World Political Ecology*. New York: Routledge.

Bryman, A. 1984. 'The debate about quantitative and qualitative research: A question of method or epistemology?' *The British Journal of Sociology* 35: 75–92.

———. 1988. *Quantity and Quality in Social Research*. London: Unwin-Hyman.

———. 2004. *Social Research Methods*. 2nd edn. Oxford: Oxford University Press.

———. 2006. 'Integrating quantitative and qualitative research: How is it done?' *Qualitative Research* 6 (1): 97–113.

———. 2008. *Social Research Methods*. 3rd edn. Oxford: Oxford University Press.

Bryman, A., and R.G. Burgess, eds. 1994. *Analyzing Qualitative Data*. London: Routledge.

Burawoy, M., et al. 1991. *Ethnography Unbound: Power and Resistance in the Modern Metropolis*. Berkeley: University of California Press.

Burgess, J. 1988. 'Exploring environmental values through the medium of small groups: 2. Illustrations of a group at work'. *Environment and Planning A* 20 (4): 457–76.

———. 1996. 'Focusing on fear: The use of focus groups in a project for the Community Forest Unit, Countryside Commission'. *Area* 28 (2): 130–5.

Burgess, J., C. Limb, and C.M. Harrison. 1988. 'Exploring environmental values through the medium of small groups: 1. Theory and practice'. *Environment and Planning A* 20 (3): 309–26.

Burgess, J., and P. Wood. 1988. 'Decoding Docklands: Place advertising and the decision-making strategies of the small firm'. In J. Eyles and D.M. Smith, eds, *Qualitative Methods in Human Geography*. Cambridge: Polity Press.

Burgess, R.G. 1982a. 'Elements of sampling in field research'. In R.G. Burgess, ed., *Field Research: A Sourcebook and Field Manual*. London: George Allen and Unwin.

———. 1982b. 'Multiple strategies in field research'. In R.G. Burgess, ed., *Field Research: A Sourcebook and Field Manual*. London: George Allen and Unwin.

———. 1982c. 'The unstructured interview as a conversation'. In R.G. Burgess, ed., *Field Research: A Sourcebook and Field Manual*. London: George Allen and Unwin.

———, ed. 1996. *Studies in Qualitative Methodology: Computing and Qualitative Research*, v. 5. London: JAI Press.

Burman, E., and I. Parker. 1993. 'Against discursive imperialism, empiricism and constructionism: Thirty-two problems with discourse analysis'. In E. Burman and I. Parker, eds, *Discourse Analytic Research: Repertoires and Readings of Texts in Action*. London: Routledge.

Burton, R.F. 1886. *The Arabian Nights*. London: Benares Edition.

Buston, K. 1997. 'NUD*IST in action: Its use and its usefulness in a study of chronic illness in young people'. *Sociological Research Online* 2 (3). http://www.socresonline.org.uk/welcome.html.

Butler, J. 2007. 'Torture and the ethics of photography'. *Environment and Planning D: Society and Space* 25: 951–66.

Butler, R. 1997. 'Stories and experiments in social inquiry'. *Organisation Studies* 18 (6): 927–48.

Butz, D. 2007. 'Sustained by the guidelines: Reflections on the limitations of standard informed consent procedures for the conduct of ethical research'. *ACME: An International E-journal for Critical Geographies* 7 (2): 239–59.

Butz, D., and K. Besio. 2004. 'The value of autoethnography for field research in transcultural settings'. *Professional Geographer* 56 (3): 350–60.

Cahill, C. 2004. 'Defying gravity: Raising consciousness through collective research'. *Children's Geographies* 2 (2): 273–86.

Cahill, C., and M. Torre. 2007. 'Beyond the journal article: Representations, audience, and the presentation of participatory action research'. In S. Kindon, R. Pain, and M. Kesby, eds, *Participatory Action Research Approaches and Methods: Connecting People, Participation and Place*, 196–205. London: Routledge.

Callon, M. 1986. 'Some elements of a sociology of translation: Domestication of the scallops and the fishermen of St Brieuc Bay'. In J. Law, ed., *Power, Action and Belief: A New Sociology of Knowledge?* London: Routledge.

Cameron, J. 1992. 'Modern-day tales of illegitimacy: Class, gender and ex-nuptial fertility'. (Department of Geography, University of Sydney, MA minor thesis).

Cameron, J., and K. Gibson. 2005. 'Participatory action research in a poststructuralist vein'. *Geoforum* 36 (3): 315–31.

Cameron, L. 2001. 'Oral history and the Freud archives: Incidents, ethics and relations'. *Historical Geography* 29: 38–44.

Campbell, D., and J. Stanley. 1966. *Experimental and Quasi-experimental Designs for Research*. Chicago: Rand McNally.

Carey, M.A. 1994. 'The group effect in focus groups: Planning, implementing and interpreting focus group research'. In J.M. Morse, ed., *Critical Issues in Qualitative Research Methods*. Thousand Oaks, CA: Sage.

Carroll, J., and J. Connell. 2000. '"You gotta love this city": The Whitlams and inner Sydney'. *Australian Geographer* 31 (2): 141–54.

Casey, M.J., et al. 1996. 'Addressing Nonpoint Source Agricultural Pollution in the Minnesota River Basin: Findings from Focus Groups Conducted with Farmers, Agency Staff, Crop Consultants and Researchers, December 1995'. Study conducted for the Minnesota Department of Agriculture. http://www.soils.umn.edu/research/mn-river/doc/fgrptweb.html.

Castree, N. 2005. 'The epistemology of particulars: Human geography, case studies and "context"'. *Geoforum* 36 (5): 541–44.

Cavendish, A.P. 1964. 'Early Greek philosophy'. In D.J. O'Connor, ed., *A Critical History of Western Philosophy*. New York: The Free Press.

Ceglowski, D. 1997. 'That's a good story, but is it really research?' *Qualitative Inquiry* 3 (2): 188–99.

Chakrabarti, R. 2004. 'Constrained spaces of prenatal care: South Asian women in New York City'. (Department of Geography, University of Illinois at Urbana-Champaign, unpublished research proposal).

Chambers, R. 1994. 'The origins and practice of participatory rural appraisal'. *World Development* 22 (7): 953–69.

Charmaz, K. 2000. 'Grounded theory: Objectivist and constructivist methods'. In N.K. Denzin and Y.S. Lincoln, *Handbook of Qualitative Research*, 2nd edn. Thousand Oaks, CA: Sage.

Chatterton, P., D. Fuller, and P. Routledge. 2007. 'Relating action to activism: Theoretical and methodological reflections'. In S. Kindon, R. Pain, and M. Kesby, eds, *Participatory Action Research Approaches and Methods: Connecting People, Participation and Place*, 216–22. London: Routledge.

Chen, P.J., and S.M. Hinton. 1999. 'Realtime interviewing using the world wide web'. *Sociological Research Online* 4 (3).

Chen, S., G. Hall, and M. Johns. 2004. 'Research paparazzi in cyberspace: The voices of the researchers'. In M. Johns, S. Chen, and G. Hall, eds, *Online Social Research: Methods, Issues and Ethics*, 157–75. New York: Peter Lang.

Cho, H., and R. LaRose. 1999. 'Privacy issues in Internet surveys'. *Social Science Computer Review* 17 (4): 421–34.

Clarke, G. 2001. 'From ethnocide to ethnodevelopment? Ethnic minorities and indigenous peoples in Southeast Asia'. *Third World Quarterly* 22 (3): 413–36.

Clifford, N., and G. Valentine, eds. 2003. *Key Methods in Geography*. London: Sage.

———. 2009. *Key Methods in Geography*. 2nd edn. London: Sage.

Cloke, P. 2006. 'Rurality and racialised others: Out of place in the countryside'. In P.J. Cloke, T. Marsden, and P.H. Mooney, eds, *Handbook of Rural Studies*. London: Sage.

Cloke, P., et al. 2000. 'Ethics, reflexivity and research: Encounters with homeless people'. *Ethics, Place and Environment* 3 (2): 133–54.

Cloke, P., et al. 2004a. *Practising Human Geography*. London: Sage.

Cloke, P., et al. 2004b. 'Talking to people'. In *Practising Human Geography*. London: Sage.

Clough, A.R., et al. 2004. 'Emerging patterns of cannabis and other substance use in Aboriginal communities in Arnhem Land, Northern Territory: A study of two communities'. *Drug and Alcohol Review* 23 (4) 381–90.

Coe, N.M., P.F. Kelly, and H.W.C. Yeung. 2007. *Economic Geography*. Malden, MA: Blackwell.

Coffey, A., and P. Atkinson. 1996. *Making Sense of Qualitative Data, Complementary Research Strategies*. Thousand Oaks, CA: Sage.

Collins, D., and R. Kearns 1998. *Avoiding the Log-jam: Exotic Forestry, Transport and Health in Hokianga*. Working Paper no. 8, Department of Geography, University of Auckland.

Colten, C. 2005. *Unnatural Metropolis: Wresting New Orleans from Nature*. Baton Rouge: Louisiana State University Press.

Connell, R.W. 1991. 'Live fast and die young: The construction of masculinity among young working-class men on the margin of the labour market'. *Australia and New Zealand Journal of Sociology* 27 (2): 141–71.

Cook, I. 1997. 'Participant observation'. In R. Flowerdew and D. Martin, eds, *Methods in Human Geography*. Harlow: Addison Wesley Longman.

———. 2000. '"Nothing can ever be the case of 'us' and 'them' again": Exploring the politics of difference through border pedagogy and student journal writing'. *Journal of Geography in Higher Education* 24 (1): 13–27.

Cook, I., et al. 2007. '"It's more than just what it is": Defetishising commodities, expanding fields, mobilising change . . .'. *Geoforum* 38: 1113–26.

Cooke, B. 2001. 'The social psychological limits of participation?' In B. Cooke and U. Kothari, eds, *Participation: The New Tyranny?* London: Zed Books.

Cooke, B., and U. Kothari. 2001. 'The case for participation as tyranny'. In B. Cooke and U. Kothari, eds, *Participation: The New Tyranny?* London: Zed Books.

Coombs, H.C. 1978. *Kulinma: Listening to Aboriginal Australians*. Canberra: Australian National University Press.

Coombs, H.C., et al. 1989. *Land of Promises: Aborigines and Development in the East Kimberley*. Canberra: Centre for Environmental Studies, Australian National University and Aboriginal Studies Press.

Cooper, A. 1994. 'Negotiating dilemmas of landscape, place and Christian commitment in a Suffolk parish'. *Transactions of the Institute of British Geographers* 19: 202–12.

———. 1995. 'Adolescent dilemmas of landscape, place and religious experience in a Suffolk parish'. *Environment and Planning D: Society and Space* 13: 349–63.

Cooper, D., and D. Herman. 1995. 'Getting the family right'. In D. Herman and C. Stychin, eds, *Legal Inversions: Lesbians, Gay Men, and the Politics of Law*, 142–62. Philadelphia: Temple University Press.

Cope, M. 2003. 'Coding qualitative data'. In N. Clifford and G. Valentine, eds, *Key Methods in Geography*, 445–59. Thousand Oaks, CA: Sage.

———. 2008. 'Patchwork neighborhood: Children's real and imagined geographies in Buffalo, NY'. *Environment and Planning A* 40: 2845–63.

Cope, M., and S. Elwood, eds. 2009. *Qualitative GIS: A Mixed-Methods Approach*. London: Sage.

Cope, M., and J.-K. Jung. 2009. 'Qualitative geographic information systems'. In N. Thrift and R. Kitchin, eds, *International Encyclopedia of Human Geography*. Amsterdam: Elsevier Science.

Corbin, J., and A. Strauss. 2007. *Basics of Qualitative Research Techniques and Procedures for Developing Grounded Theory*. 3rd edn. Thousand Oaks, CA: Sage.

Cornwall, A., and R. Jewkes. 1995. 'What is participatory research?' *Social Science and Medicine* 41: 1667–76.

Costello, L., and S. Hodge. 1998. 'Queer/clear/here: Destabilising sexualities and spaces'. In E. Stratford, ed., *Australian Cultural Geographies*, 131–52. Melbourne: Oxford University Press.

Cousins, M., and A. Houssain. 1984. *Michel Foucault*. Basingstoke: MacMillan.

Cox, K.R. 1981. 'Bourgeoise thought and the behavioural geography debate'. In R.K. Cox and R.G. Golledge, eds, *Behavioural Problems in Geography Revisited*. New York: Methuen.

Crang, M. 1997a. 'Analyzing qualitative materials'. In R. Flowerdew and D. Martin, eds, *Methods in Human Geography: A Guide for Students Doing Research*, 183–96. Harlow: Addison Wesley Longman.

———. 1997b. 'Picturing practices: Research through the tourist gaze'. *Progress in Human Geography* 21: 359–73.

———. 2003. 'Qualitative methods: Touchy, feely, look-see?' *Progress in Human Geography* 27 (4): 494–504.

———. 2005a. 'Analysing qualitative materials'. In R. Flowerdew and D. Martin, eds, *Methods in Human Geography: a Guide for Students Doing a Research Project*, 2nd edn, 218–33. Harlow: Pearson Prentice Hall.

———. 2005b, 'Qualitative methods: There is nothing outside the text?' *Progress in Human Geography* 29: (2): 225–33.

Crang, M., et al. 1997. 'Software for qualitative research: 1. Prospectus and overview'. *Environment and Planning A* 29 (5): 771–87.

Crang, M., and I. Cook. 2007. *Doing Ethnographies*. London, Thousand Oaks, CA: Sage.

Crang, P. 1994. 'It's showtime: On the workplace geographies of display in a restaurant in southeast England'. *Environment and Planning D: Society and Space* 12: 675–704.

Cresswell, T. 1999. 'Embodiment, power and the politics of mobility: The case of female tramps and hobos'. *Transactions of the Institute of British Geographers* 24 (2): 175–92.

Crush, J. 1993. 'Post-colonialism, de-colonization, and geography'. In A. Godlewska and N. Smith, eds, *Geography and Empire*. Oxford: Blackwell.

Crystal, D. 2006. *Language and the Internet*. 2nd edn. Cambridge: Cambridge University Press.

Cupples, J., and J. Harrison. 2001. 'Disruptive voices and the boundaries of respectability in Christchurch, New Zealand'. *Gender, Place and Culture* 8 (2): 189–204.

Cupples, J., and S. Kindon. 2003. 'Returning to university and writing the field'. In R. Scheyvens and D. Storey, eds, *Development Fieldwork: A Practical Guide*, 217–31. London: Sage.

Daniels, S., and D. Cosgrove. 1988. 'Introduction: Iconography and landscape'. In D. Cosgrove and S. Daniels, eds, *The Iconography of Landscape: Essays on the Symbols, Representation, Design and Use of Past Environments*. Cambridge: Cambridge University Press.

Davidson, J., L. Bondi, and M. Smith. 2005. *Emotional Geographies*. London: Ashgate.

Davies, C.A. 2008. *Reflexive Ethnography: A Guide to Researching Selves and Others*. 2nd edn. London: Routledge.

Davies, G., and C. Dwyer. 2007. 'Qualitative methods: Are you enchanted or are you alienated?' *Progress in Human Geography* 31 (2): 257–66.

Davis, C.M. 1954. 'Field techniques'. In P.E. James and C.F. Jones, eds, *American Geography: Inventory and Prospect*. Syracuse, NY: Syracuse University Press for the Association of American Geographers.

Davis, Mark. 1997. *Gangland: Cultural Elites and the New Generationalism*. Sydney: Allen and Unwin.

Davis, Mike. 1990. *City of Quartz: Excavating the Future in Los Angeles*. London: Verso.

Davison, G. 2003. 'History and archives in the 21st century'. *Australian History Association Bulletin* 55: 50–7.

Dear, M. 1988. 'The postmodern challenge: Reconstructing human geography'. *Transactions of the Institute of British Geographers* NS 13: 262–74.

de Certeau, M. 1984. *The Practice of Everyday Life*. Berkeley: University of California Press.

Deloria, V., Jr. 1988 [1969]. *Custer Died for Your Sins: An Indian Manifesto*. Norman: University of Oklahoma Press.

DeLyser, D. 2001. '"Do you really live here?" Thoughts on insider research'. *Geographical Review* 91 (1, 2): 441–53.

———. 2009. 'Writing qualitative geography'. In D. DeLyser, et al., eds., *Handbook of Qualitative Geography*. 341–358. London: Sage.

DeLyser, D., et al., eds. 2009. *The Sage Handbook of Qualitative Research in Human Geography*. London: Sage.

DeLyser, D., and P.F. Starr. 2001 'Doing fieldwork: Editors' introduction'. *The Geographical Review* 91 (1–2): iv.

Denzin, N.K. 1978. *The Research Act: A Theoretical Introduction to Sociological Methods*. New York: McGraw-Hill.

———. 2003. 'Reading and writing performance'. *Qualitative Research* 3 (2): 243–68.

———. 2008. 'Emancipatory discourses and the ethics and politics of interpretation'. In N.K. Denzin and Y.S. Lincoln, eds, *Collecting and Interpreting Qualitative Materials Research*, 3rd edn, 435–71. Los Angeles: Sage.

Denzin, N.K., and Y.S. Lincoln. 1994. 'Introduction: Entering the field of qualitative research'. In N.K. Denzin and Y.S. Lincoln, eds, *Handbook of Qualitative Research*. Thousand Oaks, CA: Sage.

———, eds. 2000. *Handbook of Qualitative Research*. 2nd edn. Thousand Oaks, CA: Sage.

———, eds. 2005. *Handbook of Qualitative Analysis*. 3rd edn. Thousand Oaks, CA: Sage.

———, eds. 2008. *Collecting and Interpreting Qualitative Materials*. 3rd edn. Los Angeles: Sage.

Derrida, J. 1976. *Of Grammatology*. tr. G. Spivak. Baltimore and London: Johns Hopkins University Press.

———. 1978. *Writing and Difference*. translated with an introduction and additional notes by A. Bass. Chicago: University of Chicago Press.

———. 1981. *Dissemination*. tr. B. Johnson. Chicago: University of Chicago Press.

de Vaus, D. 2002. *Surveys in Social Research*. 5th edn. Sydney: Allen and Unwin.

Dewalt, K.M., and B.R. Dewalt. 2002. *Participant Observation: A Guide for Fieldworkers*. Lanham, ND: Rowman Altamira.

Dey, I. 1993. *Qualitative Data Analysis: A User-Friendly Guide for Social Scientists*. London: Routledge.

Digital Research Tools (DiRT). n.d. 'Perform Qualitative Data Analysis'. http:// digitalresearchtools.pbwiki.com/Perform-Qualitative-Data-Analysis.

Dillman, D. 2007. *Mail and Internet Surveys: The Total Design Method*. 2nd edn. Hoboken, NJ: John Wiley.

Dillman, D., and D. Bowker. 2001. 'The web questionnaire challenge to survey methodologists'. In U. Reips and M. Bosnjak, eds, *Dimensions of Internet Science*, 159–78. Lengerich, Germany: Pabst Science Publishers.

Dillman, D.A., R.D. Tortora, and D. Bowker. 1998. 'Principles for Constructing Web Surveys'. sesrc (Social and Economic Sciences Research Center) Technical Report 98-50. Pullman, WA: Washington State University. http://survey.sesrc.wsu.edu/dillman/papers/websurveyppr.pdf.

Dissertation Proposal Writing Tutorial. n.d. http://www.people.ku.edu/~ebben/ tutorial_731.htm.

Dixon, D.P., and J.P. Jones III. 1996. 'For a supercalifragilisticexpialidocious scientific geography'. *Annals of the Association of American Geographers* 86: 767–79.

Doel, M.A., and D.B. Clarke. 1999. 'Dark Panopticon. Or, attack of the killer tomatoes'. *Environment and Planning D: Society and Space* 17 (4): 427–50.

Donovan, J. 1988. 'When you're ill, you've gotta carry it'. In J. Eyles and D.M. Smith, eds, *Qualitative Methods in Human Geography*. Cambridge: Polity Press.

Douglas, J.D. 1985. *Creative Interviewing*. Beverly Hills, CA: Sage.

Douglas, L., A. Roberts, and R. Thompson. 1988. *Oral History: A Handbook*. Sydney: Allen and Unwin.

Driver, F. 1988. 'Historicity and human geography'. *Progress in Human Geography* 12 (4) 479–83.

Dryzek, J.S. 2005. *The Politics of the Earth: Environmental Discourses*. 2nd edn. Oxford: Oxford University Press.

Duberman, M., M. Vicinus, and G. Chauncey. 1990. *Hidden from History: Reclaiming The Gay and Lesbian Past*. New York: Meridian Press.

Duinen, I. van, and B. van Hoven. 2003. '"We have to be extra special . . ." Gender gedifferentieerde ervaringen in de werkplaats van de Hollywood filmindustrie'. *Agora* 9 (3): 33–5.

Dumbrava, G., and A. Koronka. 2006. 'Writing for business purposes: Elements of email etiquette'. *Annals of the University of Petrosani, Economics* 6: 61–4.

Duncan, J.S. 1987. 'Review of urban imagery: Urban semiotics'. *Urban Geography* 8 (5): 473–83.

———. 1992. 'Elite landscapes as cultural (re)production: The case of Shaughnessy Heights'. In K. Anderson and F. Gale, eds, *Inventing Places: Studies in Cultural Geography*. Melbourne: Longman Cheshire.

———. 1999. 'Complicity and resistance in the colonial archive: Some issues in the method and theory of historical geography'. *Historical Geography* 27: 119–28.

Duncan, J., and N. Duncan. 1988. '(Re)reading the landscape'. *Environment and Planning D: Society and Space* 6 (2) 117–26.

Dunn, K. 1993. 'The Vietnamese concentration in Cabramatta: Site of avoidance and deprivation, or island of adjustment and participation?' *Australian Geographical Studies* 31 (2): 228–45.

———. 1995. 'The landscape as text metaphor'. In G. Dixon and D. Aitken, eds, *IAG Conference Proceedings 1993*. Melbourne: Monash Publications in Geography.

———. 1997. 'Cultural geography and cultural policy'. *Australian Geographical Studies* 35 (1) 1–11.

———. 2001. 'Representations of Islam in the politics of mosque development in Sydney'. *Tijdschrift voor Economische en Social Geografie* 92 (2): 291–308.

———. 2003. 'New cultural geographies in Australia: The social and spatial constructions of culture and citizenship'. In B.J. Garner, ed., *Geography's New Frontiers*. Gladesville: Geographical Society of New South Wales, Conference papers no. 17, 189–200.

———. 2008. 'Comparative analyses of transnationalism: A geographic contribution to the field'. *Australian Geographer* 39 (1): 1–7

Dunn, K.M., P.M. McGuirk, and H.P.M. Winchester. 1995. 'Place making: The social construction of Newcastle'. *Australian Geographical Studies* 33 (2): 149–66.

Dunn, K.M., and M. Mahtani. 2001. 'Media representations of ethnic minorities'. *Progress in Planning* 55 (3): 163–72.

Dunn, K.M., and S. Roberts. 2003. 'The social construction of an Indo-Chinese–Australian neighbourhood in Sydney: The case of Cabramatta'. In L. Wei, ed., *Enclaves to Ethoburbs*. Baltimore: John Hopkins University Press.

Durack, M. 1986 [1959]. *Kings in Grass Castles*. London: Corgi Books.

Durham, M. 2001. 'The Conservative party, New Labour and the politics of the family'. *Parliamentary Affairs* 54: 689–91.

Dyck, I. 1997. 'Dialogue with difference: A tale of two studies'. In J.P. Jones III, H.L. Nast, and S.M. Roberts, eds, *Thresholds in Feminist Geography: Difference, Methodology, Representation*. Lanham, MD: Rowman and Littlefield.

———. 1999. 'Using qualitative methods in medical geography: Deconstructive moments in a subdiscipline?' *Professional Geographer* 51 (2): 243–53.

———. 2005. 'Feminist geography, the "everyday", and local-global relations: Hidden spaces of place-making'. *The Canadian Geographer* 49 (3): 233–43.

Dyck, I., and R. Kearns. 1995. 'Transforming the relations of research: Towards culturally safe geographies of health and healing'. *Health and Place* 1 (3): 137–47.

Earickson, R., and J. Harlin. 1994. *Geographic Measurement and Quantitative Analysis*. Upper Saddle River, NJ: Prentice Hall.

Eden, S., C. Bear, and G. Walker. 2008. 'Mucky carrots and other proxies: Problematising the knowledge-fix for sustainable and ethical consumption'. *Geoforum* 39: 1044–57.

Edhlund, B. 2007. *NVivo Essentials*. Stockholm: Form and Kunskap (Lulu.com)

Edmonds, M. 1995. 'Why don't we tell them what they want to hear? Native title, politics and the intransigence of ethnography'. In J. Finlayson and D.E. Smith, eds, *Native Title: Emerging Issues for Research, Policy and Practice*, 1–8. Research Monograph no. 10. Canberra: Centre for Aboriginal Economic Policy Research, Australian National University.

Edwards, E. 2003. 'Negotiating spaces'. In J.M. Schartz and J.R. Ryan, eds, *Picturing Place Photography and the Geographical Imagination*. London and New York: I.B. Tauris.

Edwards, J.A., and M.D. Lampert, eds. 1993. *Talking Data: Transcription and Coding in Discourse Research*. Hillsdale, NJ: Lawrence Erlbaum Associates.

Eichler, M. 1988. *Nonsexist Research Methods: A Practical Guide*. Boston: Allen and Unwin.

Eliott, J.A., and I.N. Olver. 2002. 'The discursive properties of "hope": A qualitative analysis of cancer patients' speech'. *Qualitative Health Research* 12 (20): 173–93.

Ellis, C. 2004. *The Ethnographic I: A Methodological Novel about Autoethnography*. Walnut Creek, CA: AltaMira Press.

Elwood, S., et al. 2007. 'Participatory GIS: The Humboldt/West Humboldt Park Community GIS Project, Chicago, USA'. In S. Kindon, R. Pain, and M. Kesby, eds, *Participatory Action Research Approaches and Methods: Connecting People, Participation and Place*, 170–8. London: Routledge.

Ely, M. 2007. 'In-forming re-presentations'. In D.J. Clandinin, ed., *Handbook of Narrative Inquiry, Mapping a Methodology*, 567–98. Thousand Oaks, CA, and London: Sage.

England, K.V.L. 1993. 'Suburban pink collar ghettos: The spatial entrapment of women'. *Annals of the Association of American Geographers* 83 (2): 225–42.

———. 1994. 'Getting personal: Reflexivity, positionality, and feminist research'. *Professional Geographer* 46: 80–9.

Eustace, K. 1998. 'Ethnographic study of a virtual learning community'. *Internet Research: Electronic Networking Applications and Policy* 8 (1): 83–4.

Evans, B. 2006. '"I'd feel ashamed": Girls' bodies and sports participation'. *Gender, Place and Culture* 13 (5): 547–61.

Evans, M. 1988. 'Participant observation: The researcher as research tool'. In J. Eyles and D.M. Smith, eds, *Qualitative Methods in Human Geography*. Cambridge: Polity Press.

Evans, M., R. Hole, L.D. Berg, P. Hutchinson, and D. Sookraj, with the Okanagan Urban Aboriginal Health Research Collective. 2009 (in press). 'Common insights, differing methodologies: Towards a fusion of indigenous methodologies, participatory action research, and white studies in an urban Aboriginal research agenda'. *Qualitative Inquiry* 15 (7).

Evans, V., and J. Sternberg. 1999. 'Young people, politics and television current affairs in Australia'. *Journal of Australian Studies* December: 103–9.

Eyles, J. 1988. 'Interpreting the geographical world: Qualitative approaches in geographical research'. In J. Eyles and D.M. Smith, eds, *Qualitative Methods in Human Geography*. Cambridge: Polity Press.

Eyles, J., and D.M. Smith, eds. 1988. *Qualitative Methods in Human Geography*. Cambridge: Polity Press.

Fairclough, N. 1992. *Discourse and Social Change*. Cambridge: Polity Press.

———. 2003. *Analysing Discourse: Textual Analysis for Social Research*. New York and London: Routledge.

Fals-Borda, O., and M. Anisur Rahman. 1991. *Action and Knowledge: Breaking the Monopoly with Participatory Action Research*. London: Intermediate Technology Publications Ltd.

Fanon, F. 1967 [1963]. *The Wretched of the Earth*. tr. C. Farrington. Harmondsworth and London: Penguin Books.

Feinsilver, J.M. 1993. *Healing the Masses: Cuban Health Politics at Home and Abroad*. Berkeley: University of California Press.

Feitelson, E. 1991. 'The potential of mail surveys in geography: Some empirical evidence'. *Professional Geographer* 43: 190–205.

Fielding, N. 1994. 'Getting into computer-aided qualitative data analysis'. ESRC Data Archive Bulletin no. 57. http://kennedy.soc.surrey.ac.uk/caqdas/getting.htm.

———. 1995. 'Choosing the right software program'. ESRC Data Archive Bulletin no. 58. http://kennedy.soc.surrey.ac.uk/caqdas/choose.htm.

———. 1999. 'The norm and the text: Denzin and Lincoln's handbooks of qualitative method'. *British Journal of Sociology* 50 (3) 525–34.

Fielding, N., and R. Lee. 1998. *Computer Analysis and Qualitative Research*. London: Sage.

Fincher, R., and R. Panelli. 2001 'Making space: Women's urban and rural activism and the Australian state'. *Gender, Place and Culture* 8 (2): 129–48.

———. 2002. 'New patterns in the adoption and use of qualitative software'. *Field Methods* 14 (2): 197–216.

Findlay, A.M., and F.L.N. Li. 1997. 'An auto-biographical approach to understanding migration: The case of Hong Kong emigrants'. *Area* 29 (1): 34–44.

Finer, C. 2000. 'Researching in a contemporary archive'. *Social Policy and Administration* 34 (4): 434–47.

Fink, A., and J. Kosecoff. 1999. *How to Conduct Surveys: A Step-by-Step Guide*. 2nd edn. Beverly Hills, CA: Sage.

Fish, S. 1980. *Is There a Text in This Class? The Authority of Interpretive Communities*. London: Harvard University Press.

Fisher, J.F. 1990. *Sherpas: Reflections on Change in Himalayan Nepal*. Berkeley: University of California Press.

Flick, U. 1992. 'Triangulation revisited: Strategy of validation or alternative?' *Journal for the Theory of Social Behaviour* 2 (2): 175–97.

Flint, C. 2001. 'Right-wing resistance to the process of American hegemony: The changing political geography of nativism in Pennsylvania, 1920–1998'. *Political Geography* 20 (6): 763–86.

Flowerdew, R., and D. Martin, eds. 1997. *Methods in Human Geography: A Guide for Students Doing a Research Project*. Harlow: Pearson.

———. 2005. *Methods in Human Geography: A Guide for Students Doing a Research Project*. 2nd edn. Harlow: Pearson.

Flyvberg, B. 1998. *Rationality and Power: Democracy in Practice*. tr. S. Sampson. Chicago: University of Chicago Press.

———. 2006. 'Five misunderstandings about case-study research'. *Qualitative Inquiry* 12 (2): 219–45.

Foddy, W. 1993. *Constructing Questions for Interviews and Questionnaires*. Cambridge: Cambridge University Press.

Forbes, D.K. 1993. 'Multiculturalism, the Asian connection and Canberra's urban imagery'. In G. Clark, D. Forbes, and R. Francis, eds, *Multiculturalism, Difference and Postmodernism*. Melbourne: Longman Cheshire.

———. 1999. 'Globalisation, postcolonialism and new representations of the Pacific Asian metropolis'. In P. Dicken et al., eds, *The Logic(s) of Globalisation*. London: Routledge.

Forer, P., and L. Chalmers. 1987. 'Geography and information technology: Issues and impacts'. In P.G. Holland and W.B. Johnston, eds, *Southern Approaches: Geography in New Zealand*. Christchurch: New Zealand Geographical Society.

Forster, E.M., and O. Stallybrass. 2005. *Aspects of the Novel*. Harmondsworth: Penguin Classics.

Foucault, M. 1972. *The Archaeology of Knowledge*. tr. A. Sheridan Smith. New York: Pantheon.

———. 1977a. *Discipline and Punish: The Birth of the Prison*. New York: Pantheon.

———. 1977b. *Language, Counter-memory, Practice: Selected Essays and Interviews.* Ithaca: Cornell University Press.

———. 1978. *The History of Sexuality: An Introduction*, v. 1. Harmondsworth: Penguin.

———. 1980. *Power/Knowledge.* Brighton: Harvester.

———. 1981. 'The order of discourse'. In R. Young, ed., *Untying the Text: A Poststructuralist Reader*, 48–78. London: RKP.

———. 1982. 'The subject and power'. In H. Dreyfus and P. Rabinow, eds, *Beyond Structuralism and Hermeneutics.* Brighton: Harvester.

Fowler, F. 2002. *Social Survey Methods.* Thousand Oaks, CA: Sage.

Fowler, R. 1991. *Language in the News: Discourse and Ideology in the Press.* London: Routledge.

Frankenberg, R., and L. Mani. 1993. 'Crosscurrents, crosstalk: Race, "postcoloniality" and the politics of location'. *Cultural Studies* 7: 292–310.

Frankfort-Nachmaias, C., and D. Nachmaias. 1992. *Research Methods in the Social Sciences.* 4th edn. London: Edward Arnold.

Freeman, M. 2007. 'Auto-biographical understanding and narrative inquiry'. In D.J. Clandinin, ed., *Handbook of Narrative Inquiry, Mapping a Methodology*, 120–45. Thousand Oaks, CA, and London: Sage.

Freire, P. 1972a. *Cultural Action for Freedom.* Harmondsworth: Penguin.

———. 1972b. *Pedagogy of the Oppressed.* Harmondsworth: Penguin.

———. 1976. *Education: The Practice of Freedom.* London: Writers and Readers Publishing Cooperative.

Furer-Haimendorf, C. von. 1964. *The Sherpas of Nepal: Buddhist Highlanders.* London: J. Murray.

———. 1975. *Himalayan Traders: Life in Highland Nepal.* London: J. Murray.

———. 1984. *The Sherpas Transformed: Social Change in a Buddhist Society of Nepal.* New Delhi: Sterling Publishers.

Fyfe, N.R. 1992. 'Observations on observation'. *Journal of Geography in Higher Education* 16 (2): 127–33.

Fyfe, N.R., and J. Banister. 1996. 'City watching: Closed circuit television surveillance in public spaces'. *Area* 29 (1): 37–46.

Gade, D.W. 2001. 'The languages of foreign fieldwork'. *The Geographical Review* 1–2: 370–9.

Gahan, C., and M. Hannibal. 1998. *Doing Qualitative Research Using QSR NUD*IST.* London: Sage.

Gale, G.F. 1964. *A Study of Assimilation.* Adelaide: Libraries Board of South Australia.

Gale, S.J. 1996. *75 years: The Anniversary of University Geography in Australia.* Sydney: Department of Geography, University of Sydney.

Gardner, R., H. Neville, and J. Snell. 1983. 'Vietnamese settlement in Springvale'. Environmental Report no. 14. Melbourne: Monash University Graduate School of Environmental Science.

Garlick, S. 2002. 'Revealing the unseen: Tourism, art and photography'. *Cultural Studies* 16 (2): 289–305.

Geertz, C. 1973. *The Interpretation of Culture: Selected Essays.* New York: Basic Books.

———. 1980. 'Blurred genres'. *American Scholar* 49: 165–79.

———. 1984. 'Anti-anti-relativism'. *American Anthropologist* 86: 263–77.

Geographical Review. 2001. Special issue on 'doing fieldwork'. 91 (1–2).

George, A., and A. Bennett. 2005. *Case Studies and Theory Development in the Social Sciences.* Cambridge, MA: MIT Press.

George, K. 1999a. *A Place of Their Own: The Men and Women of War Service Land Settlement at Loxton after the Second World War.* Adelaide: Wakefield Press.

———. 1999b. *City Memory: A Guide and Index to the City of Adelaide Oral History Collection.* Adelaide: Corporation of the City of Adelaide.

Gerring, J. 2004. 'What is a case study and what is it good for?' *American Political Science Review* 98 (2): 341–54.

Gibbs, G.R. 2002. *Qualitative Data Analysis: Explorations with NVivo.* Buckingham: Open University Press.

Gibson, C. 2003. 'Digital divides in New South Wales: A research note on social–spatial inequality using 2001 census data on computer and Internet technology'. *Australian Geographer* 34: 239–57.

Gibson, K., J. Cameron, and A. Veno. 1999. *Negotiating Restructuring and Sustainability: A Study of Communities Experiencing Rapid Social Change.* Australian Housing and Urban Research Institute (AHURI) Working Paper. Melbourne: AHURI, Melbourne. Also available: http://www.communityeconomies.org/papers/comecon/comeconp4.pdf.

Gibson-Graham, J.K. 1994. '"Stuffed if I know!": Reflections on post-modern feminist social research'. *Gender, Place and Culture* 1 (2): 205–24.

Gilbert, M. 1994. 'The politics of location: Doing feminist research at "home"'. *Professional Geographer* 46: 90–6.

Gilbert, N., ed. 2001. *Researching Social Life*. 2nd edn. Thousand Oaks, CA: Sage.

Gill, R. 1996. 'Discourse analysis: Practical implementations'. In J.T.E. Richardson, ed., *Handbook of Qualitative Methods for Psychology and the Social Sciences*, 141–56. Leicester: British Psychological Society.

Gillham, B. 2000. *Developing a Questionnaire*. London: Continuum.

Glaser, B.G., and A.L. Strauss. 1967. *The Discovery of Grounded Theory: Strategies for Qualitative Research*. Chicago: Aldine.

Glesne, C., and A. Peshkin. 1992. *Becoming Qualitative Researchers: An Introduction*. New York: Longman.

Golafshani, N. 2003. 'Understanding reliability and validity in qualitative research'. *The Qualitative Report* 8 (4): 597–607.

Gold, J.R. 1980. *An Introduction to Behavioural Geography*. Oxford: Oxford University Press.

Gold, R.L. 1958. 'Roles in sociological field observation'. *Social Forces* 36: 219–25.

Golledge, R.G. 1997. 'On reassembling one's life: Overcoming disability in the academic environment'. *Environment and Planning D: Society and Space* 15 (4): 391–409.

———. 2005. 'Reflections on procedures for learning environments without the use of sight'. *Journal of Geography* 104 (3): 95–103.

Goss, J. 1988. 'The built environment and social theory: Towards an architectural geography'. *Professional Geographer* 40 (4): 392–403.

Goss, J.D. 1993. 'Placing the market and marketing place: Tourist advertising of the Hawaiian Islands, 1972–1992'. *Environment and Planning D: Society and Space* 11: 663–88.

Goss, J.D., and T.R. Leinbach. 1996. 'Focus groups as alternative research practice: Experience with transmigrants in Indonesia'. *Area* 28 (2): 115–23.

Gough, K. 1968. 'Anthropology and imperialism'. *Monthly Review* 19 (11): 12–27.

Gould, P. 1988. 'Expose yourself to geographic research'. In J. Eyles, ed., *Research in Human Geography*. Oxford: Blackwell.

Grbich, C. 1999. *Qualitative Research in Health: An Introduction*. London: Sage.

Green, J., M. Frauquiz, and C. Dixon. 1997. 'The myth of the objective transcription: Transcribing as a situated act'. *TESOL Quarterly* 31 (1): 172–6.

Gregory, D. 1978. *Ideology, Science and Human Geography*. London: Hutchinson.

———. 1994. 'Paradigm'. In R.J. Johnston, D. Gregory, and D.M. Smith, eds, *The Dictionary of Human Geography*. Oxford: Blackwell.

Gregory, S. 1998. 'Consuming the cool: Children's popular consumption culture' (Department of Geography, University of Auckland, MA thesis).

Griffith, D., J. Desloges, and C. Amrhein. 1990. *Statistical Analysis for Geographers*. Englewood Cliffs, NJ: Prentice Hall Engineering, Science and Mathematics.

Groves, R.M. 1990. 'Theories and methods of telephone surveys'. *Annual Review of Sociology* 16: 221–40.

Guelke, L. 1978. 'Geography and logical positivism'. In D. Herbert and R.J. Johnston, eds, *Geography and the Urban Environment*, v. 1. New York: Wiley.

Gupta, A., and J. Ferguson. 1997. *Anthropological Locations: Boundaries and Grounds of a Field Science*. Berkeley: University of California Press.

Hackel, S., and A. Reid. 2007. 'Transforming an eighteenth-century archive into a twenty-first century database: The Early Californian Population Project'. *History Compass* 5 (3): 1013–25.

Hahn, C. 2008. *Doing Qualitative Research Using Your Computer: A Practical Guide*. Thousand Oaks, CA: Sage.

Hall, B. 2005. 'In from the cold? Reflections on participatory research from 1970–2005'. *Convergence* 38 (1): 5–24.

Hall, C. 1982. 'Private archives as sources for historical geography'. In A.R.H. Baker and M. Billinge, eds, *Period and Place: Research Methods in Historical Geography*, 274–8. Cambridge: Cambridge University Press.

Hall, S. 1980. 'Encoding/decoding'. In Centre for Contemporary Cultural Studies, *Culture, Media, Language: Working Papers in Cultural Studies*, 123–38. London: Hutchinson.

———, ed. 1997. *Representation: Cultural Representations and Signifying Practices*. London: Sage.

Hallowell, N., P. Lawton, and S. Gregory. 2005. *Reflections on Research: The Realities of Doing Research in the Social Sciences*. Berkshire: Open University Press.

Hammersley, M. 1992. 'Deconstructing the qualitative–quantitative divide'. In J. Brannen, ed., *Mixing Qualitative and Quantitative Research*. Aldershot, US: Avebury.

Hammersley, M., and P. Atkinson. 1983. *Ethnography: Principles in Practice*. London: Tavistock.

Hanlon, J. 2001. 'Spaces of interpretation: Archival research and the cultural landscape'. *Historical Geography* 29: 14–25.

Haraway, D. 1991. 'Situated knowledges: The science question in feminism and the privilege of partial perspective'. In D. Haraway, ed., *Simians, Cyborgs and Women: The Reinvention of Nature*. London: Routledge.

Harley, J.B. 1992. 'Deconstructing the map'. In T. Barnes and J. Duncan, eds, *Writing Worlds: Discourse, Text and Metaphor in the Representation of Landscape*. London: Routledge.

Harper, D. 2002. 'Talking about pictures: A case for photo elicitation'. *Visual Studies* 17: 13–26.

———. 2003. 'Framing photographic ethnography: A case study'. *Ethnography* 4: 241–66.

Harris, C. 1997. *The Resettlement of British Columbia: Essays on Colonialism and Geographical Change*. Vancouver: University of British Columbia Press.

———. 2001. 'Archival fieldwork'. *Geographical Review* 91 (1–2): 328–34.

———. 2003. *Making Native Space, Colonialism, Resistance, and Reserves in British Columbia*. Vancouver: University of British Columbia Press.

Harrison, C.M., and J. Burgess. 1994. 'Social constructions of nature: A case study of conflicts over the development of Rainham marshes'. *Transactions of the Institute of British Geographers* 19 (3): 291–310.

Harrison, R.T., and D.N. Livingstone. 1980. 'Philosophy and problems in human geography: A pre-suppositional approach'. *Area* 12: 25–30.

Hartig, K.V., and K.M. Dunn. 1998. 'Roadside memorials: Interpreting new deathscapes in Newcastle, New South Wales'. *Australian Geographical Studies* 36 (1): 5–20.

Harvey, D. 1984. 'On the history and present condition of geography: An historical materialist manifesto'. *Professional Geographer* 36: 1–11.

Harvey, K. 2006. 'From bags and boxes to searchable digital collections at the Dalhousie University Archives'. *Journal of Canadian Studies* 40 (2): 120–38.

Hassini, E. 2006. 'Student–instructor communication: The role of email'. *Computers and Education* 47: 29–40.

Hawley, P. 2003. *Being Bright Is Not Enough*. 2nd edn. Springfield, IL: Charles C. Thomas.

Hay, I. 1995. 'The strange case of Dr Jekyll in Hyde Park: Fear, media and the conduct of an emancipatory geography'. *Australian Geographical Studies* 33 (2): 257–71.

———. 1996. *Communicating in Geography and the Environmental Sciences*. Melbourne: Oxford University Press.

———. 1998. 'Making moral imaginations: Research ethics, pedagogy, and professional human geography'. *Ethics, Place and Environment* 1 (1) 55–75.

———. 2003a. 'Ethical practice in geographical research'. In G. Valentine and N. Clifford, eds, *Key Methods in Geography* London: Sage.

———. 2003b. 'From "millennium" to "profiles": Geography's oral histories across the Tasman'. *Proceedings of the 22nd Conference of the New Zealand Geographical Society*, 6–11 July, University of Auckland, New Zealand, 5–6.

———. 2006. *Communicating in Geography and the Environmental Sciences*. 3rd edn. Melbourne: Oxford University Press.

Hay, I., and D. Bass. 2002. 'Making news in geography and environmental management'. *Journal of Geography in Higher Education* 26 (1): 129–42.

Hay, I., D. Bochner, and C. Dungey. 2006. *Making the Grade. A Guide to Successful Communication and Study*. 3rd edn. Melbourne: Oxford University Press.

Hay, I., A. Hughes, and M. Tutton. 2004. 'Monuments, memory and marginalisation in Adelaide's Prince Henry Gardens'. *Geografiska Annaler B* 86 (3): 200–15.

Hay, I., and M. Israel. 2001. '"Newsmaking geography": communicating geography through the media'. *Applied Geography* 21 (2): '107–25.

Head, L. 2000. *Second Nature: The History and Implications of Australia as Aboriginal Landscape*. New York: Syracuse University Press.

Head, L., and P. Muir. 2006. 'Suburban life and the boundaries of nature: Resilience and rupture in Australian backyards'. *Transactions of the Institute of British Geographers* 31 (4): 505–24

Heath, A.W. 1997. 'The proposal in qualitative research'. *The Qualitative Report* 3 (1). http://www.nova.edu/ssss/QR/QR3-1/heath.html.

Heathcote, R.L. 1975. *Australia*. London: Longman.

Heidegger, M. 1996 [1927]. *Being and Time*. tr. J. Stambaugh. Albany: State University of New York Press.

Henry, M., and L.D. Berg. 2006. 'Geographers performing nationalism and hetero-masculinity'. *Gender, Place and Culture* 13 (6): 629–45.

Herlihy, P.H. 2003. 'Participatory research mapping of indigenous lands in Darien, Panama'. *Human Organization* 62 (4): 315–31.

Herlihy, P.H., and G. Knapp. 2003. 'Maps of, by, and for the peoples of Latin America'. *Human Organization* 62 (4): 303–14.

Herman, R.D.K. 1999. 'The Aloha State: Place names and the anti-conquest of Hawai'i'. *Annals of the Association of American Geographers* 89 (1): 76–102.

Herman, T., and D. Mattingly. 1999. 'Community, justice, and the ethics of research: Negotiating reciprocal research relations'. In J.D. Proctor and D.M. Smith, eds, *Geography and Ethics: Journeys in a Moral Terrain*. London: Routledge.

Herod, A. 1993. 'Gender issues in the use of interviewing as a research method'. *Professional Geographer* 45 (3): 305–17.

———. 1999. 'Reflections on interviewing foreign elites: Praxis, positionality, validity and the cult of the insider'. *Geoforum* 30: 313–27.

Hesse-Biber, S., and P. Leavy. 2004. *Approaches to Qualitative Research: A Reader on Theory and Practice*. Oxford: Oxford University Press.

Hessey, R. 1997. 'I, Caesar'. *Sydney Morning Herald*, Metro Section, 14–20 February.

Hewson, C., et al. 2003. *Internet Research Methods*. London: Sage.

Hibbard, M., M.B. Lane, and K. Rasmussen. 2008. 'The split personality of planning: Indigenous peoples and planning for land and resource management'. *Journal of Planning Literature* 23 (2): 136–51.

Hickey, S., and G. Mohan, eds. 2004. *Participation: From Tyranny to Transformation*. London: Zed Books.

Hiebert, W., and D. Swan. 1999. 'Positively fit: A case study in community development and the role of participatory research'. *Community Development Journal* 34: 356–64.

Hinchliffe, S., et al. 'Software for qualitative research: 2. Some thoughts on "aiding" analysis'. *Environment and Planning A* 29 (5): 1109–24.

Hirst, P.A. 1979. *On Law and Ideology*. London: MacMillan.

Hoad, N. 2000. 'Arrested development or the queerness of savages: Resisting evolutionary narrative of difference'. *Postcolonial Studies* 3 (2): 133–58.

Hodge, R., and G. Kress. 1980. *Social Semiotics*. Cambridge: Polity Press.

Hoggart, K., L. Lees, and A. Davies. 2002. *Researching Human Geography*. London: Arnold.

Holbrook, B., and P. Jackson. 1996. 'Shopping around: Focus group research in North London'. *Area* 28 (2): 136–42.

Holland, P. 1991. 'Poetry and landscape in New Zealand'. *New Zealand Journal of Geography* 92: 8–9.

Holland, P.G., and R.P. Hargreaves. 1991. 'The trivial round, the common task: Work and leisure on a Canterbury Hill Country run in the 1860s and 1870s'. *New Zealand Geographer* 47 (1): 19–25.

Holland, P., E. Pawson, and T. Shatford. 1991. 'Qualitative resources in geography'. *New Zealand Journal of Geography* 92 (special issue).

Holliday, A. 2007. *Doing and Writing Qualitative Research*. 2nd edn. Thousand Oaks, CA, and London: Sage.

Holt-Jensen, A. 1988. 'Multiple meanings: Shopping and the cultural politics of identity'. *Environment and Planning A* 27: 1913–30.

Hook, D. 2001. 'Discourse, knowledge, materiality, history: Foucault and discourse analysis'. *Theory and Psychology* 11 (4): 521–47.

hooks, b. 1990. *Yearning: Race, Gender, and Cultural Politics*. Boston: South End Press.

Hoppe, M.J., et al. 1995. 'Using focus groups to discuss sensitive topics with children'. *Evaluation Review* 19 (1): 102–14.

Horowitz, I.L. 1967. 'The rise and fall of Project Camelot'. In I. Horowitz, *The Rise and Fall of Project Camelot: Studies in the Relationship between Science and Practical Politics*. Cambridge, MA: MIT Press.

Horvath, R.J. 1971. 'The Detroit Geographical Expedition and Institute experience'. *Antipode* 3 (1): 73–85.

Howitt, R. 1992. 'The political relevance of locality studies: A remote Antipodean viewpoint'. *Area* 24 (1): 73–81.

———. 1997. 'Getting the scale right: The geopolitics of regional agreements'. *Northern Analyst* 2: 15–17.

———. 2001. *Rethinking Resource Management: Justice, Sustainability and Indigenous Peoples*. London: Routledge.

———. 2002a. 'Worlds turned upside down: Inclusionary research in Australia'. *Proceedings of the Association of American Geographers Conference*, Los Angeles, March. http://www.es.mq.edu.au/~rhowitt/AAG2002.panel.htm.

————. 2002b. 'Decolonizing research: Ethical and methodological issues'. *Proceedings of the Institute of Australian Geographers Conference*, Australian National University, Canberra.

————. 2002c. 'Scale and the other: Levinas and geography'. *Geoforum* 33: 299–313.

————. 2003. 'Scale'. In J. Agnew, K. Mitchell, and G. Toal, eds, *A Companion to Political Geography*. Oxford: Blackwell.

Howitt, R., G. Crough, and B. Pritchard. 1990. 'Participation, power and social research in Central Australia'. *Australian Aboriginal Studies* 1: 2–10.

Howitt, R., and S. Jackson. 1998. 'Some things do change: Indigenous rights, geographers and geography in Australia'. *Australian Geographer* 29 (2): 155–73.

Howitt, R., and S. Suchet-Pearson. 2003. 'Ontological pluralism in contested cultural landscapes'. In K. Anderson, M. Domosh, S. Pile, and N. Thrift, eds, *Handbook of Cultural Geography*. London: Sage.

Huggan, G. 1995. 'Decolonizing the map'. In B. Ashcroft, G. Griffiths, and H. Tiffin, eds, *The Post-colonial Studies Reader*. London: Routledge.

Hughes, A. 1999. 'Constructing economic geographies from corporate interviews: Insights from a cross-country comparison of retailer–supplier relationships'. *Geoforum* 30: 363–74.

Hume-Cook, G., et al. 2007. 'Uniting people with place using participatory video in Aotearoa/New Zealand'. In S. Kindon, R. Pain, and M. Kesby, eds, *Participatory Action Research Approaches and Methods: Connecting People, Participation and Place*, 160–9. London: Routledge.

Hutchins, M. 1993. *Talking History: A Short Guide to Oral History*. Wellington: Bridget Williams Books and Historical Branch of Internal Affairs.

Imrie, R., and C. Edwards. 2007. 'The geographies of disability: Reflections on the development of a sub-discipline'. *Geography Compass* 1 (3): 623–40.

Institute of International Studies, University of California, Berkeley. 2001. 'Dissertation Proposal Workshop'. http://globetrotter.berkeley.edu/DissPropWorkshop

Isaacman, A., P. Lalu, and T. Nygren. 2005. 'Digitization, history, and the making of a postcolonial archive of southern Africa liberation struggles: The Aluka project'. *Africa Today* 52 (2): 54–77.

Israel, M., and I. Hay. 2006. *Research Ethics for Social Scientists: Between Ethical Conduct and Regulatory Compliance*. London: Sage.

Ivanitz, M. 1999. 'Culture, ethics and participatory methodology in cross-cultural research'. *Australian Aboriginal Studies* 2: 46–58.

Jackson, J. 1990. '"I am a field note": Fieldnotes as a symbol of professional identity'. In R. Sanjek, ed., *Fieldnotes: The Makings of Anthropology*. Ithaca, NY: Cornell University Press.

Jackson, P. 1983. 'Principles and problems of participant observation'. *Geografiska Annaler* 65B: 39–46.

————. 1993. 'Changing ourselves: A geography of position'. In R.J. Johnston, ed., *The Challenge for Geography*. Oxford: Basil Blackwell.

————. 1994. 'Black male: Advertising and the cultural politics of masculinity'. *Gender, Place and Culture* 1 (1): 49–59.

————. 1999. 'Constructions of culture, representations of race: Edward Curtis' "way of seeing"'. In K. Anderson and F. Gale, eds, *Cultural Geographies*. Melbourne: Longman.

————. 2001. 'Making sense of qualitative data'. In M. Limb and C. Dwyer, eds, *Qualitative Methodologies for Geographers*. London: Oxford University Press.

Jackson, P., and B. Holbrook. 1995. 'Multiple meanings: Shopping and the cultural politics of identity'. *Environment and Planning A* 27: 1913–30.

Jackson, P., and J. Penrose, eds. 1993. *Constructions of Race, Place and Nation*. London: University College Press.

Jackson-Lears, T. 1985. 'The concept of cultural hegemony: Problems and possibilities'. *American Historical Review* 90: 567–93.

Jacobs, J.M. 1992. 'Culture of the past and urban transformation: The Spitalfield Market redevelopment in East London'. In K.J. Anderson and F. Gale, eds, *Inventing Places: Studies in Cultural Geography*. Melbourne: Longman Cheshire.

————. 1993. '"Shake 'im this country": The mapping of the Aboriginal sacred in Australia—The case of Coronation Hill'. In P. Jackson and J. Penrose, eds, *Constructions of Race, Place and Nation*. London: University College Press.

————. 1996. *Edge of Empire: Postcolonialism and the City*. London: Routledge.

————. 1999. 'The labour of cultural geography'. In E. Stratford, ed., *Australian Cultural Geographies*. Melbourne: Oxford University Press.

James, N., and H. Busher. 2006. 'Credibility, authenticity and voice: Dilemmas in online interviewing'. *Qualitative Research* 6 (3): 403–20.

Jarrett, R.L. 1994. 'Living poor: Family life among single parent African American women'. *Social Problems* 41 (1): 30–49.

Jay, N. 1981. 'Gender and dichotomy'. *Feminist Studies* 7: 38–56.

Jiang, H. 2003. 'Stories remote sensing images can tell: Integrating remote sensing analysis with ethnographic research in the study of cultural landscapes'. *Human Ecology* 31: 215–32.

Jick, T.D. 1979. 'Mixing qualitative and quantitative methods: Triangulation in action'. *Administrative Science Quarterly* 24 (December): 602–11.

Johnsen, S., J. May, and P. Cloke. 2008. 'Imag(in)ing "homeless places": Using auto-photography to (re) examine the geographies of homelessness'. *Area* 40 (2): 194–207.

Johnson, A. 1996. '"It's good to talk": The focus group and the sociological imagination'. *The Sociological Review* 44 (3): 517–38.

Johnson, J.T., et al. 2007. 'Creating anti-colonial geographies: Embracing indigenous peoples' knowledges and rights'. *Geographical Research* 45 (2): 117–20.

Johnson, J.T., R.P. Louis, and A. Pramono. 2005. 'Facing future: Encouraging critical cartographic literacies in indigenous communities'. ACME: *An International E-Journal for Critical Geographies* 4 (1): 80–98.

Johnson, J.T., and B. Murton. 2007. 'Replacing native science: Indigenous voices in contemporary constructions of nature'. *Geographical Research* 45 (2): 121–29.

Johnson, R.B. 1997. 'Examining the validity structure of qualitative research'. *Education* 118 (2): 282–90.

Johnston, R.J. 1978. *Multivariate Statistical Analysis in Geography: A Primer on the General Linear Model.* New York: Longman.

———. 1979. *Geography and Geographers: Anglo-American Human Geography Since 1945.* London: Edward Arnold.

———. 1983. *Philosophy and Human Geography: An Introduction to Contemporary Approaches.* London: Edward Arnold.

———. 2000. 'On disciplinary history and textbooks: Or where has spatial analysis gone?' *Australian Geographical Studies* 38 (2): 125–37.

Johnston, R.J., et al., eds. 2000. *The Dictionary of Human Geography.* Oxford: Blackwell.

Johnston, L. 2001. '(Other) bodies and tourism studies'. *Annals of Tourism Research* 28 (1): 180–201.

Jones, A. 1992. 'Writing feminist educational research: Am "I" in the text?'. In S. Middleton and A. Jones, eds, *Women and Education in Aotearoa.* Wellington: Bridget Williams Books.

Jones, J.P. III, H.J. Nast, and S.M. Roberts, eds. 1997. *Thresholds in Feminist Geography: Difference, Methodology, Representation.* Lanham, MD: Rowman and Littlefield.

Jorgensen, J. 1971. 'On ethics and anthropology'. *Current Anthropology* 12 (3): 321–56.

Judd, C.M., E.R. Smith, and L.H. Kidder. 1991. *Research Methods in Social Relations.* 6th edn. Sydney: Holt, Rinehart and Winston.

Jung, J.-K. 2009. 'Computer-aided qualitative GIS: A software-level integration of qualitative research and GIS. In M. Cope and S. Elwood, eds, *Qualitative GIS: A Mixed-Methods Approach*, 115–35. London: Sage.

Katz, C. 1994. 'Playing the field: Questions of fieldwork in geography'. *Professional Geographer* 46 (1): 67–72.

Kearns, R. 1987. 'In the shadow of illness: A social geography of the chronically mentally disabled in Hamilton, Ontario'. (Department of Geography, McMaster University, PhD dissertation).

———. 1991a. 'Talking and listening: Avenues to geographical understanding'. *New Zealand Journal of Geography* 92: 2–3.

———. 1991b. 'The place of health in the health of place: The case of the Hokianga special medical area'. *Social Science and Medicine* 33: 519–30.

———. 1997. 'Constructing (bi)cultural geographies: Research on, and with, people of the Hokianga District'. *New Zealand Geographer* 52: 3–8.

Kearns, R.A., D.C.A. Collins, and P.M. Neuwelt. 2003. 'The walking school bus: Extending children's' geographies'. *Area* 35: 285–92.

Kearns, R., and I. Dyck. 2005. 'Culturally safe research'. In D. Wepa, ed., *Cultural Safety in Aotearoa New Zealand*, 79–88. Auckland: Pearson/Prentice Hall.

Kearns, R.A., C.J. Smith, and M.W. Abbott. 1991. 'Another day in paradise? Life on the margins in urban New Zealand'. *Social Science and Medicine* 33: 369–79.

Keen, J., and T. Packwood. 1995. 'Case study evaluation'. *British Medical Journal* 311 (7002): 444–8.

Kelle, U. 1995. 'An overview of computer-aided methods in qualitative research'. In U. Kelle, ed., *Computer-Aided Qualitative Data Analysis: Theory, Methods and Practice.* London: Sage.

———. 1997a. 'Capabilities for theory building and hypothesis testing in software for computer aided qualitative data analysis'. *The Data Archive Bulletin* 65 (10). http://www.data-archive.ac.uk.

———. 1997b, 'Theory building in qualitative research and computer programs for the management of textual data'. *Sociological Research Online* 2 (2). http://www.socresonline.org.uk/ welcome.html.

Kellehear, A. 1993. *The Unobtrusive Researcher: A Guide to Methods.* Sydney: Allen and Unwin.

Kenny, A. 2005. 'Interaction in cyberspace: An online focus group'. *Methodological Issues in Nursing Research* 49 (4): 414–22.

Kesby, M. 2000. 'Participatory diagramming: Deploying qualitative methods through an action research epistemology'. *Area* 32 (4): 423–535.

———. 2005. 'Retheorizing empowerment-through-participation as a performance in space: Beyond tyranny to transformation'. *Signs: Journal of Women in Culture and Society* 30: 2037–65.

———. 2007. 'Spatialising participatory approaches: The contribution of geography to a mature debate'. *Environment and Planning A* 39 (12): 2813–31.

Kesby, M., S. Kindon, and R. Pain. 2005. '"Participatory" diagramming and approaches'. In R. Flowerdew and D. Martin, eds, *Methods in Human Geography*, 2nd edn, 144–66. London: Pearson.

———. 2007. 'Participation as a form of power: Retheorising empowerment and spatialising participatory action research'. In S. Kindon, R. Pain, and M. Kesby, eds, *Participatory Action Research Approaches and Methods: Connecting People, Participation and Place*, 19–25. London: Routledge.

Kidder, L.H., C.M. Judd, and E.R. Smith. 1986. *Research Methods in Social Relations.* New York: CBS Publishing.

Kindon, S. 1995. 'Exploring empowerment methodologies with women and men in Bali'. *New Zealand Geographer* 51 (1): 10–12.

———. 1998. 'Of mothers and men: Challenging gender and community myths in Bali, Indonesia'. In I. Guijt and M. Kaul Shah, eds, *The Myth of Community: Gender Issues in Participatory Development.* London: Intermediate Technology Publications.

———. 2003. 'Participatory video in geographic research: A feminist practice of looking?' *Area* 35 (2): 142–53.

———. 2009. 'Participation'. In S. Smith, R. Pain, S. Marston, and J.P. Jones III, eds, *The Handbook of Social Geography.* London: Sage.

Kindon, S., and J. Cupples. 2003. 'Anything to declare: The politics and practicalities of leaving the field'. In R. Scheyvens and D. Storey, eds, *Development Fieldwork: A Practical Guide.* London: Sage.

Kindon, S., and A. Latham. 2002. 'From mitigation to negotiation: Ethics and the geographical imagination in Aotearoa/New Zealand'. *New Zealand Geographer* 58 (1): 14–22.

Kindon, S., and C. Odell. 1990. 'Mejors unidas que solas (better together than alone)'. (unpublished report for Acosta Women's Association [AMA], San Ignacio de Acosta, Costa Rica).

Kindon, S., R. Pain, and M. Kesby, eds. 2007a. *Participatory Action Research Approaches and Methods: Connecting People, Participation and Place.* London: Routledge.

———. 2007b. 'Participatory action research: Origins, approaches and methods'. In S. Kindon, R. Pain, and M. Kesby, eds, *Participatory Action Research Approaches and Methods: Connecting People, Participation and Place*, 9–18. London: Routledge.

Kirby, S., and I. Hay. 1997. '(Hetero)sexing space: Gay men and "straight" space in Adelaide, South Australia'. *Professional Geographer* 49 (3): 295–305.

Kirk, J., and M. Miller. 1986. *Reliability and Credibility in Qualitative Research.* Beverly Hills, CA: Sage.

Kitchin, R., and N. Tate. 2000. *Conducting Research in Human Geography: Theory, Methodology and Practice.* London: Longman.

Kitchin, R., and N. Thrift, eds. 2009. *International Encyclopedia of Human Geography.* Amsterdam: Elsevier Science.

Kitzinger, J. 1994. 'The methodology of focus groups: The importance of interaction between research participants'. *Sociology of Health and Illness* 16 (1): 103–21.

Kivell, L. 1995. 'Sex-gender and race: Constructing a harvest workforce'. (Department of Geography, University of Auckland, MA thesis).

Kluckhohn, F.R. 1940. 'The participant observer technique in small communities'. *American Journal of Sociology* 46: 331–43.

Kneale, J. 2001. 'Working with groups'. In M. Limb and C. Dwyer, eds, *Qualitative Methodologies for Geographers*, 136–50. London: Arnold.

Kneale, P. 1999. *Study Skills for Geography Students.* London: Arnold.

Knigge, L., and M. Cope. 2006. 'Grounded visualization: Integrating the analysis of qualitative and quantitative data through grounded theory and visualization'. *Environment and Planning A* 38: 2021–37.

Kobayashi, A. 1994. 'Coloring the field: Gender, "race", and the politics of fieldwork'. *Professional Geographer* 46 (1): 73–80.

Kolakowski, L. 1972. *Positivist Philosophy: From Hume to the Vienna Circle*. London: Penguin.

Kong, L. 1998. 'Refocussing on qualitative methods: Problems and prospects for research in a specific Asian context'. *Area* 30 (1): 79–82.

———. 1999. 'Cemeteries and columbaria, memorials and mausoleums: Narrative and interpretation in the study of deathscapes in geography'. *Australian Geographical Studies* 37 (1): 1–10.

König, T. n.d. '*CAQDAS—A primer: New methods for the analysis of media content*'. http://www.lboro.ac.uk/research/mmethods/research/software/ caqdas_primer.html#what.

Kothari, U. 2001. 'Power, knowledge and social control in participatory development'. In B. Cooke and U. Kothari, eds, *Participation: The New Tyranny?*. London: Zed Books.

Krippendorff, K. 2003. *Content Analysis: An Introduction to Its Methodology*. Thousand Oaks, CA: Sage.

Krueger, R. 1994. *Focus Groups: A Practical Guide for Applied Research*. 2nd edn. Thousand Oaks, CA: Sage.

———. 1998. *Analyzing and Reporting Focus Group Results*. Focus Group Kit 6. Thousand Oaks, CA: Sage.

Krug, S. 2000. *Don't Make Me Think! A Common Sense Approach to Web Usability*. Indianapolis: Que.

Kunstler, H. 1994. *The Geography of Nowhere*. New York: Free Press.

Kurtz, M. 2001. 'Situating practices: The archives and the file cabinet'. *Historical Geography* 29: 26–37.

Kusenbach, M. 2002. 'Up close and personal: Locating the self in qualitative research'. *Qualitative Sociology* 25 (1): 149–52.

Kwan, M.P. 2002. 'Feminist visualization: Reenvisioning GIS as a method in feminist geographic research'. *Annals of the Association of American Geographers* 92 (4): 645–61.

Kwan, M.P., and L. Knigge. 2006. 'Doing qualitative research using GIS: An oxymoronic endeavor?' *Environment and Planning A* 38: 1999–2002.

Labovitz, S., and R. Hagedorn. 1981. *Introduction to Social Research*. 3rd edn. Sydney: McGraw-Hill.

Lane, M.B., et al. 2003. 'Sacred land, mineral wealth, and biodiversity at Coronation Hill, Northern Australia: Indigenous knowledge and SIA'. *Impact Assessment and Project Appraisal* 21 (2): 89–98.

Lane, R. 1997. 'Oral histories and scientific knowledge in understanding environmental change: A case study in the Tumut region, NSW'. *Australian Geographical Studies* 35 (2): 195–205.

Lane, R., and G. Waitt. 2001. 'Authenticity in tourism and Native title: Place, time and spatial politics in the East Kimberley'. *Social and Cultural Geography* 2 (4): 381–404.

La Pelle, N. 2004. 'Simplifying qualitative data analysis using general purpose software tools'. *Field Methods* 16: 85–108.

Latham, A. 2000. 'Urban renewal, heritage planning and the remaking of an inner-city suburb: A case study of heritage planning in Auckland, New Zealand'. *Planning Practice and Research* 15 (4): 285–98.

———. 2003. 'Research, performance, and doing human geography: Some reflections on the diary-photography, diary-interview method'. *Environment and Planning A* 35: 1993–2017.

Latham, A., and D.P. McCormack. 2007. 'Digital photography and web-based assignments in an urban field course: Snapshots from Berlin'. *Journal of Geography in Higher Education* 31 (2): 241–56.

Latimer, B. 1998. 'Masculinity, place and sport: Rugby Union and the articulation of the "new man" in Aotearoa/New Zealand'. (Department of Geography, University of Auckland, MA thesis).

Laurier, E., and C. Philo. 2006. 'Cold shoulders and napkins handed: Gestures of responsibility'. *Transactions of the Institute of British Geographers* NS 31: 193– 207.

Law, J. 2004. *After Method: Mess in Social Science Research*. London: Routledge.

Law, L. 2000. *Sex Work in Southeast Asia: The Place of Desire in a Time of AIDS*. London and New York: Routledge.

Lawrence, J. 2007. 'Placing the lived experience(s) of TB in a refugee community in Auckland, New Zealand'. (University of Auckland, New Zealand, PhD thesis [Geography]).

Lawrence, R., and M. Adams. 2005. 'First Nations and the politics of indigeneity: Australian perspectives on indigenous peoples, resource management and global rights'. *Australian Geographer* 36 (2): 265–72.

Lawson, V. 1995. 'The politics of difference: Examining the quantitative/qualitative dualism in post-structuralist feminist research'. *Professional Geographer* 47 (4): 449–57.

Le Doeff, M. 1987. 'Women and philosophy'. In T. Moi, ed., *French Feminist Thought: A Reader*. Oxford: Basil Blackwell.

Lee, R. 1992. 'Teaching qualitative geography: A JGHE written symposium'. *Journal of Geography in Higher Education* 16 (2): 123–6.

———. 2000. *Unobtrusive Methods in Social Research*. Buckingham: Open University Press.

Lee, R.M., and N.G. Fielding. 2004. 'Tools for qualitative data analysis'. In M.A. Hardy and A. Bryman, eds, *Handbook of Data Analysis*. Thousand Oaks, CA: Sage.

Lees, L. 1994. 'Rethinking gentrification: Beyond the positions of economics or culture'. *Progress in Human Geography* 18 (2): 137–50.

————. 2004. 'Urban geography: Discourse analysis and urban research'. *Progress in Human Geography* 28 (1): 101–7.

Lefebvre, H. 1991. *The Production of Space*. Oxford: Blackwell.

Le Heron, R., et al. 2001. 'Global supply chains and networking: A critical perspective on learning challenges in the New Zealand dairy and sheepmeat commodity chains'. *Journal of Economic Geography* 1 (4): 439–56.

Leurs, R. 1997. 'Critical reflections on rapid and participatory rural appraisal'. *Development in Practice* 7 (3): 290–3.

Levinas, E. 1969. *Totality and Infinity: An Essay on Exteriority*. tr. A. Lingis. Pittsburgh: University of Pittsburgh Press.

Levitas, R. 1997. *The Inclusive Society? Social Exclusion and New Labour*. Basingstoke: MacMillan.

Lewin, K. 1946. 'Action research and minority problems'. *Journal of Social Issues* 1–2: 34–6.

Lewins, A., and C. Silver. 2006. 'Choosing a CAQDAS package: A working paper'. 5th edn. CAQDAS Networking Project. http://caqdas.soc.surrey.ac.uk/ ChoosingLewins&SilverV5July06.pdf.

————. 2007. *Using Software in Qualitative Research: A Step-by-Step Guide*. London: Sage.

Lewis, P.F. 1979. 'Axioms for reading the landscape: Some guides on the American scene'. In D.W. Meinig, ed., *The Interpretation of Ordinary Landscapes: Geographical Essay*, 11–33. New York: Oxford University Press.

Lewis, R.B. 2004. 'NVivo 2.0 and ATLAS.ti 5.0: A comparative review of two popular qualitative data-analysis programs'. *Field Methods* 16 (4): 439–64.

Ley, D. 1974. *The Black Inner City as Frontier Outpost*. Monograph no. 7. Washington: Association of American Geographers.

Liamputtong, P. 2007. *Researching the Vulnerable: A Guide to Sensitive Research Methods*. London: Sage.

Liebow, E. 1967. *Tally's Corner: A Study of Negro Streetcorner Men*. Boston: Little Brown.

Limb, M., and C. Dwyer, eds. 2001. *Qualitative Methodologies for Geographers: Issues and Debates*. New York: Oxford University Press.

Lincoln, Y., and E. Guba. 1981. *Effective Evaluation*. San Francisco: Jossey-Bass.

————. 1985. *Naturalistic Inquiry*. Beverly Hills, CA: Sage.

————. 2000. 'Paradigmatic controversies, contradictions, and emerging confluences'. In N.K. Denzin and Y.S. Lincoln, eds, *Handbook of Qualitative Research*, 2nd edn. Thousand Oaks, CA: Sage.

————. 2002. 'Judging the quality of case study reports'. In M. Huberman and M. Miles, eds, *The Qualitative Researcher's Companion*. London: Sage.

Lindsay, J.M. 1997. *Techniques in Human Geography*. London: Routledge.

Lindsey, D., and R.A. Kearns. 1994. 'The writing's on the wall: Graffiti, territory and urban space'. *New Zealand Geographer* 50: 7–13.

Linstead, S., and R. Grafton-Small. 1990. 'Organisational bricolage'. In B. Turner, ed., *Organisational Symbolism*. Berlin: Walter de Gruyter.

Livingstone, D.N. 2005. 'Science, text and space: Thoughts on the geography of reading'. *Transactions of the Institute of British Geographers* 30 (4): 391–401.

Lloyd, G. 1984. *The Man of Reason: 'Male' and 'Female' in Western Philosophy*. London: Methuen.

Lockwood, M., et al. 2007. *Strengths and Challenges of Regional NRM Governance: Interviews with Key Players and Insights from the Literature*. Report no. 4 of the Land and Water Australia Project, Pathways to Good Practice in Regional NRM Governance. Hobart: University of Tasmania.

Longhurst, R. 1995. 'The geography closest in—The body . . . the politics of pregnability'. *Australian Geographical Studies* 33: 214–23.

————. 1996. 'Refocusing groups: Pregnant women's geographical experiences of Hamilton, New Zealand/Aotearoa'. *Area* 28 (2): 143–9.

————. 2000. '"Corporeographies" of pregnancy: "Bikini babes"'. *Environment and Planning D: Society and Space* 18: 453–72.

————. 2001. *Bodies: Exploring Fluid Boundaries*. London: Routledge.

————. 2005. 'Fat bodies: Developing geographical research agendas'. *Progress in Human Geography* 29 (3): 247–259.

Louis, R.P. 2007. 'Can you hear us now? Voices from the margin: Using indigenous methodologies in geographic research'. *Geographical Research* 45 (2): 130–39.

Lowenthal, D., and H. Prince. 1965. 'English landscape tastes'. *Geographical Review* 47 (4): 449–57.

Lukinbeal, C. 2004. 'The map that precedes the territory: An introduction to essays in cinematic geography'. *GeoJournal* 59: 247–51.

Lumsden, J. 2005. 'Guidelines for the design of online questionnaires'. National Research Council of Canada. Published as NRC/ERC-1127. 9 June, NRC48231. http://iit-iti.nrc-cnrc.gc.ca/iit-publications-iti/docs/NRC-48231.pdf.

Lunt, P., and S. Livingstone. 1996. 'Rethinking the focus group in media and communications research'. *Journal of Communication* 46 (2): 79–98.

Lutz, C.A., and J.L. Collins. 1993. *Reading National Geographic*. Chicago: University of Chicago Press.

McBride, B. 1999. 'The (post)colonial landscape of Cathedral Square: Urban redevelopment and representation in the "Cathedral City"'. *New Zealand Geographer* 55 (1): 3–11.

McClean, R., L.D. Berg, and M.M. Roche. 1997. 'Responsible geographies: Co-creating knowledges in Aotearoa'. *New Zealand Geographer* 53 (2): 9–15.

Maccoby, E., and N. Maccoby. 1954. 'The interview: A tool of social science'. In G. Lindzey, ed., *Handbook of Social Psychology*. Cambridge, MA: Addison-Wesley.

McCoyd, J.L.M., and T.S. Kerson. 2006. 'Conducting intensive interviews using e-mail: A serendipitous comparative opportunity'. *Qualitative Social Work* 5 (3): 389–406.

McCracken, J. 1991. 'Looking at photographs'. *New Zealand Journal of Geography* 92: 12–14.

McDowell, L. 1992a. 'Multiple voices: Speaking from outside and inside the project'. *Antipode* 24: 56–72.

———. 1992b. 'Valid games? A response to Erica Schoenberger'. *Professional Geographer* 44 (2): 212–15.

———. 1998. 'Illusions of power: Interviewing local elites'. *Environment and Planning A* 30: 2121–32.

———. 2005. 'The men and the boys: Bankers, burger makers and barmen'. In B. van Hoven and K. Horschelmann, eds, *Spaces of Masculinity*, 19–30. London: Routledge.

McDowell, L., and G. Court. 1994. 'Performing work: Bodily representations in merchant banks'. *Environment and Planning D: Society and Space* 12: 727–50.

McFarlane, T., and I. Hay. 2003. 'The battle for Seattle: Protest and popular geopolitics in *The Australian* newspaper'. *Political Geography* 22: 211–32.

McGregor, G. 1994. *EcCentric Visions: Re-constructing Australia*. Waterloo, ON: Wilfrid Laurier University Press.

McGuirk, P.M. 2002. 'Producing the capacity to govern in global Sydney: A multiscaled account'. *Journal of Urban Affairs* 25: 201–23.

McGuirk, P.M., and D. Rowe. 2001. '"Defining moments" and refining myths in the making of place identity: The Newcastle Knights and the Australian Rugby League grand final'. *Australian Geographical Studies* 39 (1): 52–66.

McIlwaine, C., and C. Moser. 2003. 'Poverty, violence and livelihood security in urban Colombia and Guatemala'. *Progress in Development Studies* 3 (2): 113–30.

McIntyre, J. 2007. 'Guide to writing a research proposal'. http://www.education.uts.edu.au/research/degrees/guide.html.

McKay, D. 2002. 'Negotiating positionings: Exchanging life stories in research interviews'. In P. Moss, ed., *Feminist Geography in Practice*. Oxford: Blackwell.

McKendrick, J.H. 1996. *Multi-method Research in Population Geography: A Primer to Debate*. Manchester: University of Manchester, Population Geography Research Group.

Mackenzie, S. 1989. *Visible Histories: Women and Environments in a Post-war British City*. Montreal: McGill-Queen's University Press.

McLafferty, S. 2003. 'Conducting questionnaire surveys'. In N. Clifford and G. Valentine, eds, *Key Methods in Geography*, 87–100. London: Sage.

McManus, P.A. 2000. 'Beyond Kyoto? Media representations of an environmental issue'. *Australian Geographical Studies* 38 (3): 306–19.

McNay, L. 1994. *Foucault: A Critical Introduction*. Cambridge: Polity Press.

McNeill, D. 1998. 'Writing the new Barcelona'. In T. Hall and P. Hubbard, eds, *The Entrepreneurial City: Geographies of Politics, Regime and Representation*. London: John Wiley.

Madge, C. 2007. 'Developing a geographer's agenda for online research ethics'. *Progress in Human Geography* 31: 654–74.

Maguire, P. 1987. *Doing participatory research: A feminist approach*. Amherst: Centre for International Education, University of Massachusetts.

Manalo, E., and J. Trafford. 2004. *Thinking to Thesis. A Guide to Graduate Success at All Levels*. Auckland: Pearson Education.

Mangabeira, W.C., R.M. Lee, and N.G. Fielding. 2004. 'Computers and qualitative research: Adoption, use, and representation'. *Social Science Computer Review* 22 (2): 167–78.

Mann, C. 2000. *Internet Communication and Qualitative Research: A Handbook for Researching Online*. London: Sage.

Mann, C., and F. Stewart. 2002. 'Internet interviewing'. In J.F. Gubrium and J.A. Holstein, eds, *Handbook of Interview Research: Context and Method*, 603–27. Thousand Oaks, CA: Sage.

Manning, K. 1997. 'Authenticity in constructivist inquiry: Methodological considerations without prescription'. *Qualitative Inquiry* 3 (1): 93–104.

Manzo, L., and N. Brightbill. 2007. 'Towards a participatory ethics'. In S. Kindon, R. Pain, and M. Kesby, eds, *Participatory Action Research Approaches and Methods: Connecting People, Participation and Place*, 33–40. London: Routledge.

Marcus, G.E., and M.M.J. Fisher. 1986. *Anthropology as Cultural Critique: An Experimental Moment in the Human Sciences*. Chicago: University of Chicago Press.

Markwell, K. 1997. 'Dimensions of a nature-based tour'. *Annals of Tourism Research* 24 (1): 131–55.

———. 2002. 'Mardi Gras tourism and the construction of Sydney as an international gay and lesbian city'. *GLQ* 8 (1–2): 81–100.

Martin, D. 2007. 'Bureaucratizing ethics: Institutional review boards and participatory research'. ACME: *An International E-journal for Critical Geographies* 6 (3): 319–28.

Mason, J. 2004. *Qualitative Researching*. 2nd edn. London: Sage.

Mason, K., and T. Zanish-Belcher. 2007. 'Raising the archival consciousness: How women's archives challenge traditional approaches to collecting and use—What's in a name?' *Library Trends* 56 (2): 344–59.

Massey, D. 1993. 'Power-geometry and a progressive sense of place'. In J. Bird, B. Curtis, G. Robertson, and L. Tickner, eds, *Mapping the futures*. London: Routledge.

———. 1994. *Space, Place and Gender*. Cambridge: Polity Press.

Massey, D., and R. Meegan, eds. 1985. *Politics and Method—Contrasting Studies in Industrial Geography*. London: Methuen.

Matless, D., J. Oldfield, and A. Swain. 2007. 'Encountering Soviet geography: Oral histories of British geographical studies of the USSR and eastern Europe, 1945–1991' *Social and Cultural Geography* 8 (3): 353–72.

Matthews S., Burton L., Detwiler J., 2001. 'Viewing people and places: Conceptual and methodological issues in coupling geographic information analysis and ethnographic research'. Paper presented at the GIS and Critical Geographic Research conference, 25 February, New York. Copy available from Dr. Matthews, Sociology and Crime, Law and Justice, Pennsylvania State University, University Park, CA. Paper presented to the Institute of Australian Geographers conference, 27 September to 1 October, Sydney.

Matthews, H., M. Limb, and M. Taylor. 1998 'The geography of children: Some ethical and methodological considerations for project and dissertation work'. *Journal of Geography in Higher Education* 22: 311–24.

May, T. 2001. *Social Research: Issues, Methods and Process*. 3rd edn. Philadelphia: Open University.

Maya People of Southern Belize, Toledo Maya Cultural Council, and Toledo Alcaldes Association. 1997. *Maya Atlas: The Struggle to Preserve Maya Land in Southern Belize*. Berkeley, CA: North Atlantic Books.

Mayhew, R. 2003. 'Researching historical geography'. In A. Rogers and H. Viles, eds, *The Student's Companion to Geography*, 2nd edn. Oxford: Blackwell.

Mays, N., and C. Pope. 1997. 'Rigour and qualitative research'. *British Medical Journal* 310 (6997): 109–13.

Mee, K.J. 1994. 'Dressing up the suburbs: Representations of western Sydney'. In K. Gibson and S. Watson, eds, *Metropolis Now: Planning and the Urban in Contemporary Australia*. Sydney: Pluto Press.

Mee, K., and R. Dowling. 2003. 'Reading *Idiot Box*: Film reviews intertwining the social and cultural'. *Social and Cultural Geography* 4 (2): 185–215.

Meho, L.I. 2006. 'E-mail interviewing in qualitative research: A methodological discussion'. *Journal of the American Society for Information Science and Technology* 57 (10): 1284–95.

Meijering, L. 2006. 'Making a place of their own: Rural intentional communities in northwest Europe'. *Netherlands Geographical Studies* 349. http://irs.ub.rug.nl/ppn/297985914.

Merlan, F. 1996. 'Formulations of claim and title: A comparative discussion'. In J. Finlayson and A. Jackson-Nakano, eds, *Heritage and Native Title: Anthropological and Legal Perspectives*, 165–77. Canberra: Australian Institute of Aboriginal and Torres Strait Islander Studies.

Merton, R.K. 1987. 'The focussed interview and focus groups: Continuities and discontinuities'. *Public Opinion Quarterly* 51 (4): 550–66.

Meth, P. 2003. 'Entries and omissions: Using solicited diaries in geographical research'. *Area* 35 (2): 195–205.

Miles, M.B., and A.M. Huberman. 1984. *Qualitative Data Analysis: A Sourcebook of New Methods*. Beverly Hills, CA: Sage.

———. 1994. *Qualitative Data Analysis: An Expanded Sourcebook*. 2nd edn. Thousand Oaks, CA: Sage.

Mills, S. 1997. *Discourse*. London and New York: Routledge.

Minichiello, V., et al. 1990. *In-depth Interviewing: Researching People*. Melbourne: Longman Cheshire.

———. 1995. *In-depth Interviewing: Principles, Techniques, Analysis*. 2nd edn. Melbourne: Longman Cheshire.

Mishler, E. 1986. *Research Interviewing: Context and Narrative*. Cambridge, MA: Harvard University Press.

Mitchell, H., R.A. Kearns, and D.C.A. Collins. 2007. 'Nuances of neighbourhood: Children's perceptions of the space between home and school in Auckland, New Zealand'. *Geoforum* 38: 614–27.

Mohammad, R. 1999. 'Marginalisation, Islamism and the production of the "Other's" "Other"'. *Gender, Place and Culture* 6 (3): 221–40.

Mohanty, C.T. 1991. 'Cartographies of struggle: Third world women and the politics of feminism'. In C. Mohanty, A. Russo, and L. Torres, eds, *Third World Women and the Politics of Feminism*. Bloomington: University of Indiana Press.

Monk, J., and S. Hanson. 1982. 'On not excluding half of the human in human geography'. *Professional Geographer* 34: 11–23.

Monk, J., P. Manning, and C. Denman. 2003. 'Working together: Feminist perspectives on collaborative research and action'. *ACME: An International E-Journal for Critical Geographies* 2 (1): 91–106.

Morgan, D.L., ed. 1993. *Successful Focus Groups: Advancing the State of the Art*. Newbury Park: Sage.

———. 1996. 'Focus groups'. *Annual Review of Sociology* 22: 129–52.

———. 1997. *Focus Groups as Qualitative Research*. 2nd edn. Thousand Oaks, CA: Sage.

Morgan, D.L., and R.A. Krueger. 1993. 'When to use focus groups and why'. In D.L. Morgan, ed., *Successful Focus Groups: Advancing the State of the Art*. Newbury Park: Sage.

Moser, C.A., and G. Kalton. 1983. *Survey Methods in Social Investigation*. London: Heinemann.

Moss, P. 1995. 'Reflections on the "gap" as part of the politics of research design'. *Antipode* 27 (1): 82–90.

———. 2001. *Placing Autobiography in Geography*. Syracuse, NY: Syracuse University Press.

———. 2002. *Feminist Geography in Practice*. Oxford: Blackwell.

Mosse, D. 1994. 'Authority, gender and knowledge: Theoretical reflections on the practice of participatory rural appraisal'. *Development and Change* 25: 497–526.

Mostyn, B. 1985. 'The content analysis of qualitative research data: A dynamic approach'. In M. Brown, J. Brown, and D. Canter, eds, *The Research Interview: Uses and Approaches*. London: Academic Press.

mrs c kinpaisby-hill. 2008. 'Publishing from participatory research'. In A. Blunt, ed., *Publishing in Geography: A Guide for New Researchers*, 45–7. London: Wiley-Blackwell.

mrs kinpaisby. 2008. 'Taking stock of participatory geographies: Envisioning the communiversity'. *Transactions of the Institute of British Geographers* 33: 292–9.

Muller, S. 1999. 'Myths, media and politics: Implications for koala management decisions in Kangaroo Island, South Australia'. Paper presented to the Institute of Australian Geographers conference, 27 September to 1 October, Sydney.

Mullings, B. 1999. 'Insider or outsider, both or neither: Some dilemmas of interviewing in a cross-cultural setting'. *Geoforum* 30: 337–50.

Munt, S.R. 1998. 'Sisters in exile: The lesbian nation'. In A. Rosa, ed., *New Frontiers of Spaces, Bodies and Gender*, 3–19. London and New York: Routledge.

Murdoch, J. 2006. *Post-structuralist Geography*. London: Sage.

Myers, G. 1998. 'Displaying opinions: Topics and disagreement in focus groups'. *Language in Society* 27: 85–111.

Myers, G., T. Klak, and T. Koehl. 1996. 'The inscription of difference: News coverage of the conflicts in Rwanda and Bosnia'. *Political Geography* 15 (1): 21–46.

Myers, G., and P. Macnaghten. 1998. 'Rhetorics of environmental sustainability: Commonplaces and places'. *Environment and Planning A* 30: 333–53.

Nader, L. 1974. 'Up the anthropologist—Perspective gained from studying up'. In D. Hymes, *Reinventing Anthropology*, 284–311. New York: Vintage Books.

Nagar, R. 2000. '*Mujhe Jawab Do!* (answer me!): Women's grass-roots activism and social spaces in Chitrakoot (India)'. *Gender, Place and Culture* 7 (4): 341–62.

Nast, H. 1994. 'Opening remarks: Women in the field'. *Professional Geographer* 46 (1): 54–66.

National Council on Public History. 2003. 'NCPH ethics guidelines'. http://ncph.org/ethics.html.

National Research Foundation (South Africa). n.d. 'Proposal writing resources'. http://www.nrf.ac.za/methods/proposals.htm.

Nelson, C., P. Treichler, and L. Grossberg. 1992. 'Cultural studies: An introduction'. In L. Grossberg, C. Nelson, and P. Treichler, eds, *Cultural Studies*. New York: Routledge.

Nelson, S. 2003. 'It's I mean like uh disrespectful'. *Times Higher Educational Supplement* 28 March: 16.

Nettlefold, P.A., and E. Stratford. 1999. 'The production of climbing landscapes-as-texts'. *Australian Geographical Studies* 37 (2): 130–41.

Neuendorf, K. 2001. *The Content Analysis Guidebook*. Thousand Oaks, CA: Sage. (See online accompaniment at http://academic.csuohio.edu/kneuendorf/content).

NHMRC (National Health and Medical Research Council). 2003. *Values and Ethics—Guidelines for Ethical Conduct in Aboriginal and Torres Strait Islander Health Research*. Canberra: NHMRC.

Nicholson, B. 2000. 'Something there is . . .' In K. Reed-Gilbert, ed., *The Strength of Us as Women: Black Women Speak*, 27–30. Canberra: Ginninderra Press.

Nicholson, H. 2002. 'Telling travelers' tales: The world through home movies'. In T. Creswell and D. Dixon, eds, *Engaging Film: Geographies of Mobility and Identity*. Lanham, MD: Rowman and Littlefield.

Nietschmann, B.Q. 1973. *Between Land and Water: The Subsistence Ecology of the Miskito Indians, Eastern Nicaragua*. New York: Seminar Press.

———. 1979. *Caribbean Edge: The Coming of Modern Times to Isolated People and Wildlife*. Indianapolis: Bobbs-Merrill.

———. 1987. 'The third world war'. *Cultural Survival Quarterly* 11 (3): 1–16.

———. 1995. 'Defending the Miskito reefs with maps and GPS: Mapping with sail, scuba, and satellite'. *Cultural Survival Quarterly* 18 (4): 34–7.

———. 1997. 'Protecting indigenous coral reefs and sea territories, Miskito Coast, RAAN, Nicaragua'. In S. Stevens, ed., *Conservation through Cultural Survival: Indigenous Peoples and Protected Areas*, 193–224. Washington: Island Press.

———. 2001. 'The Nietschmann syllabus: A vision of the field'. *Geographical Review* 91 (1–2): 175–84.

Nietzsche, F.W. 1969. *On the Genealogy of Morals*. tr. W. Kaufmann and R.J. Hollingdale. New York: Vintage.

Nijoux, D., and J. Myers. 2006. 'Interviews, solicited diaries and photography: "New" ways of accessing everyday experiences of place'. *Graduate Journal of Asia-Pacific Studies* 4 (1): 44–64.

Nolan, N. 2003. 'The ins and outs of skateboarding and transgression in public space in Newcastle, Australia'. *Australian Geographer* 34 (3): 311–27.

Nunavut Research Institute and Inuit Tapiriit Kanatami. n.d. 'Negotiating research relationships with Inuit communities: A guide for researchers'. http://www.nri.nu.ca/pdf/06-068%20ITK%20NRR%20 booklet.pdf.

Oakley, A. 1981. *From Here to Maternity: Becoming a Mother*. Harmondsworth: Penguin.

O'Brien, K. 1993. 'Improving survey questionnaires through focus groups'. In D.L. Morgan, ed., *Successful Focus Groups: Advancing the State of the Art*. Newbury Park: Sage.

O'Connell-Davidson, J., and D. Layder. 1994. *Methods, Sex and Madness*. London: Routledge.

O'Connor, H., and C. Madge. 2003. '"Focus groups in cyberspace": Using the Internet for qualitative research'. *Qualitative Market Research: An International Journal* 6 (2): 133–43.

Ogborn, M. 2003. 'Knowledge is power: Using archival research to interpret state formation'. In A. Blunt et al., eds, *Cultural Geography in Practice*, 9–20. London: Arnold.

———. 2006. 'Finding historical data'. In N. Clifford and G. Valentine, eds, *Key Methods in Geography*, 101–15. London: Sage.

Oliver, W.H. 2002. *Looking for the Phoenix*. Wellington: Bridget Williams Books.

Olver, I.N., J.A. Eliott, and J. Blake-Mortimer. 2002. 'Cancer patients' perceptions of Do Not Resuscitate orders'. *PsychoOncology* 11 (3): 181–7.

O'Neill, P.M. 2001. 'Financial narratives of the modern corporation'. *Journal of Economic Geography* 1: 181–99.

Opie, A. 1992. 'Qualitative research, appropriation of the other and empowerment'. *Feminist Review* 40: 52–69.

Oppenheim, N. 1992. *Questionnaire Design, Interviewing and Attitude Measurement*. London: Pinter.

Oritz, S.M. 1994. 'Shopping for sociability in the mall'. *Research in Community Sociology* 4 (supplement): 183–99.

Ortner, S.B. 1978. *Sherpas through Their Rituals*. Cambridge: Cambridge University Press.

———. 1989. *High Religion: A Cultural and Political History of Sherpa Buddhism*. Princeton, NJ: Princeton University Press.

———. 1999. *Life and Death on Mt. Everest: Sherpas and Himalayan Mountaineering*. Princeton, NJ: Princeton University Press.

Osumanu, I. 2007. 'Environmental concerns of poor households in low-income cities: The case of the Tamale Metropolis, Ghana'. *Geojournal* 68: 343–55.

Padfield, M., and I. Procter. 1996. 'The effect of the interviewer's gender on the interviewing process: A comparative enquiry'. *Sociology* 30: 355–66.

Pain, R. 2003. 'Social geography: On action-oriented research'. *Progress in Human Geography* 27 (5): 677–85.

Pain, R., and P. Francis. 2003. 'Reflections on participatory research'. *Area* 35 (1): 46–54.

Panelli, R. 2004. *Social Geographies*. London: Sage.

Panofsky, E. 1957. *Meaning in the Visual Arts*. New York: Doubleday Anchor.

Parfitt, J. 2005. 'Questionnaire design and sampling'. In R. Flowerdew and D. Martin, eds, *Methods in Human Geography: A Guide for Students Doing a Research Project*. Harlow: Longman.

Park, P. 1993. 'What is participatory research? A theoretical and methodological perspective'. In P. Park, M. Brydon-Miller, B. Hall, and T. Jackson, eds, *Voices of Change: Participatory Research in the United States and Canada*. Westport, CT: Bergin and Garvey.

Parkes, M., and R. Panelli. 2001. 'Integrating catchment ecosystems and community health: The value of participatory action research'. *Ecosystem Health* 7 (2): 85–106.

Parnwell, M. 2003. 'Consulting the poor in Thailand: Enlightenment or delusion?' *Progress in Development Studies* 3 (2): 99–112.

Parr, H. 1998. 'Mental health, ethnography and the body'. *Area* 30: 28–37.

Patton, M.Q. 1990. *Qualitative Evaluation and Research Methods*. 2nd edn. Beverly Hills: Sage.

———. 2002. *Qualitative Evaluation and Research Methods*. 3rd edn. Beverly Hills: Sage.

Pavlovskaya, M. 2004. 'Other transitions: Multiple economies of Moscow households in the 1990s'. *Annals of the Association of American Geographers* 94 (2): 329–51.

Pawson, E. 1991. 'Monuments, memorials and cemeteries: Icons in the landscape'. *New Zealand Journal of Geography* 92: 26–7.

Pawson, E., and E. Teather. 2002. 'Geographical expeditions: Assessing the benefits of a student-driven fieldwork method'. *Journal of Geography in Higher Education* 26 (3): 275–89.

Peace, R. 1998. 'CAQDAS/NUD*IST: Computer assisted qualitative data analysis software/nonnumerical, unstructured data. Indexing, searching and theorising—a geographical perspective'. In E. Bliss, ed., *Proceedings of the Second Joint Conference, Institute of Australian Geographers and New Zealand Geographical Society*, January 1997, 382–5. Hamilton: New Zealand Geographical Society.

Peake, L. (on behalf of Red Thread Women's Development Programme). 2000. *Women Researching Women: Methodology Report and Research Projects on the Study of Domestic Violence and Women's Reproductive Health in Guyana*. Georgetown, Guyana: Interamerican Bank.

Pearson, L.J. 1996. 'Place re-identification: The "Leisure Coast" as a partial representation of Wollongong'. (School of Geography, University of New South Wales, BSc Honours thesis).

Pearson, L.J., and K.M. Dunn. 1999. 'Reidentifying Wollongong: Dispossession of the local citizenry'. Proceedings of the Australian University Tourism and Hospitality Education 1999 National Research Conference, Adelaide.

Perry, P. 1969. 'Twenty-five years of New Zealand historical geography'. *New Zealand Geographer* 25 (2): 93–105.

Philip, L.J. 1998. 'Combining quantitative and qualitative approaches to social research in human geography—An impossible mixture?' *Environment and Planning A* 30: 261–76.

Phillips, L., and Jørgensen, M.W. 2002. *Discourse Analysis as Theory and Method*. London: Sage.

Phillips, N., and C. Hardy. 2002. *Discourse Analysis: Investigating Processes of Social Construction*. Thousand Oaks, CA: Sage.

Pickerill, J. 2009. 'Finding common ground? Spaces of dialogue and the negotiation of indigenous interests in environmental campaigns in Australia'. *Geoforum* 40: 66–79.

Pickles, J. 1992. 'Texts, hermeneutics and propaganda maps'. In T.J. Barnes and J.S. Duncan, eds, *Writing Worlds*. London: Routledge.

Pickles, K. 2002. 'Kiwi icons and the re-settlement of New Zealand as colonial space'. *New Zealand Geographer* 58 (2): 5–16.

Pile, S. 1992. 'Oral history and teaching qualitative methods'. *Journal of Geography in Higher Education* 16 (2): 135–43.

Pinkerton, A., and K. Dodds. 2009. 'Radio geopolitics: Broadcasting, listening and the struggle for acoustic spaces'. *Progress in Human Geography* 33 (1): 10–27.

pla notes. December 2003. London: International Institute for Environment and Development.

Platt, J. 1988. 'What can case studies do?' *Studies in Qualitative Methodology* 1: 1–23.

————. 1992. '"Case study" in American methodological thought'. *Current Sociology* 40: 17-48.

Poland, B.D. 1995. 'Transcription quality as an aspect of rigor in qualitative research'. *Qualitative Inquiry* 1 (3): 290–310.

Ponga, M. 1998. 'I Nga ra o Mua: (Re)constructions of symbolic layers in Te Poho-o-Hinemihi Marae'. (Department of Geography, University of Auckland, MA thesis).

Popper, K. 1959. *The Logic of Scientific Discovery*. London: Hutchinson.

Popping, R. 2000. *Computer-assisted text analysis*. New Technologies for Social Research, Research Methods Series. Thousand Oaks, CA, and London: Sage.

Porteous, D.J. 1985. 'Smellscape'. *Progress in Human Geography* 9: 356–78.

Potter, J. 1996. 'Discourse analysis and constructionist approaches: Theoretical background'. In J.T.E. Richardson, ed., *Handbook of Qualitative Methods for Psychology and the Social Sciences*. Leicester: British Psychological Society.

Powell, J.M. 1973 *Yeomen and bureaucrats: The Victorian Crown Lands Commission, 1878–79*. Melbourne: Oxford University Press.

————. 1988. *An Historical Geography of Modern Australia: The Restive Fringe*. Cambridge: Cambridge University Press.

Powell, R.C. 2008. 'Becoming a geographical scientist: Oral histories of Arctic fieldwork'. *Transactions of the Institute of British Geographers* 33 (4): 548–65.

Pratt, G. 1999. 'From registered nurse to registered nanny: Discursive geographies of Filipina domestic workers in Vancouver, B.C.'. *Economic Geography* 75 (3): 215–36.

————. 2000. 'Participatory action research'. In R. Johnston, D. Gregory, G. Pratt, and M. Watts, eds, *Dictionary of Human Geography*, 4th edn. Oxford: Blackwell.

————. 2002. 'Studying immigrants in focus groups'. In P. Moss, ed., *Feminist Geography in Practice: Research and Methods*. Oxford: Blackwell.

Pratt, G., in collaboration with the Philippine Women's Centre. 1999. 'Is this Canada? Domestic workers' experiences in Vancouver, B.C.'. In J. Momson, ed., *Gender, Migration and Domestic Service*. London: Routledge.

Pratt, G., in collaboration with the Philippine Women's Centre of B.C. and Ugnayan Ng Kabataang Pilipino sa Canada/Filipino Canadian Youth Alliance. 2007. 'Working with migrant communities: Collaborating with the Kalayaan Centre in Vancouver, Canada'. In S. Kindon, R. Pain, and M. Kesby, eds, *Participatory Action Research Approaches and Methods: Connecting People, Participation and Place*, 95–103. London: Routledge.

Pretty, J., et al. 1995. *Participatory Learning and Action: A Trainer's Guide*. London: International Institute for Environment and Development.

Professional Geographer. 1994. 46 (1: special issue on women in the field).

Prosser, J., and A. Loxley. 2008. *Introducing Visual Methods*. ESRC National Centre for Research Methods Review Paper, NCRM/010.

Przeworski, A., and F. Salmon. 1995. 'The art of writing proposals: Some candid suggestions for applicants to Social Science Research Council competitions'. http://fellowships.ssrc.org/art_of_writing_proposals.

Pulvirenti, M. 1997. 'Unwrapping the parcel: An examination of culture through Italian home ownership'. *Australian Geographical Studies* 35: (1): 32–9.

Punch, K.F. 2006. *Developing Effective Research Proposals*. 2nd edn. London: Sage.

Quanchi, M. 2006. 'Photography and history in the Pacific Islands'. *The Journal of Pacific History* 41 (2): 165–73.

Radicati Group. 2005. *Taming the Growth of Email—An ROI Analysis*. White Paper by the Radicati Group, Palo Alto, CA.

Radley, A., D. Hodgetts, and A. Cullen. 2005. 'Visualizing homelessness: A study in photography and estrangement'. *Journal of Community and Applied Social Psychology* 15 (4): 273–95.

Raitz, K.B. 2001. 'Field observations, archives, and exploration'. *Geographical Review* 91 (1 and 2): 121–31.

Reason, P., and J. Rowan, eds. 1981. *Human Inquiry: A Sourcebook of New Paradigm Research*. Chichester: John Wiley and Sons.

Reed, M., and D. Harvey. 1992. 'The new science and the old: Complexity and realism in the social sciences'. *Journal for the Theory of Social Behaviour* 22 (4): 353–80.

Reinharz, S. 1992. *Feminist Methods in Social Research*. New York: Oxford University Press.

Rettie, R. 2005. 'Exploiting freely available software for social research'. *Social Research Update* 48.

Reynolds, G. 2008. *presentationzen: Simple Ideas on Presentation, Design and Delivery*. Berkeley, CA: New Riders.

Reynolds, H. 1998. *This Whispering in Our Hearts*. Sydney: Allen and Unwin.

Richards, L. 1990. *Nobody's Home: Dreams and Realities in a New Suburb*. Melbourne: Oxford University Press.

———. 1997. 'Computers and qualitative analysis'. In *The International Encyclopedia of Education*. Oxford: Elsevier Science.

———. 2005. *Handling Qualitative Data, A Practical Guide*. London: Sage.

Richards, L., and T. Richards. 1995. 'Using hierarchical categories in qualitative data analysis'. In U. Kelle, ed., *Computer-Aided Qualitative Data Analysis: Theory, Methods and Practice*. London: Sage. http://www.qsr.com.au/otherinfo/papers/hierarchies.html.

Richardson, L., and E.A. St Pierre. 2005. 'Writing: A method of inquiry'. In N.K. Denzin and Y.S. Lincoln, eds, *Collecting and Interpreting Qualitative Materials Research*, 3rd edn, 473–99. Los Angeles: Sage.

Riley, M., and D. Harvey. 2007. 'Editorial: Talking landscapes: On oral history and the practice of geography'. *Social and Cultural Geography* 8 (4): 391–415.

Rimmer, P.J., and S. Davenport. 1998. 'The geographer as itinerant: Peter Scott in flight, 1952–1996'. *Australian Geographical Studies* 36 (2): 123–42.

Rinaldi, A.H. 1994. 'The Net: User guidelines and netiquette'. Academic/Institutional Support Services, Florida Atlantic University. http://www.wifak.uni-wuerzburg.de/wilan/sysgroup/texte/netiquet/netiquet.txt.

Ritchie, J., and J. Lewis. 2003. *Qualitative Research Practice*. London: Sage.

Rivera, M. 1997. 'Various definitions of geography'. http://www2.westga.edu/~geograph/define.html.

Roberts, J., and G. Sainty. 1996. *Listening to the Lachlan*. Potts Point, NSW: Sainty and Associates.

Robertson, B.M. 1994. *Oral History Handbook*. Adelaide: Oral History Association of Australia.

———. 2006. *Oral History Handbook*. 5th edn. Adelaide: Oral History Association of Australia SA Branch Inc.

Robinson, G. 1998. *Methods and Techniques in Human Geography*. Chichester: John Wiley and Sons.

Rodaway, P. 1994. *Sensuous Geographies: Body, Sense, Place*. London: Routledge.

Rofe, M.W. 2000. 'Gentrification within Australia's "problem city": Inner Newcastle as a zone of residential transition'. *Australian Geographical Studies* 38 (1): 54–70.

———. 2004. 'From "problem city" to "promise city": Gentrification and the revitalisation of Newcastle'. *Australian Geographical Studies* 42 (2): 193–206.

———. 2007. 'Urban revitalisation and masculine memories: Towards a critical awareness of gender in the postindustrial landscape'. *Australian Planner* 44 (2): 26–33.

Rofe, M.W., and H.P.M. Winchester. 2003. 'Masculine scripting and the mythology of motorcycling'. *Journal of Interdisciplinary Gender Studies* 7 (1 and 2): 161–79.

———. 2007. 'Lobethal the *Valley of Praise*: Inventing tradition for the purposes of place making in rural South Australia'. In R. Jones and B.J. Shaw, eds, *Loving a Sunburned Country? Geographies of Australian Heritage*, 133–50. London: Ashgate.

Rogerson, P.A. 2001. *Statistical Methods for Geography*. Thousand Oaks, CA: Sage.

Rose, C. 1988. 'The concept of reach and the anglophone minority in Quebec'. In J. Eyles and D. Smith, eds, *Qualitative Methods in Human Geography*. Cambridge: Polity Press.

Rose, D.B. 1996a. 'Histories and rituals: Land claims in the Territory'. In B. Attwood, ed., *In the Age of Mabo: History, Aborigines and Australia*, 35–52. Sydney: Allen and Unwin.

———. 1996b. *Nourishing Terrains: Australian Aboriginal Views of Landscape and Wilderness*. Canberra: Australian Heritage Commission.

———. 1999. 'Indigenous ecologies and an ethic of connection'. In N. Low, ed., *Global Ethics and Environment*. London: Routledge.

Rose, G. 1993. *Feminism and Geography*. Minneapolis: University of Minnesota Press.

———. 1996. 'Teaching visualised geographies: Towards a methodology for the interpretation of visual materials'. *Journal of Geography in Higher Education* 20 (3): 281–94.

———. 1997. 'Situating knowledges: Positionality, reflexivities and other tactics'. *Progress in Human Geography* 21: 305–20.

———. 2001. *Visual Methodologies. An Introduction to the Interpretation of Visual Materials*. London: Sage.

———. 2003. 'Family photographs and domestic spacings: A case study'. *Transactions of the Institute of British Geographers* 28: 5–18.

———. 2007. *Visual Methodologies: An Introduction to the Interpretation of Visual Materials*. 2nd edn. London: Sage.

———. 2008. 'Using photographs as illustrations in human geography'. *Journal of Geography in Higher Education* 32 (1): 151–60.

Routledge, P. 2001. 'Within the river: Collaboration and methodology'. *Geographical Review* 91 (1 and 2): 113–20.

———. 2002. 'Travelling east as Walter Kurtz: Identity, performance and collaboration in Goa, India'. *Environment and Planning D: Society and Space* 20 (4): 477–96.

Rowles, G.D. 1978. *Prisoners of Space: Exploring the Geographical Experience of Older People*. Boulder, CO: Westview Press.

Rowley, C.D. 1970. *The Destruction of Aboriginal Society: Aboriginal Policy and Practice, Volume I*. Canberra: Australian National University Press.

———. 1971a. *Outcasts in White Australia: Aboriginal Policy and Practice, Volume II*. Canberra: Australian National University Press.

———. 1971b. *The Remote Aborigines: Aboriginal Policy and Practice, Volume III*. Canberra: Australian National University Press.

Rowse, T. 2000. *Obliged to Be Difficult: Nugget Coombs' Legacy in Indigenous Affairs*. Cambridge: Cambridge University Press.

Ruby, J. 1995. *Secure the Shadow; Death and Photography in America*. Cambridge, MA: MIT Press.

Ruddick, S. 2004. 'Activist geographies: Building possible worlds'. In P. Cloke, P. Crang, and M. Goodwin, eds, *Envisioning Human Geographies*. London: Arnold.

Rugendyke, B. 2005. 'W(h)ither development geography in Australia?' *Geographical Research* 43 (3): 306–18.

Ruming, R., K. Mee, and P.M. McGuirk. 2004. 'Questioning the rhetoric of social mix: Courteous community or hidden hostility?' *Australian Geographical Studies* 42 (2): 234–48.

Ryan, G.W. 2004. 'Using a word processor to tag and retrieve blocks of text'. *Field Methods* 16: 109–30.

Sackett, H. 2005. 'Nothing is true but change: Archaeology, time and landscape in the writing of Lewis Grassic Gibbon'. *Scottish Archaeological Journal* 27 (1): 13–29.

Said, E. 1978. *Orientalism*. New York: Vintage Books.

———. 1993. *Culture and Imperialism*. New York: Vintage Books.

Sanders, R. 2007. 'Developing geographers through photography: Enlarging concepts'. *Journal of Geography in Higher Education* 31 (1): 181–95.

Sanderson, E., with Holy Family Settlement Research Team, R. Newport, and *Umaki* Research Participants. 2007. 'Participatory cartographies: Reflections from research performances in Fiji and Tanzania'. In S. Kindon, R. Pain, and M. Kesby, eds, *Participatory Action Research Approaches and Methods: Connecting People, Participation and Place*, 122–31. London: Routledge.

Sanderson, E., and S. Kindon. 2004. 'Progress in participatory development: Opening up the possibility of knowledge through progressive participation'. *Progress in Development Studies* 4 (2): 114–26.

Sanjek, R., ed. 1990. *Fieldnotes: The Makings of Anthropology*. Ithaca, NY: Cornell University Press.

Sarantakos, S. 1993. *Social Research*. South Melbourne: MacMillan.

———. 2005. *Social Research*. 3rd edn. Melbourne: Palgrave.

Sauer, C.O. 1925. 'The morphology of landscape'. *University of California Publications in Geography* 2: 19–54.

———. 1941. 'Foreword to historical geography'. *Annals of the Association of American Geographers* 31 (1): 1–24.

Sausseur, F. de. 1983. *Course in General Linguistics*. London: Duckworth.

Sayer, A. 1992. *Method in Social Science: A Realist Approach*. 2nd edn. London: Routledge.

———. 2000. *Realism and Social Science*. Newbury Park: Sage.

Sayer, A., and K. Morgan. 1985. 'A modern industry in a declining region: Links between method, theory and policy'. In D. Massey and R. Meegan, eds, *Politics and Method: Contrasting Studies in Industrial Geography*. London: Methuen.

Schaffer, K. 1988. *Women and the Bush: Forces of Desire in the Australian Cultural Tradition*. Cambridge: Cambridge University Press.

Schein, R.H. 1997. 'The place of landscape: A conceptual framework for interpreting an American scene'. *Annals of the Association of American Geographers* 87 (4): 660–80.

Scheuermann, L., and G. Taylor. 1997. 'Netiquette'. *Internet Research: Electronic Networking Applications and Policy* 7 (4): 269–73.

Schoenberger, E. 1991. 'The corporate interview as a research method in economic geography'. *Professional Geographer* 43 (2) 180–9.

———. 1992. 'Self-criticism and self-awareness in research: A reply to Linda McDowell'. *Professional Geographer* 44 (2): 215–18.

Schollmann, A., H.C. Perkins, and K. Moore. 2000. 'Intersecting global and local influences in urban place promotion: The case of Christchurch, New Zealand'. *Environment and Planning D: Society and Space* 32: 55–76.

School of Global Studies. 1998. *Essay Format and Essay Writing for Geography Students*. Palmerston North: School of Global Studies, Massey University.

Schwartz, J., and R. Ryan, eds. 2003. *Picturing Place, Photography and the Geographical Imagination*. London: I.B. Tauris.

Scollon, R., and S. Wong Scollon. 2003. *Discourses in Place: Language in the Material World*. London: Routledge.

Scott, K., J. Park, C. Cocklin, and G. Blunden. 1997. *A Sense of Community: An Ethnography of Rural Sustainability in the Mangakahia Valley, Northland*. Occasional Publication 33. Auckland: Department of Geography, University of Auckland.

Scott, K., J. Park, and C. Cocklin. 2000. 'From "sustainable rural communities" to "social sustainability": Giving voice to diversity in Mangakahia Valley'. *New Zealand Journal of Rural Studies* 16 (4): 433–46.

Scribner, J.P. 2003. 'Teacher learning in context: The special case of a rural high school'. *Education Policy Analysis Archives* 11 (12). http://epaa.asu.edu/epaa/v11n12.

Seale, C.F. 2002. 'Computer assisted analysis of qualitative data'. In J.F. Gubrium and J.A. Holstein, eds, *Handbook of Interview Research*, 651–70. Thousand Oaks, CA: Sage.

Secor, A. 2003. 'Citizenship in the city: Identity, community and rights among women migrants to Istanbul'. *Urban Geography* 24 (2): 147–68.

———. 2004. '"There is an Istanbul that belongs to me": Citizenship, space, and identity in the city'. *Annals of the Association of American Geographers* 94 (2): 352–68.

Seebohm, K. 1994. 'The nature and meaning of the Sydney Mardi Gras in a landscape of inscribed social relations'. In R. Aldrich, ed., *Gay Perspectives II: More Essays in Australian Gay Culture*. Sydney: Department of Economic History with the Australian Centre for Gay and Lesbian Research, University of Sydney.

Sekaran, U. 1992. *Research Methods for Business: A Skill Building Approach*. 2nd edn. New York: John Wiley and Sons.

Shapcott, M., and P. Steadman. 1978. 'Rhythms of urban activity'. In T. Carlstein, D. Parkes, and N. Thrift, eds, *Human Activity and Time Geography*. New York: John Wiley and Sons.

Sharp, J. 2009. 'Geography and gender: What belongs to feminist geography? Emotion, power and change'. *Progress in Human Geography* 33 (1): 74–80.

Shaw, G., and D. Wheeler. 1994. *Statistical Techniques in Geographical Analysis*. New York: Halsted Press.

Shaw, W.S. 2000. 'Ways of whiteness: Harlemising Sydney's Aboriginal Redfern'. *Australian Geographical Studies* 38 (3): 291–305.

———. 2007. *Cities of Whiteness*. Oxford: Blackwell.

Sheehan, K., and S. McMillan. 1999. 'Response variation in e-mail surveys: An exploration'. *Journal of Advertising Research* 29: 45–54.

Sheridan, G. 2001. 'Dennis Norman Jeans: Historical geographer and landscape interpreter extraordinaire'. *Australian Geographical Studies* 39 (1): 96–106.

Shopes, L. 2000. 'International review boards have a chilling effect on oral history'. *Perspectives OnLine*. http://www.theaha.org/perspectives/issues/2000/009/009vie.cfm9.

Shurmer-Smith, P. 2002. *Doing Cultural Geography*. London: Sage.

Sibley, D., and B. van Hoven. 2009. 'The contamination of personal space: Boundary construction in a prison environment'. *Area* 41 (2): 198–206.

Sides, C.H. 1999. *How to Write and Present Technical Information*. 3rd edn. Phoenix: Oryx Press.

Silverman, D. 1991. *Interpreting Qualitative Data: Methods for Analysing Talk, Text and Interaction*. Thousand Oaks, CA: Sage.

———. 1993. *Interpreting Qualitative Data: Methods for Analysing Talk, Text and Interaction*. 2nd edn. London: Sage.

Singer, A., and L. Woodhead. 1988. *Disappearing World: Television and Anthropology*. London: Boxtree and Granada Television.

Smith, A. 1994. *New Right Discourses on Race and Sexuality*. Cambridge, MA: Cambridge University Press.

Smith, D.A. 2003. 'Participatory mapping of community lands and hunting yields among the Buglé of western Panama'. *Human Organization* 62 (4): 332–41.

Smith, D.M. 1977. *Human Geography: A Welfare Approach*. London: Edward Arnold.

Smith, K. 2003. 'Pushing the boundaries: The exclusion of disability rights groups from political influence in Victoria'. *Australian Geographer* 34 (3): 345–54.

Smith, L.T. 1999. *Decolonizing Methodologies: Research and Indigenous Peoples*. Dunedin and London: University of Otago Press and Zed Books.

Smith, S.J. 1981. 'Humanistic method in contemporary social geography'. *Area* 15: 355–8.

———. 1988. 'Constructing local knowledge: The analysis of self in everyday life'. In J. Eyles and D.M. Smith, eds, *Qualitative Methods in Human Geography*. Cambridge: Polity Press.

———. 1994. 'Soundscape'. *Area* 26: 232–40.

Solís, P. 2009. 'Preparing competitive research grant proposals'. In M. Solem, K. Foote, and J. Monk, eds, *Aspiring Academics: A Resource Book for Graduate Students and Early Career Faculty*. Upper Saddle River, NJ: Pearson/Prentice Hall.

Sparke, M. 1998. 'A map that roared and an original atlas: Canada, cartography and the narration of a nation'. *Annals of the Association of American Geographers* 88 (3): 463–95.

Spate, O.H.K., and A. Learmonth. 1967. *India and Pakistan: A General and Regional Geography*. London: Methuen.

Spinks, N., B. Baron Wells, and M. Meche. 1999. 'Netiquette: A behavioral guide to electronic business communication'. *Corporate Communications: An International Journal* 4 (3): 145–55.

Spradley, J.P. 1980. *Participant Observation*. New York: Holt, Rinehart and Wilson.

Srinivas, M.N., A.M. Shah, and E.A. Ramaswamy. 1979. *The Fieldworker and the Field: Problems and Challenges in Sociological Investigation*. New Delhi: Oxford University Press.

Stacey, J. 1988. 'Can there be a feminist ethnography?' *Women's Studies International Forum* 11: 21–7.

Stake, R. 1995. *The Art of Case Study Research*. Thousand Oaks, CA: Sage.

Staller, K. 2002. 'Musings of a skeptical software junkie and the HyperRESEARCH fix'. *Qualitative Social Work* 1 (4): 473–87.

Stanczak, G., ed. 2007. *Visual Research Methods: Image, Society and Representation*. Thousand Oaks, CA, and London: Sage.

Stanley, L., and S. Wise. 1993. *Breaking Out Again: Feminist Ontology and Epistemology*. 2nd edn. London: Routledge.

Stanton, N. 1996. *Mastering Communication*. 3rd edn. London: MacMillan.

Starkey, A., (video recording) and K. George (interviews). 2003. *Balfour's City Site, 1910–2003*. Adelaide: Corporation of the City of Adelaide.

Stenson, K., and P. Watt. 1999. 'Governmentality and "the death of the social"?: A discourse analysis of local government texts in south-east England'. *Urban Studies* 36 (3): 189–201.

Stevens, S. 1993. *Claiming the High Ground: Sherpas, Subsistence, and Environmental Change in the Highest Himalaya*. Berkeley: University of California Press.

———. 1997. 'Consultation, co-management, and conflict in Sagarmatha (Mt Everest) National Park, Nepal'. In S. Stevens, ed., *Conservation through Cultural Survival: Indigenous Peoples and Protected Areas*, 63–97. Washington: Island Press.

———. 2001. 'Fieldwork as commitment'. *Geographical Review* 91 (1–2): 66–73.

———. 2003. 'Tourism and deforestation in the Mt Everest region of Nepal'. *Geographical Journal* 169 (3): 255–77.

———. 2004. 'Imperialism, Sharwas (Sherpas), and protected areas in the Chomolungma (Sagarmatha/ Mt Everest) Region of Nepal'. Paper presented to the Association of American Geographers Centennial Meeting, Philadelphia, March.

Stevens, S., and M.N. Sherpa. 1993. 'Indigenous peoples and protected areas: New approaches to conservation in highland Nepal'. In L.S. Hamilton, D.P. Bauer, and H.F. Takeuchi, eds, *Parks, Peaks, and People*. Honolulu, HI: East-West Center, Program on Environment.

Stewart, D.W., and P.N. Shamdasani. 1990. *Focus Groups: Theory and Practice*. Newbury Park: Sage.

Stewart, D., P. Shamdasani, and D. Rook. 2007. *Focus Groups: Theory and Practice*. 2nd edn. Thousands Oaks, CA: Sage.

Stewart, E.J. 2009. 'Comparing resident attitudes toward tourism: Community-based cases from Arctic Canada'. (Department of Geography, University of Calgary, unpublished PhD thesis).

Stewart, E.J., and D. Draper. 2009. 'Reporting back research findings: A case study of community-based tourism research in northern Canada'. *Journal of Ecotourism* 8 (2): 128–43.

Stratford, E. 1997. 'Memory work in geography and environmental studies: Some suggestions for teaching and research'. *Australian Geographical Studies* 35 (2): 208–21.

———. 1998. 'Public spaces, urban youth and local government: The skateboard culture in Hobart's

Franklin Square'. In R. Freestone, ed., *20th Century Urban Planning Experience: Proceedings of the 8th International Planning History Conference*. Sydney: University of New South Wales.

———, ed. 1999. *Australian Cultural Geographies*. Melbourne: Oxford University Press.

———. 2001. 'The Millennium Project on Australian Geography and Geographers: An introduction'. *Australian Geographical Studies* 39 (1): 91–5.

———. 2002. 'On the edge: A tale of skaters and urban governance'. *Social and Cultural Geography* 3 (2): 193–206.

———. 2008. 'Islandness and struggles over development: A Tasmanian case study'. *Political Geography* 27 (2): 160–75.

Stratford, E., and A. Harwood. 2001. 'The regulation of skating in Australia: An overview and commentary on the Tasmanian case'. *Urban Policy and Research* 19 (2): 61–76.

Strauss, A., and J. Corbin. 1990. *Basics of Qualitative Research: Grounded Theory, Procedures and Techniques*. Newbury Park: Sage.

Strunk, William, Jr, and E.B. White. 1959. *The Elements of Style*. New York: MacMillan.

Stychin, C.F. 2003. *Governing Sexuality. The Changing Politics of Citizenship and Law Reform*. Oxford and Portland, OR: Hart Publishing.

Suchet, S. 2002. '"Totally wild"? Colonising discourses, indigenous knowledges and managing wildlife'. *Australian Geographer*: 33 (3): 141–57.

Sudman, S., and N.M. Bradburn. 1982. *Asking Questions: A Practical Guide to Questionnaire Design*. San Francisco: Jossey-Bass.

Sue, V., and L. Ritter. 2007. *Conducting Online Surveys*. London: Sage.

Sundberg, J. 2004. 'Masculinist epistemologies and the politics of fieldwork in Latin Americanist geography'. *Professional Geographer* 55 (2): 180–90.

Swanson, K. 2007. '"Bad mothers" and "delinquent children": Unravelling anti-begging rhetoric in the Ecuadorian Andes'. *Gender, Place and Culture* 14 (6): 703–20.

———. 2008. 'Witches, children and Kiva-the-research dog: Striking problems encountered in the field'. *Area* 40 (1): 55–64.

Sweet, C. 2001. 'Designing and conducting virtual focus groups'. *Qualitative Market Research: An International Journal* 4 (3): 130–5.

Swenson, J.D., W.F. Griswold, and P.B. Kleiber. 1992. 'Focus groups: Method of inquiry/intervention'. *Small Group Research* 23 (4): 459–74.

Symonds, J.A. 1896. *A Problem of Modern Ethics*. London: privately printed.

Symposium on Computing and Qualitative Geography. 1995. University of Durham, 11–12 July. http://www.helsinki.fi/neu/lists/qual-software/0034.html.

Szili, G., and M.W. Rofe. 2007. 'Greening port misery: The "green face" of waterfront redevelopment in Port Adelaide, South Australia'. *Urban Policy and Research* 25 (3): 363–84.

Tashakkori, A., and C. Teddlie. 1998. *Mixed Methodology: Combining Qualitative and Quantitative Approaches*. Thousand Oaks, CA: Sage.

Taylor, C. 1984. 'Foucault on freedom and truth'. *Political Theory* 12 (2): 152–83.

Tesch, R., ed. 1989. 'Computer software and qualitative analysis: A reassessment'. In G. Blank, E. Brent, and J.L. McCartney, eds, *New Technology in Sociology: Practical Applications in Research and Work*. New Brunswick, NJ: Transaction Books.

———. 1990. *Qualitative Research: Analysis Types and Software Tools*. Basingstoke: Falmer Press.

Thatcher, J., C. Waddell, and M. Burks. 2002. *Constructing Accessible Websites*. Birmingham, UK: Glasshaus.

Thiesmeyer, L., ed. 2003. *Discourse and Silencing. Representation and the Language of Displacement*. Amsterdam and Philadelphia: John Benjamins Publishing.

Thomas, M. 2004. 'Pleasure and propriety: Teen girls and the practice of straight space'. *Environment and Planning D: Society and Space* 22 (5): 773–89.

Thomas, W., and F. Znaniecki. 1918. *The Polish Peasant in Europe and America*. New York: Dover Publications.

Thomas-Slayter, B. 1995. 'A brief history of participatory methodologies'. In R. Slocum, L. Wichart, D. Rocheleau, and B. Thomas-Slayter, eds, *Power, Process and Participation: Tools for Change*. London: Intermediate Technology Publications.

Thompson, P. 2000. *The Voice of the Past: Oral History*. 3rd edn. New York: Oxford University Press.

Thompson, S. 1994. 'Suburbs of opportunity: The power of home for migrant women'. In K. Gibson and S. Watson, eds, *Metropolis Now: Planning and the Urban in Contemporary Australia*. Sydney: Pluto.

Thrift, N. 1996. *Spatial Formations*. Thousand Oaks, CA: Sage.

———. 2000. 'Dead or alive?' In I. Cook, D. Crouch, S. Naylor, and J. Ryan, eds, *Cultural Turns/Geographical Turns*, 1–6. London: Sage.

Thrift, N., and S. French. 2002. 'The automatic production of space'. *Transactions of the Institute of British Geographer* 27: 309–35.

Tomsen, S., and K. Markwell. 2007. *When the Glitter Settles: Safety and Hostility at and around Gay and Lesbian Public Events*. Survey Report, Centre for Cultural Industries and Practices, University of Newcastle, New South Wales.

Tonkiss, F. 1998. 'Analysing discourse'. In C. Seale, ed., *Researching Society and Culture*, 245–60. London: Sage.

Townsend, J., with U. Arrevillaga, J. Bain, S. Cancino, S. Frenk, S. Pacheco, and E. Perez. 1995. *Women's Voices from the Rainforest*. London: Routledge.

Treby, E., I. Hewitt, and A. Shaw. 2006. 'Embedding "disability and access" into the geography curriculum'. *Teaching in Higher Education* 11 (4): 413–25.

Tremblay, M.A. 1982. 'The key informant technique: A non-ethnographic application'. In R.G. Burgess, ed., *Field Research: A Sourcebook and Field Manual*. London: Allen and Unwin.

Tri-Council (Medical Research Council of Canada, National Science and Engineering Research Council of Canada, Social Sciences and Humanities Research Council of Canada). 2005. 'Policy statement: Ethical conduct for research involving humans'. http://pre.ethics.gc.ca/english/policystatement/policystatement.cfm.

Tuan, Y.F. 1977. *Space and Place: The Perspective of Experience*. Minneapolis: University of Minnesota Press.

———. 1991. 'Language and the making of place: A narrative–descriptive approach'. *Annals of the Association of American Geographers* 81 (4): 684–96.

Tufte, E. 2003. *The Cognitive Style of PowerPoint*. Cheshire: Graphics Press LLC.

Tully, J. 1995. *Strange Multiplicity: Constitutionalism in an Age of Diversity*. Cambridge: Cambridge University Press.

Turner, F.J. 1920. *The Frontier in American History*. New York: Henry Holt.

Urry, J. 2002. *The Tourist Gaze*. 2nd edn. London: Sage.

Valentine, G. 1989. 'The geography of women's fear'. *Area* 21 (4): 385–90.

———. 1993. '(Hetero)sexing space: Lesbian perceptions and experiences of everyday spaces'. *Environment and Planning D: Society and Space* 11 (4): 395–413.

———. 1997. 'Tell me about . . . : Using interviews as a research methodology'. In R. Flowerdew and D. Martin, eds, *Methods in Human Geography: A Guide for Students Doing a Research Project*. Harlow: Longman.

———. 2002. 'People like us: Negotiating sameness and difference in the research process'. In P. Moss, ed., *Feminist Geography in Practice: Research and Methods*. Oxford: Blackwell.

———. 2003. 'Geography and ethics: In pursuit of social justice—Ethics and emotions in geographies of health and disability research'. *Progress in Human Geography* 27 (3): 375–80.

Valentine, G., T. Skelton, and R. Butler. 2003. 'Coming out and outcomes: Negotiating lesbian and gay identities with, and in, the family'. *Environment and Planning D: Society and Space* 21: 479–99.

van Hoven, B. 2003. 'Analysing qualitative data using CAQDAS'. In N. Clifford and G. Valentine, eds, *Key Methods in Geography*, 461–76. London: Sage.

van Hoven, B., and A. Poelman. 2003. 'Using computers for qualitative data analysis: An example using NUD*IST'. *Journal of Geography in Higher Education* 27 (1): 113–20.

van Hoven B., and D. Sibley. 2008. '"Just duck": The role of vision in the production of prison spaces'. *Environment and Planning D: Society and Space* 26 (6): 1001–17.

van Hoven-Iganski, B. 2000. *Made in the GDR. The Changing Geographies of Women in the Post-socialist Rural Society in Mecklenburg-Westpomerania*. Utrecht: KNAG/NGS.

Van Selm, M., and N. Jankowski. 2006. 'Conducting online surveys'. *Quality and Quantity* 40: 435–56.

VanWynsberghe, R., and S. Khan. 2007. 'Redefining case study'. *International Journal of Qualitative Methods* 6 (2): 2–10.

Ventresca, M., and J. Mohr. 2005. 'Archival research methods'. In J. Baum, ed., 'The Blackwells companion to organizations'. Blackwell Reference Online. http://www.blackwellreference.com/subscriber/tocnode?id=g9780631216940_chunk_g978063121694040.

Wadsworth, Y. 1998. 'What is participatory action research?' Action Research International Paper 2. http://www.scu.edu.au/schools/gecm/ar/ari/p-ywadsworth98.html.

Wainwright, J. 2008. *Decolonizing Development: Colonial Power and the Maya*. Oxford: Blackwell Press.

Waitt, G. 1997. 'Selling paradise and adventure: Representations of landscape in the tourist advertising of Australia'. *Australian Geographical Studies* 35 (1): 47–60.

————. 1999. 'Naturalizing the primitive: A critique of marketing Australia's indigenous peoples as hunter gatherers'. *Tourism Geographies* 1 (2): 142–63.

Waitt, G., et al. 2000. *Introducing Human Geography: Globalisation, Difference and Inequality*. Sydney: Pearson Education Australia.

Waitt, G., and L. Head. 2002. 'Postcards and frontier mythologies: Sustaining views of the Kimberley as timeless'. *Environment and Planning D: Society and Space* 20: 319–44.

Waitt, G., and P.M. McGuirk. 1996. 'Marking time: Tourism and heritage representation at Millers Point, Sydney'. *Australian Geographer* 27 (1): 11–29.

Waitt, G., and K. Markwell. 2006. *Gay Tourism: Culture and Context*. Binghamton, NY: Haworth Press.

Wakefield, S. 2007. 'Reflective action in the academy: Exploring praxis in critical geography using a "food movement" case study'. *Antipode* 39 (2): 331–54.

Walmsley, D.S., and G.I. Lewis. 1984. *Human Geography: Behavioural Approaches*. New York: Longman.

Walsh, B., and T. Lavalli. 1996. 'Comparative review of NUD*IST, ATLAS/ti, Folio Views'. *Microtimes* 162. http://www.microtimes.com/162/research.html.

Wang, C., and M.A. Burris. 1997. 'Photovoice: Concept, methodology, and use for participatory needs assessment'. *Health Education and Behaviour* 24: 369–87.

Ward, B. 1972. *What's Wrong with Economics?* London: MacMillan.

Ward, R. 1958. *The Australian Legend*. Melbourne: Oxford University Press.

Ward, V.M., J.T. Bertrand, and L.F. Brown. 1991. 'The comparability of focus group and survey results: Three case studies'. *Evaluation Review* 15 (2): 266–83.

Warde, A. 1989. 'Recipes for a pudding: A comment on locality'. *Antipode* 21 (3): 274–81.

Wearing, B. 1984. *The Ideology of Motherhood: A Study of Sydney Suburban Mothers*. Sydney: Allen and Unwin.

Webb, B. 1982. 'The art of note-taking'. In R.G. Burgess, ed., *Field Research: A Sourcebook and Field Manual*. London: Allen and Unwin.

Webb, E.J., et al. 1966. *Unobtrusive Measures: Non-reactive Research in the Social Sciences*. Chicago: Rand McNally.

Weinstein, D., and M. Weinstein. 1991. 'Georg Simmel: Sociological flâneur bricoleur'. *Theory, Culture and Society* 8: 151–68.

Weis, T. 2000. 'Beyond peasant deforestation: Environment and development in rural Jamaica'. *Global Environmental Change* 10: 299–305.

Weitzman, E.A. 2000. 'Software and qualitative research'. In N. Denzin and Y. Lincoln, eds, *Handbook of Qualitative Research*, 2nd edn, 803–20. Thousand Oaks, CA: Sage.

Weitzman, E., and M. Miles. 1995. *Computer Programs for Qualitative Data Analysis: Software Sourcebook*. Thousand Oaks, CA: Sage.

Welsh, I. 1996. *Trainspotting*. London: W.W. Norton.

Western, J.C. 1981. *Outcast Cape Town*. Minneapolis: University of Minnesota Press.

Wetherell, M., S. Taylor, and S.J. Yates. 2001. *Discourse Theory and Practice: A Reader*. Thousands Oaks, CA: Sage.

White, P., and P.A. Jackson. 1995. '(Re)theorising population geography'. *International Journal of Population Geography* 1: 111–23.

Whyte, W.F. 1982. 'Interviewing in field research'. In R.G. Burgess, ed., *Field Research: A Sourcebook and Field Manual*. London: Allen and Unwin.

Whyte, W.H. 1943. *Street Corner Society*. Chicago: University of Chicago Press.

Widdowfield, R. 2000. 'The place of emotions in academic research'. *Area* 32 (2): 199–208.

Wiles, J. 2003. 'Daily geographies of caregivers: Mobility, routine, scale'. *Social Science and Medicine* 57: 1307–25.

Wilkinson, A. 1999. 'New Labour and Christian socialism'. In G.R. Taylor, ed., *The Impact of New Labour*, 37–50. Basingstoke: Macmillan.

Williams, C.C., and J. Round. 2007. 'Re-thinking the nature of the informal economy: Some lessons from Ukraine'. *International Journal of Urban and Regional Research* 31 (2): 425–41.

Williams, G., et al. 2003. 'Enhancing pro-poor governance in eastern India: Participation, politics and action'. *Progress in Development Studies* 3 (2): 159–78.

Williams, K., and C. Johnstone. 2000. 'The politics of the selective gaze: Closed circuit television and the policing of public space'. *Crime, Law and Society* 34: 183–210.

Williams, M. 1992. 'Archives in geographical research'. In A. Rogers, H. Viles, and A. Goudie, eds, *The Student's Companion to Geography*. Oxford: Blackwell.

Williamson, J.E. 1978. *Decoding Advertisements: Ideology and Meaning in Advertising.* London: Marion Boyars.

Wilson, A.G. 1972. 'Theoretical geography: Some considerations'. *Transactions of the Institute of British Geographers* 7: 31–44.

Wilson, K., and E.J. Peters. 2005. '"You can make a place for it": Remapping urban First Nations spaces of identity'. *Environment and Planning D: Society and Space* 23: 395–413.

Winchester, H.P.M. 1992. 'The construction and deconstruction of women's role in the urban landscape'. In F. Gale and K. Anderson, eds, *Inventing Places: Studies in Cultural Geography.* Melbourne: Longman Cheshire.

———. 1996. 'Ethical issues in interviewing as a research method in human geography'. *Australian Geographer* 27 (1): 117–31.

———. 1999. 'Interviews and questionnaires as mixed methods in population geography: The case of lone fathers in Newcastle, Australia'. *Professional Geographer* 51 (1): 60–7.

———. 2000. 'Qualitative research and its place in geography'. In I. Hay, ed., *Qualitative Research Methods in Human Geography.* Melbourne: Oxford University Press.

Winchester, H.P.M., and L.N. Costello. 1995. 'Living on the street: Social organisation and gender relations of Australian street kids'. *Environment and Planning D: Society and Space* 13: 329–48.

Winchester, H.P.M., and K.M. Dunn. 1999. 'Cultural geographies of film: Tales of urban reality'. In K.J. Anderson and F. Gale, eds, *Inventing Places: Studies in Cultural Geography.* Melbourne: Addison Wesley Longman.

Winchester, H.P.M., K.M. Dunn, and P.M. McGuirk. 1997. 'Uncovering Carrington'. In J. Moore, J. Ostwald, and A. Chawner, eds, *Hidden Newcastle: The Invisible City and the City of Memory,* 174–81. Arlington, VA: Gadfly Media.

Winchester, H.P.M., L. Kong, and K.M. Dunn. 2003. *Landscapes: Ways of Imagining the World.* Harlow: Pearsons.

Winchester, H.P.M., P.M. McGuirk, and K. Everett. 1999. 'Celebration and control: Schoolies Week on the Gold Coast, Queensland'. In E. Teather, ed., *Embodied Geographies: Spaces, Bodies and Rites of Passage.* London: Routledge.

Winchester, H.P.M., and M.W. Rofe. 2005. 'Christmas in the "Valley of Praise": Intersections of the rural idyll, heritage and community in Lobethal, South Australia'. *Journal of Rural Studies* 21: 265–79.

Wolcott, H. 1990. 'On seeking—and rejecting—validity in qualitative research'. In E. Eisner and A. Peshkin, eds, *Qualitative Inquiry in Education: The Continuing Debate.* New York: Teachers College Press.

———. 2001. *Writing up Qualitative Research.* 2nd edn. Newbury Park: Sage.

Wolf, D., ed. 1996. *Feminist Dilemmas in Fieldwork.* Boulder, CO: Westview Press.

Wolf, E. 1982. *Europe and the People without History.* Berkeley: University of California Press.

Wood, L.A., and R.O. Kroger. 2000. *Doing Discourse Analysis: Methods for Studying Action in Talk and Text.* Thousand Oaks, CA: Sage.

Wood, L., and S. Williamson. 1996. *Consultants' Report on Franklin Square: Users, Activities and Conflicts.* Hobart: UNITAS Consulting.

Wood, N., M. Duffy, and S.J. Smith. 2007. 'The art of doing (geographies of) music'. *Environment and Planning D: Society and Space* 25 (5): 867–89.

Wood, N., and S. Smith. 2004. 'Instrumental routes to emotional geographies'. *Social and Cultural Geography* 5: 533–48.

Wooldridge, S.W. 1955. 'The status of geography and the role of fieldwork'. *Geography* 40: 73–83.

Wrigley, E.A. 1970. 'Changes in the philosophy of geography'. In R.J. Chorley and P. Haggett, eds, *Frontiers in Geographical Teaching.* London: Methuen.

Wrigley, N., et al. 2004. 'The Leeds "food deserts" intervention study: What the focus groups reveal'. *International Journal of Retail and Distribution Management* 32 (2): 123–36.

Wylie, J.W. 2007. *Landscape.* New York: Routledge.

Yin, R. 2003. *Case Study Research: Design and Methods.* Los Angeles: Sage.

Young, L., and H. Barratt. 2001. 'Adapting visual methods: Action research with Kampala street children'. *Area* 33 (2): 141–52.

Zeigler, D.J., S.D. Brunn, and J.H. Johnson. 1996. 'Focusing on Hurricane Andrew through the eyes of the victims'. *Area* 28 (2) 124–9.

Zelinsky, W. 1973. *Cultural Geography of the United States.* New York: Prentice Hall.

———. 2001. 'The geographer as voyeur'. *Geographical Review* 91 (1–2): 1–8.

Zimmerer, K.S., and T.J. Bassett, eds. 2003. *Political Ecology: An Integrative Approach to Geography and Environment–Development Studies.* New York: Guilford Press.

Index